ENHANCED BOILING HEAT TRANSFER

ENHANCED BOILING HEAT TRANSFER

John R. Thome

Engineering Consultant
Rome, Italy

HEMISPHERE PUBLISHING CORPORATION
A member of the Taylor & Francis Group

New York Washington Philadelphia London

ENHANCED BOILING HEAT TRANSFER

1 2 3 4 5 6 7 8 9 0 B R B R 8 9 8 7 6 5 4 3 2 1 0 9

The editors were Susan E. Zinninger and Lisa McCullough.
The typesetter was WorldComp.
Cover design by Reneé Winfield.
Braun-Brumfield, Inc. was printer and binder.

Library of Congress Cataloging-in-Publication Data

Thome, John R.
Enhanced boiling heat transfer / by John R. Thome.
p. cm.

1. Heat engineering. 2. Ebullition. 3. Heat—Transmision.
I. Title.
TJ260.T495 1989
621.402′5—dc20 89-36432
ISBN 0-89116-745-5

CONTENTS

PREFACE ix

1 INTRODUCTION 1

2 FUNDAMENTALS OF BOILING HEAT TRANSFER ON PLAIN SURFACES 4

2.1 Introduction 4
2.2 Pool Boiling 5
2.3 Boiling on Tube Bundles 14
2.4 Boiling Inside Tubes 17
2.5 Nomenclature 24
References 26

3 HISTORICAL DEVELOPMENT OF ENHANCED BOILING SURFACES 28

3.1 Introduction 28
3.2 Early Developments 28
3.3 Novel Enhancement Techniques 39
3.4 Development of Enhanced Surfaces for Intube Boiling 53
3.5 Nomenclature 58
References 59

4 COMMERCIAL ENHANCED BOILING TUBES 64

4.1 Introduction 64
4.2 Externally Finned Tubes 64
4.3 Modified Externally Finned Tubes 66
4.4 Porous Layer-Coated Tubes 70
4.5 Internally Finned Tubes 71
4.6 Doubly Enhanced Tubes 72
4.7 Tube Inserts 75
References 77

5 BOILING NUCLEATION ON ENHANCED BOILING SURFACES 78

5.1 Introduction 78
5.2 Visualization Studies 79
5.3 Nucleation Superheat Experiments 83
5.4 Onset of Nucleate Boiling 88
References 90

6 MECHANISMS OF ENHANCED NUCLEATE POOL BOILING 91

6.1 Introduction 91
6.2 Heat Transfer Mechanisms and Processes 91
6.3 Fundamental Experimental Studies 98
6.4 Miscellaneous Effects on Boiling Performance 117
6.5 Nomenclature 122
References 123

7 PERFORMANCE COMPARISONS OF ENHANCED BOILING TUBES 125

7.1 Water 127
7.2 Refrigerants 128
7.3 Cryogens 138
7.4 Hydrocarbons 141
7.5 Alcohols 144
7.6 Solvents (Acetone) 147
7.7 Dielectrics (Fluorinert FC-72) 148
7.8 Summary 148
7.9 Nomenclature 149
References 150

8 NUCLEATE POOL BOILING CORRELATIONS 152

8.1 Introduction 152
8.2 Low Finned Tubes 152

8.3 Porous Coatings 155
8.4 Modified Externally Finned Tubes 161
8.5 Conclusions 170
8.6 Nomenclature 171
References 173

9 ENHANCED BOILING OF MIXTURES 175

9.1 Vapor-Liquid Phase Equilibria 175
9.2 Mixture Physical Properties 177
9.3 Boiling Nucleation 180
9.4 Boiling Heat Transfer in Mixtures 183
9.5 Boiling of Refrigerant-Oil Mixtures 195
9.6 Nomenclature 201
References 201

10 ENHANCED FLOW BOILING 204

10.1 Intube Enhanced Boiling 204
10.2 Enhanced Boiling on Tube Bundles 254
10.3 Nomenclature 260
References 262

11 APPLICATION OF ENHANCED BOILING TUBES IN THE CHEMICAL PROCESSING INDUSTRIES 266

11.1 Introduction 266
11.2 Advantages of Enhanced Boiling Tubes 267
11.3 Heat Exchanger Configurations 268
11.4 Process Applications 271
11.5 Special Thermal Design Considerations 280
11.6 Mechanical Design Considerations 288
11.7 Fouling 289
11.8 Operating Considerations 297
11.9 Economic Considerations and Surface Selection 298
11.10 Summary 302
11.11 Nomenclature 302
References 303

12 DESIGN CASE HISTORIES 305

12.1 Introduction 305
12.2 Fluorocarbon Refrigeration Services 306
12.3 Refinery and Chemical Plant Refrigeration Services 313
12.4 Reboilers 315
12.5 Nomenclature 321
References 321

13 BOILING IN PLATE-FIN HEAT EXCHANGERS 323

13.1 Introduction 323
13.2 Types of Plate-Fins 326
13.3 Two-Phase Flow Regimes 326
13.4 Onset of Nucleate Boiling 329
13.5 Flow Boiling Studies 332
13.6 Plate-Fin Boiling Correlations 340
13.7 Nomenclature 348
References 349

INDEX 353

PREFACE

Enhanced boiling is one of the most exciting and dynamic areas of thermal engineering. Although single-phase heat transfer provides only a marginal opportunity for improvement above and beyond the addition of extended surface area, some enhanced boiling surfaces produce up to 100 times the performance of a conventional, plain surface under certain conditions. Thus in this sense enhanced boiling surfaces can be thought of as "superconductors" of heat. Enhanced boiling is a subject still ripe for significant advances, such as in innovative development of new enhancement geometries and their manufacture. Key fundamental problems regarding the physical processes and phenomena remain to be solved, providing an incentive to the creative investigator to explore and then to explain how and why these enhancement geometries work as outstandingly as they do. Thus enhanced boiling is still an emerging science and technology.

In the past 20 years there has been extensive research and development in enhanced boiling heat transfer. The application of enhanced boiling surfaces in heat exchangers has become widely, although not extensively, established, and their use in other services, such as in heat pipes and for cooling electronic chips in the electronics industry, is widening. Therefore, it is the right time for an extensive survey of the subject, first to bring organization to the massive and exponentially growing amount of literature available and second to provide thermal design information for the application of enhanced boiling surfaces in practice.

The book is organized as follows. After a short introduction to the subject in Chapter One, Chapter Two provides a brief, basic review of conventional boiling heat transfer for nonspecialists. The historical development of enhanced boiling surfaces together with the corresponding advancements in nucleation theory influencing their invention are presented in Chapter Three. In Chapter Four, many types of commercially available enhanced boiling tubes are described. Boiling nucleation on enhanced boiling surfaces is surveyed in Chapter Five, and in Chapter Six the thermal mechanisms responsible for the tremendous heat transfer coefficients of enhanced boiling surfaces and the geometric factors affecting their performance are discussed. In Chapter Seven, the boiling performances of many enhancement geometries are compared for a wide range of fluids to help in the selection of suitable surfaces for a particular application and to use as a "yardstick" for the evaluation of new enhancement geometries. A complete survey of correlations available for predicting enhanced nucleate pool boiling coefficients is presented in Chapter Eight. Chapter Nine surveys the enhanced boiling of mixtures. Section 10-1 of Chapter Ten reviews the experimental studies on enhanced boiling inside tubes and includes an extensive review of the thermal design correlations available. Enhanced shell-side boiling is treated in Section 10-2 of Chapter Ten, and practical application guidelines and case histories are covered in Chapters Eleven and Twelve. Evaporation in plate-fin heat exchangers is an important form of enhanced boiling, and thus two-phase flow patterns and heat transfer in these units are reviewed in Chapter Thirteen.

One novel feature presented for the first time is an enhanced boiling surface selection aid for heat exchangers. Utilizing a specially developed graph together with boiling data available throughout the book, one can quickly determine the merits of using plain, low-finned, highly enhanced, or doubly augmented tubes in a particular application. In addition, photographs never previously published of the boiling nucleation and liquid film formation processes, inside boiling enhancements are shown. Other highlights are an extensive survey of special thermal design considerations and constraints involving enhanced boiling heat exchangers; a review of energy conservation interventions that use enhanced boiling tubes in ethylene plants, refineries, and the like; a survey of the effects of oil on both intube and shell-side boiling of refrigerants; a photographic sequence of the fouling and cleaning of a reboiler's low-finned tube bundle together with a review of the benefits of using low-finned tubes in highly fouling boiling services; and a survey of the thermal design correlations for plate-fin heat exchangers.

Efficient utilization of energy is still an elusive goal of the engineering community and humankind as a whole. Enhanced boiling should play a principal role in our efforts because of the ability of these special geometries to transfer megawatts of heat at boiling temperature differences on the order of only 1.0 K. One should consider that every industrialized country in the world has a "renewable" and incredibly large natural resource in the form of nonpolluting energy that can be extracted from petrochemical facilities, refineries, fertilizer plants, and so forth by use of enhanced boiling technology in conjunction with "pinch" technology, vapor recompression cycles, and the like. Consequently, another goal of this book is to stimulate further the application of enhanced boiling surfaces in the refrigeration, petroleum, and chemical processing industries by presentation of special thermal guidelines for the design and operation of enhanced boiling heat exchangers, a survey of possible process interventions, and a description of several case histories of actual operating units. In summary, it is my goal to provide a useful treatise for both the research engineer and the heat transfer practitioner.

I would like especially to thank my wife, Carla, and my two sons, Luca and Alessandro, for their moral support and sacrifice, which made the writing of this book possible. I am grateful to the many friends and acquaintances who provided photographs for the book. I would also like to thank William Begell and Florence Padgett of Hemisphere Publishing Corporation for their encouragement and assistance.

John R. Thome

CHAPTER

ONE

INTRODUCTION

Enhanced boiling is one of the *high-technology* areas of heat transfer. In single-phase heat transfer the augmentation of performance is incremental, depending almost exclusively on the provision of additional heat transfer surface area with fins, corrugations, and the like. In contrast, enhanced boiling surfaces can in some instances produce heat transfer coefficients as much as 100 times larger than those of a comparable plain, smooth surface that is boiling at the same wall superheat. Thus a significant leap forward in technology has been achieved that may change many conventional practices in heat transfer design and heat exchange equipment for years to come.

Enhanced boiling heat transfer is of great importance to evaporation processes in the refrigeration and air-conditioning industries, the hydrocarbon and chemical industries, the microelectronics industries, the growing heat pipe industry, and others. In refrigeration systems, for instance, enhanced boiling tubes are widely used to augment performance of flooded evaporators. In refineries and chemical processing plants, enhanced boiling tubes can be utilized for reboilers on distillation towers in nonfouling services and in evaporators of cascade refrigeration systems. In the electronics industry, enhanced boiling is a promising new technology for cooling high–power density components. In the air separation and gas processing industries, plate-fin geometries are utilized to augment boiling heat transfer and to obtain compact designs. In heat pipes, enhanced boiling surfaces improve performance in the evaporation zone.

The principal economic benefits derived from the application of enhanced boiling technology in heat exchangers are a reduction in size and cost of new units and a decrease in energy-related operating costs when retubing existing units with enhanced boiling tubes. The primary advantage to electronic components is the augmentation of cooling rates, which lowers the start-up and operational temperatures of the power-dissipating components and in turn increases their service life and reliability.

Numerous ways to augment boiling heat transfer coefficients have been developed. Active techniques include rotation or vibration of the heated surface, periodic wiping of the heated surface or suction at its surface, ultrasonic vibration of the evaporating fluid, application of electrostatic fields to the fluid, and the introduction of trace additives to the fluid. Most of these active methods are difficult to implement outside the laboratory or to justify economically. Passive techniques include roughening the heat transfer surface; placing tiny nonwetting spots (such as Teflon or epoxy) on the heated surface; pitting a surface with a corrosive chemical; using integral fins or rolling, splitting, or knurling integral finned tubes to form complex geometries; and coating the heated surface with a porous layer. Also, dendritic types of scaling have been used to augment boiling performance. Only several of these methods are practical on a large scale and economically advantageous compared to conventional, unmodified surfaces such as ordinary drawn tubing. Various other passive techniques have also been developed that do not require modification of the heated surface. These include twisted metallic tapes, coiled wire inserts, and inlet vortex generators, all of which are primarily installed inside conventional shell-and-tube heat exchangers to improve below-design boiling performances. Also, tubes have been externally wrapped with various materials such as wire and wire screens.

As can be expected for any new technology, further developments in enhancement techniques, boiling performances, and manufacturing methods and novel applications to thermal processes are forthcoming. This area of heat transfer is especially exciting for the heat transfer specialist in that he or she has the opportunity to significantly improve the thermal performance of a heat transfer process by utilizing his or her own creativity rather than following standard, ordinary design procedures.

The ultimate goal of enhanced boiling research and development is to obtain the optimal combination of surface and boiling performance for each general area of application together with accurate and reliable thermal design methodology. The optimal surface, in this case, does not necessarily mean the one with the best boiling performance but rather the best surface when such factors as unit cost, quality of manufacture, uniformity in performance, and service life are taken into account. The biggest impediment to widespread application of enhanced boiling surfaces is the scarcity of well-documented thermal design methods for use by the engineering practitioner. This is one aspect of enhanced boiling heat transfer in which extensive research and development are urgently required to obtain the necessary correlating equations and then to implement them in widely used design software. The principal design parameters that are needed for enhanced boiling surfaces are:

1. the nucleation superheat, which denotes the minimum operational wall superheating permissible to sustain boiling;
2. the maximum or critical heat flux, which is the largest heat flux that can be applied while remaining in the nucleate boiling or wetted wall regime; and
3. the boiling heat transfer coefficient as a function of wall superheat or heat flux for the particular operating conditions of the enhanced boiling surface.

Each of these parameters is a function of numerous geometric factors, the type of fluid, and the flow conditions. These include the particular type of boiling enhancement being used, the operating pressure, whether there is subcooling of the bulk liquid, the local vapor quality (in flow boiling), and mass diffusion effects (in mixtures).

Chapter Two treats boiling heat transfer on plain surfaces. Chapters Three to Nine deal primarily with the fundamentals of enhanced nucleate pool boiling. Chapters Ten to Thirteen cover flow boiling inside tubes, outside tube bundles, and inside plate-fin heat exchangers. Practical design and operation information and a survey of applications of enhanced boiling heat exchangers are addressed principally in Chapters Eleven and Twelve.

CHAPTER

TWO

FUNDAMENTALS OF BOILING HEAT TRANSFER ON PLAIN SURFACES

2-1 INTRODUCTION

The thermal design of conventional evaporators is based on our knowledge of boiling on plain, smooth surfaces. Plain, drawn tubing is the typical surface in heat exchangers, with boiling either on its inside or outside surface. Because conventionally designed evaporators with drawn tubing are the reference against which enhanced boiling tube evaporators are compared, a review of the fundamentals of boiling heat transfer on plain, smooth surfaces is presented.

The general topics covered in this chapter are nucleate pool boiling, intube boiling, and boiling on the outside of tube bundles. The aim is to provide some background on these processes for nonspecialists in boiling heat transfer. Another objective is to furnish some of the heat transfer correlations for conventional boiling that are used later in the text to determine the relative augmentation in heat transfer performances obtained by enhanced boiling surfaces under various circumstances.

Boiling heat transfer is distinct from single-phase forced convection in that the boiling heat transfer coefficient is a strong function of the temperature difference between the heated wall and the bulk liquid. During boiling, a large quantity of the heat is transported from the heated wall as latent heat in vapor bubbles in addition to the sensible heat added to the liquid. Another distinguishing characteristic of boiling heat transfer is the effect of the microsurface geometry on the boiling process. Therefore, boiling heat transfer is a much more complex process than the more widely studied single-phase forced convection process, and hence much more effort is required to obtain a basic understanding of the involved phenomena, heat transfer mechanisms, and predictive methods.

The subject of boiling heat transfer is traditionally divided into two parts: pool boiling and flow boiling. Pool boiling represents the situation in which boiling occurs on a single heated surface, usually a tube or a disk, in a large pool of otherwise quiescent liquid (the only motion is caused by the boiling process itself). As its name implies, flow boiling refers to the situation in which there is bulk motion of the liquid past the heated surface (or surfaces) either by natural circulation of the liquid or by the driving action of a pump. Flow in the natural circulation regime is provided by the driving head produced by a column of liquid in the inlet piping. Pool boiling by itself is of limited practical significance as a process, but it is studied extensively to gain insight into the more complex problem of flow boiling.

The reader who requires further information about conventional boiling heat transfer is referred to Rohsenow (1973) for a review of the general principles of boiling, Van Stralen and Cole (1979) for an exhaustive survey of the fundamentals of nucleate boiling, and Collier (1981) for an extensive review of flow boiling inside tubes. For boiling on the outside of horizontal tube bundles, the reader is referred to the article by Palen, Yarden, and Taborek (1972).

2-2 POOL BOILING

Pool Boiling Curve

Pool boiling heat transfer is most easily explained with reference to the pool boiling curve, which represents the functional dependence of the heat flux leaving the heated wall on the temperature difference between the surface of the heated wall and the surrounding bulk liquid. Figure 2-1 depicts the pool boiling curve and delineates the various boiling regimes occurring when the wall temperature is the independently controlled variable. The wall superheat is plotted along the abscissa and is defined as the difference between the surface temperature of the wall and the saturation temperature of the liquid (when the bulk liquid is not at saturation but is subcooled, the bulk temperature is used in place of the saturation temperature). The heat flux passing from the heated wall into the fluid is plotted along the ordinate.

The pool boiling curve in Fig. 2-1 for boiling on the outside of a typical plain, smooth tube in water is divided into four distinct heat transfer regimes: single-phase natural convection, nucleate pool boiling, transition boiling, and film boiling. As the wall superheat is raised from an initial value of zero, the regime of single-phase natural convection is in effect until nucleation occurs and the first vapor bubbles form on the heated wall. The wall temperature must reach a minimum value (point A) above the saturation temperature of the liquid for boiling to initiate. Hence natural convection can occur up to this point, even though the temperature of the liquid in the thermal boundary layer at the heated wall is higher than the saturation temperature. Liquid in this state is said to be superheated.

Once boiling nucleation occurs and bubbles begin to grow and depart from various sites on the heated wall, the mechanism of heat transfer is augmented by the boiling process, and the heat flux passing through the heated wall increases to the higher value at point B. The regime of nucleate pool boiling has now been attained. This regime is

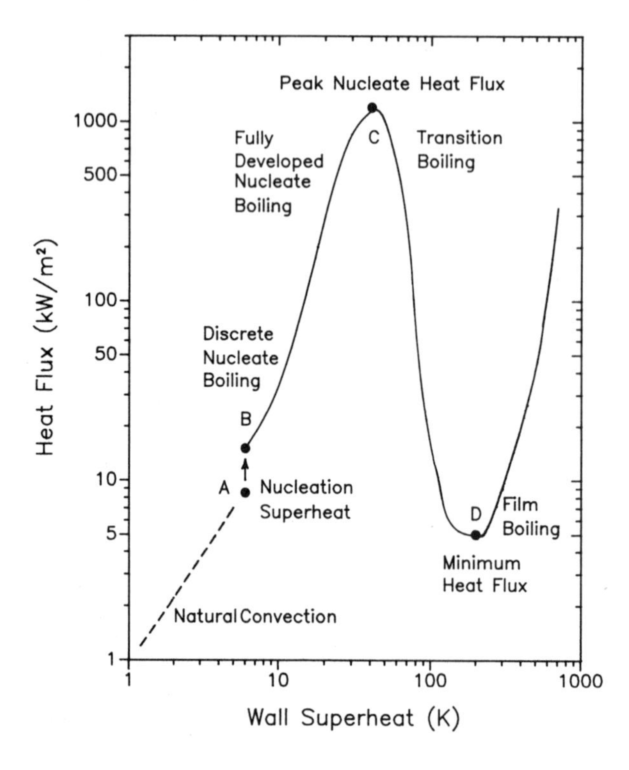

Figure 2-1 Boiling curve and boiling regimes for a horizontal, steam-heated tube.

characterized by discrete bubbles growing and departing from the heated wall at numerous nucleation sites at low heat fluxes, often called the discrete boiling region, and by large patches and slugs of vapor leaving the surface at medium and high heat fluxes, the latter of which is commonly referred to as fully developed nucleate pool boiling. These boiling regimes are depicted schematically in Fig. 2-2. Heat can leave the surface either as sensible heat in the form of superheated liquid or as latent heat in vapor bubbles. Here the principal mechanisms of heat transfer are (1) liquid-phase convection, augmented by the agitating action of the vapor bubbles and the cyclic stripping of the thermal boundary layer by departing bubbles (the later process is

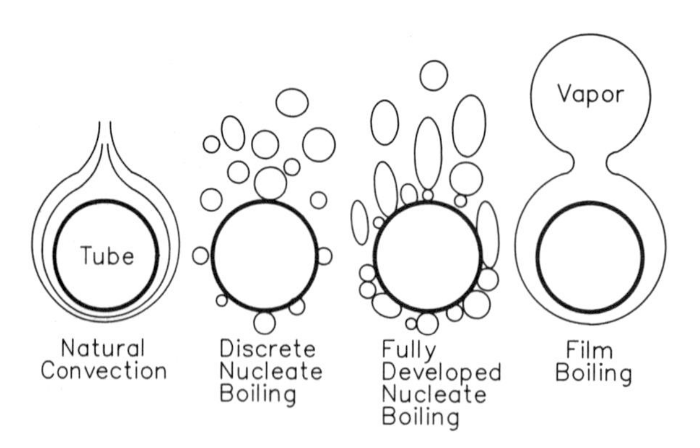

Figure 2-2 Depiction of the nucleate pool boiling regimes.

depicted in Fig. 2-3), and (2) microlayer evaporation of the thin liquid film trapped between the growing bubble and the heated surface.

As the wall superheat continues to rise, a maximum in the heat flux is attained (Point C in Fig. 2-1). This point is referred to as the peak nucleate heat flux or sometimes as the maximum heat flux or the departure from nucleate boiling. This maximum is thought to occur as a result of a hydrodynamic instability in the vapor jets leaving the heated wall, which in turn causes a vapor film to form over portions of the heated wall. As the wall superheat is raised yet higher, the regime of transition boiling is initiated. This regime is characterized by only partial and intermittent contact of the fluid with the heated wall, which significantly reduces the rate at which heat can be transferred from the heated wall to the fluid. This regime is bounded at higher wall superheats by point D, the minimum heat flux. At this superheat, the liquid is no longer able to come in contact with the heated wall because of the rapid rate of vapor generation in the vapor film covering the surface.

The last regime of the pool boiling curve is film boiling, which is bounded at lower heat fluxes by the minimum heat flux and at its maximum by the melting temperature of the wall material or the maximum temperature of the heating source. This regime is characterized by a stable vapor film that blankets the heated wall. Heat thus must be conducted or convected through this vapor film to the free liquid-vapor interface before being transported away into the bulk. This process is inefficient and results in small heat transfer coefficients and large wall superheats. Because of this poor thermal performance, film boiling is avoided whenever possible in thermal design.

Boiling Nucleation

Bubbles growing and departing from a heated wall in boiling originate from small grooves, holes, or other cavities in the surface. A surface that appears to be smooth to the naked eye actually looks like a "lunar landscape" when viewed under an electron microscope. Cavities of various sizes, shapes, and depths are present in the surface, as illustrated schematically in Fig. 2-4. These cavities trap vapor and gases, which then serve as the nuclei for the formation of vapor bubbles. The diameters of the cavities acting as nucleation sites are on the order of 0.5 to 10 μm.

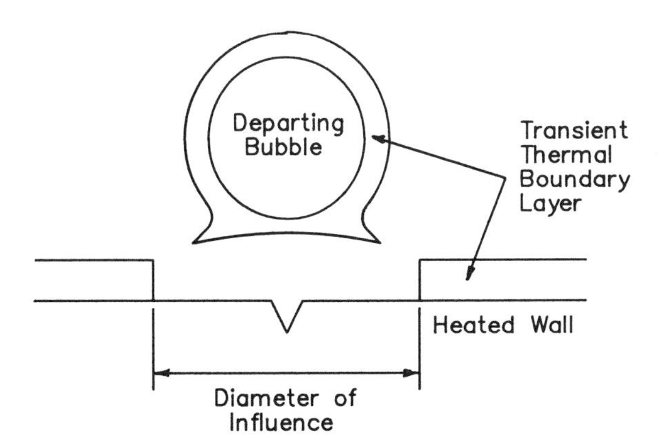

Figure 2-3 Cyclical thermal boundary layer stripping mechanism.

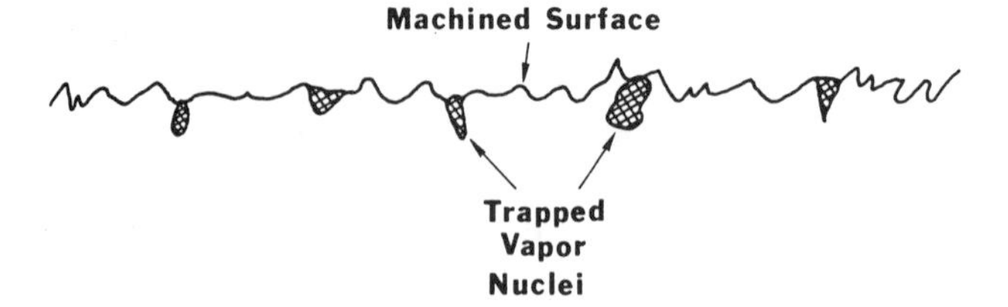

Figure 2-4 Magnified cross-section of a plain, smooth metallic surface.

For these vapor nuclei to be stable, they must be in thermal and mechanical equilibrium with the heated wall and the immediately surrounding liquid. For thermal equilibrium to exist, the temperature of the vapor in the nuclei must be the same as the temperature of the heated wall and liquid. If, instead, the temperature of the liquid is less than that of the vapor, then the vapor in the nuclei will tend to condense, and the vapor nuclei will decrease in size and perhaps cease to exist altogether. On the other hand, if the surrounding liquid is hotter than both the vapor in the nuclei and the saturation temperature, then the vapor nuclei will tend to grow and form vapor bubbles on the heated wall.

For mechanical equilibrium to exist, the pressure difference between the inside and the outside of the vapor nucleus must be balanced by the surface tension forces at the liquid-vapor-metal interface, as illustrated in Fig. 2-5. Because the pressure inside the nucleus is greater than the ambient pressure outside, the saturation temperature of the vapor inside the nucleus is higher than the bulk saturation temperature. Therefore, the temperature of the ambient liquid must be greater than its saturation temperature (i.e., the liquid must be superheated) for thermal equilibrium to exist. The nucleation superheat required for thermal and mechanical equilibrium to exist for a vapor nucleus of radius R trapped in a heated surface is

$$\Delta T_{\text{sat}} = \frac{2\sigma}{R(dP/dT)_{\text{sat}}} \tag{2-1}$$

The radius R in general is unknown, and hence the nucleation superheat is difficult to predict for a particular surface. As an example of typical values, Table 2-1 lists nucleation superheats measured by Thome, Shakir, and Mercier (1982) and by Shakir and Thome (1986) for various pure fluids. The large difference between the ethanol values for the disk and those for the tube is primarily a reflection of different surface roughnesses and thus cavity sizes.

Peak Nucleate Heat Flux

The largest heat flux that can be transferred in the nucleate pool boiling regime is the peak nucleate heat flux. Zuber (1959) explained this peak flux as being the result of a Helmholz instability that can occur at the vapor-liquid interface of a rising vapor jet. In nucleate boiling at large heat fluxes, vapor jets can be formed by the coalescence of rapidly growing and departing bubbles from the same boiling site. The instability is

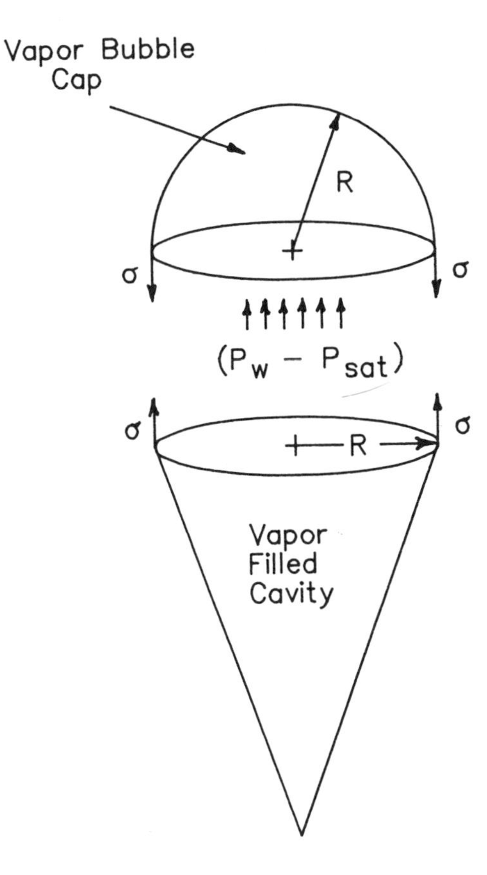

Figure 2-5 Mechanical equilibrium of a vapor nucleus at the mouth of a conical cavity.

Table 2-1 Measured nucleation superheats

Fluid/surface combination	Pressure, bar	Superheat, K
Nitrogen/copper disk	1.11	1.6
Argon/copper disk	1.25	2–3
Water/copper disk	1.01	6–8
Water/copper tube	1.01	4–8
Ethanol/copper disk	1.01	23–24
Ethanol/copper tube	1.01	6–10
n-Propanol/copper tube	1.01	8
Methanol/copper tube	1.01	9–10
Benzene/copper tube	1.01	10–11

caused by the pressure gradient that forms across the interface as the velocity of the rising vapor in the jet increases (i.e., the higher the vapor velocity, the lower the local pressure in the vapor jet). At a critical vapor velocity, the pressure gradient across the interface is sufficient to overcome the surface tension forces that act to hold the interface intact. The vapor jet then becomes unstable and is thought to prevent liquid from continuously wetting the heated wall. The resulting expression for the peak nucleate heat flux is

$$q_{max} = 0.149\rho_v^{1/2}\Delta h_v[g(\rho_L - \rho_v)\sigma]^{1/4} \tag{2-2}$$

Equation (2-2) can be corrected to predict the peak nucleate heat flux for horizontal tubes of diameters typically used in shell-and-tube heat exchangers by replacing the constant 0.149 with the empirical value 0.131 obtained earlier by Kutateladze (1948). Note, however, that the above expression is only valid for a single tube, not a tube bundle.

Minimum Heat Flux

Point D in Fig. 2-1 represents the minimum heat flux, which denotes the boundary between the film boiling regime and the transition boiling regime in pool boiling. Zuber (1959) postulated that, as the wall superheat is decreased in the film boiling regime, the vapor generation rate at some point becomes too small to maintain a stable vapor film on the heated wall. Assuming that this is caused by a Taylor instability, which occurs when there is an imbalance between the surface tension and gravity and inertia forces acting on an interface, Zuber developed an analytic expression for the minimum heat flux involving an empirical constant. Berenson (1960) improved on this theory and, together with experimental data, obtained the following expression for the minimum heat flux:

$$q_{min} = 0.09\rho_v\Delta h_v\left[\frac{g(\rho_L - \rho_v)\sigma}{(\rho_L + \rho_v)^2}\right]^{1/4} \tag{2-3}$$

Heat Transfer Correlations

Natural convection. The natural convection curve in Fig. 2-1 can be adequately predicted by using the well-known correlation of Churchill and Chu (1975) for single, isolated, horizontal cylinders with an isothermal wall condition:

$$\text{Nu}_D = 0.36 + 0.518\text{Ra}_D^{1/4}\left[1 + \left(\frac{0.559}{\text{Pr}_L}\right)^{9/16}\right]^{-4/9} \tag{2-4}$$

where the Rayleigh number based on the tube diameter is defined as

$$\text{Ra}_D \equiv \frac{g\beta_L\Delta T D_o^3}{\nu_L a_L} \tag{2-5}$$

and the average Nusselt number over the surface of the cylinder is defined as

$$\text{Nu}_D \equiv \frac{\alpha_{nc}D_o}{\lambda_L} \tag{2-6}$$

The recommended range of application of Eq. (2-4) is

$$10^{-4} \leq \mathrm{Ra}_D \leq 10^9 \tag{2-7}$$

The fluid properties should be evaluated at the mean temperature between the heating wall and the liquid.

Nucleate pool boiling. For the nucleate pool boiling regime, the Mostinski (1963) correlation is simple to apply and gives reasonable results for a wide range of pure fluids and over a wide range of reduced pressures. The correlating expression is

$$\alpha_{\mathrm{nb}} = 0.00417 q^{0.7} P_c^{0.69} F_P \tag{2-8}$$

where α_{nb} is the nucleate pool boiling heat transfer coefficient, q is the heat flux, P_{c} is the critical pressure of the fluid, and F_P is the pressure correction factor, which is calculated as

$$F_P = 1.8 P_{\mathrm{r}}^{0.17} + 4 P_{\mathrm{r}}^{1.2} + 10 P_{\mathrm{r}}^{10} \tag{2-9}$$

where P_{r} is the reduced pressure. For nucleate pool boiling, the convention is to evaluate the fluid properties at the saturation temperature.

More recently, Stephan and Abdelsalam (1980) developed accurate nucleate pool boiling correlations for several classes of pure fluids. These correlations are based on the physical properties of the fluid rather than on the reduced pressure. Their correlation for organic fluids is given as

$$\frac{\alpha_{\mathrm{nb}} d}{\lambda_{\mathrm{L}}} = 0.0546 \left[\left(\frac{\rho_{\mathrm{v}}}{\rho_{\mathrm{L}}} \right)^{1/2} \left(\frac{q d}{\lambda_{\mathrm{L}} T_{\mathrm{sat}}} \right) \right]^{0.67} \left(\frac{\Delta h_{\mathrm{v}} d^2}{a_{\mathrm{L}}^2} \right)^{0.248} \left(\frac{\rho_{\mathrm{L}} - \rho_{\mathrm{v}}}{\rho_{\mathrm{L}}} \right)^{-4.33} \tag{2-10}$$

where the bubble departure diameter d is determined by the following expression:

$$d = 0.0146 \beta \left[\frac{2\sigma}{g(\rho_{\mathrm{L}} + \rho_{\mathrm{v}})} \right]^{1/2} \tag{2-11}$$

These investigators utilized a fixed value of 35° for the contact angle of all fluids irrespective of their actual values. This method has been verified independently by Sardesai, Palen, and Thome (1986) for numerous hydrocarbons and by Thome and Shakir (1987) and Bajorek (1988) for various alcohols and solvents, all for boiling on the outside of a single horizontal copper tube.

Figure 2-6 shows a comparison of the Mostinski (1963) and the Stephan and Abdelsalam (1980) correlations with data obtained by Uhlig and Thome (1985) for acetone boiling on a horizontal copper tube (19.05 mm in diameter). In this instance the Stephan and Abdelsalam correlation gives a good overall fit to the data, and the Mostinski correlation gives reasonable accuracy at large heat fluxes.

Surface roughness can have a significant effect on nucleate pool boiling, especially at low reduced pressures. The boiling curve tends to move to the right as the surface

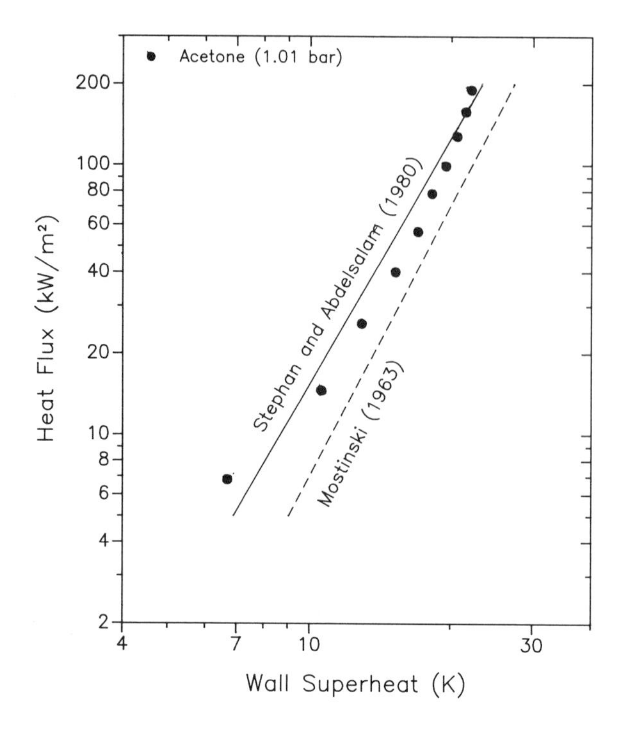

Figure 2-6 Comparison of correlating expressions for nucleate pool boiling of acetone.

becomes smoother. Neither of the above correlations takes roughness of the surface into account. They do model the boiling performance of drawn tubing quite well, however. The relationship between the surface roughness and the boiling heat transfer coefficient is an area requiring further development because previous attempts to correlate the two are not acceptable for general application (surface roughness as a means of boiling augmentation is discussed in more detail in Chapter Three, Sec. 3-1).

Subcooled nucleate pool boiling. When the bulk liquid is not at its saturation temperature but at a lower temperature, subcooled boiling may exist on the heated wall if the wall temperature is sufficiently higher than the saturation temperature to sustain boiling. Figure 2-7 [from Hui and Thome (1985)] depicts the effect of liquid subcooling on the nucleate pool boiling curve of benzene boiling on a vertical brass disk. The boiling curve moves to the right as subcooling increases. Because the subcooled heat transfer coefficient is defined as

$$\alpha_{sub} \equiv \frac{q}{T_w - T_b} \tag{2-12}$$

it decreases as the bulk liquid temperature T_b decreases. Some texts erroneously define the subcooled heat transfer coefficient with the expression for saturated boiling, that is with T_{sat} in place of T_b in Eq. (2-12). This definition is unrealistic, however; as the subcooling increases T_w will eventually decrease to T_{sat}, at which point the heat transfer coefficient will incorrectly become infinitely large. Figure 2-7 demonstrates that even

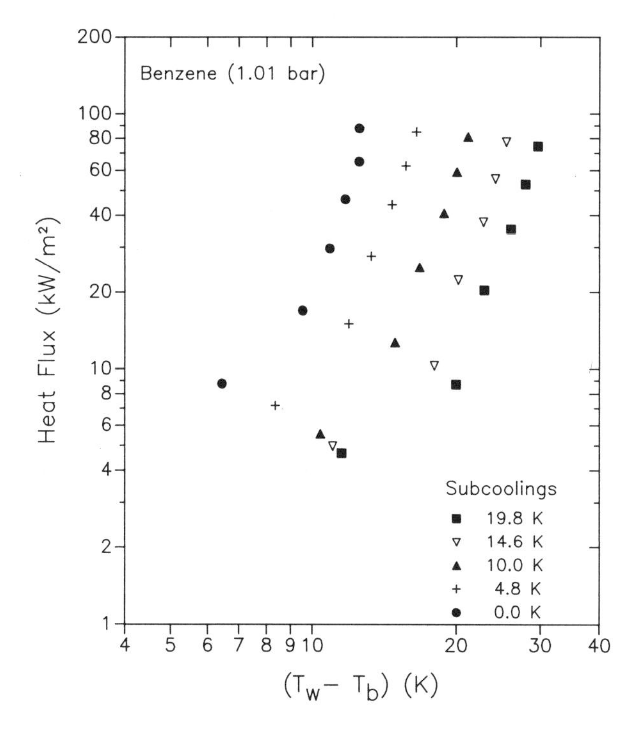

Figure 2-7 Subcooled boiling of benzene on a vertical heated disk (25.4 mm in diameter) at 1.01 bar.

small levels of subcooling can substantially affect the heat transfer process and significantly reduce the boiling heat transfer coefficient.

Film boiling. At large temperature differences boiling occurs in the film boiling regime, which has many similarities to film condensation. Bromley (1950), in fact, utilized the well-known Nusselt equation for film condensation on the outside of a horizontal cylinder to predict film boiling on horizontal tubes. He replaced the thermal conductivity and kinematic viscosity of the liquid with those of the vapor and changed Nusselt's lead constant from 0.725 to 0.62 to correlate film boiling data. The Bromley correlation for film boiling on horizontal tubes is

$$\mathrm{Nu}_D = 0.62\left[\frac{(\rho_\mathrm{L}-\rho_\mathrm{v})\,g\Delta h_\mathrm{v}' D_\mathrm{o}^3}{\nu_\mathrm{v}\lambda_\mathrm{v}(T_\mathrm{w}-T_\mathrm{sat})}\right]^{1/4} \tag{2-13}$$

where the average Nusselt number for the tube surface is defined as

$$\mathrm{Nu}_D \equiv \frac{\alpha_\mathrm{fb} D_\mathrm{o}}{\lambda_\mathrm{v}} \tag{2-14}$$

The corrected latent heat of vaporization (as for film condensation) is approximated as

$$\Delta h_v' = \Delta h_v \left[1 + 0.34 \left(\frac{c_{Pv} \Delta T}{\Delta h_v} \right) \right] \tag{2-15}$$

The nondimensional term in Eq. (2-15) is often called the Jacob number; it represents the ratio of the sensible heat capacity of the vapor to the latent heat of vaporization. The liquid properties are evaluated at the saturation temperature, whereas the vapor properties are evaluated at an intermediate vapor film temperature of

$$T_f = T_{sat} + \frac{\Delta T}{2} \tag{2-16}$$

Because of the large wall superheats and the relatively low heat transfer coefficients, thermal radiation from the heated wall can become an important component of the heat transfer process. The radiation contribution to heat transfer reduces the contribution of convection through the vapor film, so that the radiant heat transfer coefficient cannot simply be added to that of convection. Bromley (1950) proposed the following approximation for combining the contributions of convection and radiation to calculate the heat transfer coefficient for film boiling at large temperature differences:

$$\alpha = \alpha_{fb} + \frac{3}{4} \alpha_{rad} \tag{2-17}$$

where α_{fb} refers to the film boiling heat transfer coefficient [from Eq. (2-13)] and α_{rad} refers to the radiative heat transfer coefficient calculated from a suitable radiation expression.

2-3 BOILING ON TUBE BUNDLES

Boiling on the outside of horizontal tube bundles is a common industrial practice. In spite of their extensive use as evaporators in the refrigeration and petrochemical processing industries, there is still a large amount of uncertainty about the basic physical phenomena controlling the process, which is reflected in the marginal accuracy of the correlations used for the thermal design of these units. Heat transfer coefficients for a tube bundle are usually larger than those for nucleate pool boiling on a single tube under the same conditions. The increase in the bundle boiling heat transfer coefficient is primarily the result of the additional contribution of turbulent forced convection of the liquid-phase over the tubebank. At large heat fluxes, however, a tube bundle tends to perform worse than a single tube because the upper rows of tubes become inundated with vapor.

Figure 2-8 depicts the heat transfer coefficient contours in a tube bundle obtained experimentally by Cornwell, Duffin, and Schuller (1980) for refrigerant-113 boiling at 1.01 bar on a 241-tube bundle with ¾-inch (19-mm) diameter tubes on a 1.0-inch (25.4-mm) square tube pitch. Each tube was 1.0 inches (25.4 mm) long and electrically heated and represented a "slice" of a normal evaporator's tube bundle. These contours illustrate a large variation in the boiling heat transfer coefficient with position in the

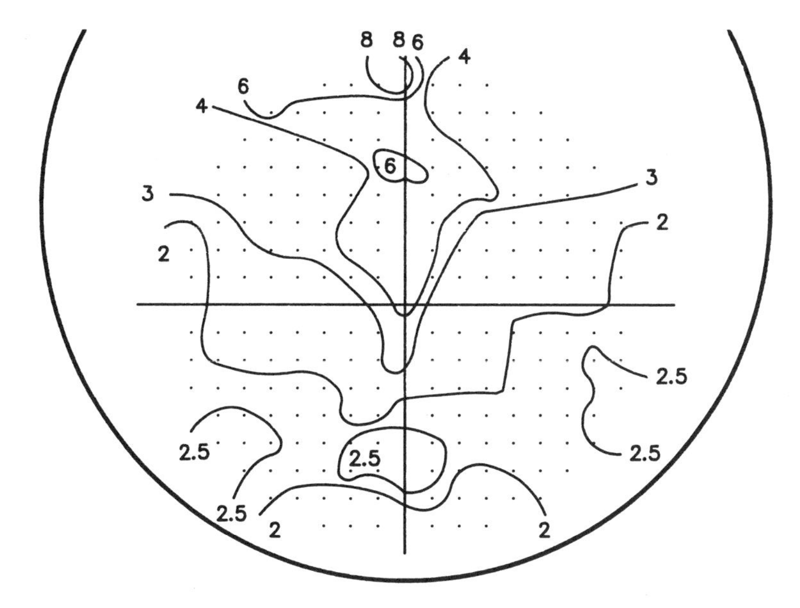

Figure 2-8 Iso–heat transfer coefficient [kW/(m^2·K)] contours in a kettle reboiler evaporating refrigerant-113 at 20 kW/m^2.

tube bundle (although the variation was typically less than shown here). The bottom values are similar to those obtained for nucleate pool boiling on single horizontal tubes.

The increase in the boiling heat transfer coefficient from bottom to top in the bundle can be qualitatively explained by two factors:

1. The volumetric expansion of the fluid during its evaporation greatly increases the liquid flow rate past the tubes with increasing height in the bundle and thus augments the contribution of liquid convection to the process.
2. The large void fraction of vapor in the flow creates a thin liquid film on the outside of many of the tubes, which produces high heat transfer coefficients analogous to those for film condensation.

Bundle Boiling Correlation

An "ideal" thermal design method for boiling on horizontal tube bundles would incorporate the spatial variation in the boiling heat transfer coefficient in the bundle. Brisbane, Grant, and Whalley (1980), for example, made some progress toward this goal by applying intube boiling concepts (discussed in Sec. 2-4) to circulation through the bundle. The conventional approach, however, is to use a superposition model for calculating the average boiling heat transfer coefficient for a tube bundle. Palen (1983), for instance, recommends the following simple expression, which algebraically sums the contributions of boiling and convection:

$$\alpha_{\mathrm{b}} = \alpha_{\mathrm{nb}} F_{\mathrm{b}} F_{\mathrm{c}} + \alpha_{\mathrm{nc}} \tag{2-18}$$

where the nucleate pool boiling heat transfer coefficient is calculated from the Mostinski (1963) correlation given in Eqs. (2-8) and (2-9), F_b is the bundle boiling factor to account for the two-phase flow augmentation, F_c is the mixture boiling correction factor, and α_{nc} is the natural convection coefficient for the tube bundle for liquid flow only. The bundle boiling factor is reported to vary typically from 1 to 3 depending primarily on the heat flux, bundle configuration, and bundle diameter. Palen (1983) recommends using 1.5 as a conservative estimate when no specific information is available.

The mixture boiling correction factor in Eq. (2-18) is equal to 1.0 when boiling pure fluids but may drop to as low as 0.1 for evaporating mixtures with wide boiling ranges (a method for predicting it is given in Chapter Nine's section on nucleate pool boiling on smooth surfaces). The natural convection heat transfer coefficient is determined by using a suitable correlation for the tube bundle. Otherwise, Palen (1983) recommends using a conservative value of 250 W/(m^2·K).

Maximum Bundle Heat Flux

The maximum heat flux for a tube bundle is less than the peak nucleate heat flux for a single tube in pool boiling. This is primarily the result of two different phenomena:

1. At high heat fluxes and large bundle diameters, so much vapor can be generated that the upper tubes in the bundle become blanketed with rising vapor, effectively forming an insulation layer that renders these tubes thermally ineffective.
2. The liquid circulation through the bundle may not be sufficient as the result of poor heat exchanger layout or geometric factors (such as a small tube pitch) and thus may prevent liquid from reaching and wetting the upper tube rows.

Therefore, the maximum heat flux for a bundle is an important parameter that should be checked during thermal design.

Palen and Small (1964) studied this problem and found that the maximum heat flux is a strong function of bundle geometry. They developed the following relationship between the peak nucleate heat flux [Eq. (2-2) with Kutateladze's constant] and the maximum heat flux for horizontal tube bundles:

$$q_{b,\,max} = q_{max}\,\phi_b \tag{2-19}$$

where the bundle correction factor is

$$\phi_b = 2.2\left(\frac{\pi D_b L}{A}\right) \tag{2-20}$$

Here D_b is the outside diameter of the tube bundle, L is the length of the tube bundle, and A is the total heat transfer surface area of the bundle. The ratio of these three parameters is equivalent to the external envelope surface area of the tube bundle divided by its heat transfer surface area. This ratio increases to 1.0 for a single tube; thus Eqs.

(2-19) and (2-20) do not tend toward the single tube peak nucleate heat flux value for small tube bundles but toward a value 2.2 times larger. Therefore, the maximum recommended design value for the bundle correction factor is 1.0; bundles should not be designed with larger values.

Equation (2-20) can be simplified to the following convenient expression:

$$\phi_b = \frac{KL}{D_o(D_b/L)^{1.1}} \tag{2-21}$$

where $K = 4.12$ for square tube layout
$K = 3.56$ for triangular pitch layout

At large operational heat fluxes a square tube layout is therefore preferable. The minimum value for the validity of this bundle correction factor is 0.1. If the calculated value is less than this, then vapor release channels should be examined to facilitate vapor flow out of the bundle. This method has been compared against the operation of many different reboilers in service in the petrochemical processing industry by Palen, Yarden, and Taborek (1972) and is reported to give reliable, although conservative, values.

Designing horizontal evaporators with the use of the maximum bundle heat flux values only can lead to an unsafe design. The slope of the bundle's boiling curve is sometimes steep, and an increase in the available wall superheat at initially clean conditions or other such changes can take the process into the film boiling regime. Thus it is wise at large heat fluxes to calculate the wall superheat corresponding to the bundle's maximum heat flux and to compare this value to the design condition. If the process is in film boiling, the operating heat duty will be much less than expected. The wall superheat at the maximum heat flux is readily obtained by solving Eqs. (2-18), (2-19), and (2-20) or (2-21) together with the definition of the heat transfer coefficient:

$$\alpha \equiv \frac{q}{T_w - T_{sat}} \tag{2-22}$$

2-4 BOILING INSIDE TUBES

The subject of intube boiling is normally divided into two separate geometries: vertical tubes and horizontal tubes. The boiling characteristics of both these orientations are significantly different from one another, so that different design correlations are used for each of them. Boiling inside vertical tubes is utilized in thermosyphon reboilers and various types of steam generators, and boiling in horizontal tubes is common to refrigeration system evaporators and the convection sections of steam generators.

The two most important types of heat transfer regimes in flow boiling are nucleate boiling and convective boiling. In the former, the nucleate boiling contribution is dominant and is characterized by vapor bubbles growing and departing from the heated tube wall. In convective boiling, heat is transferred by convection across a liquid film coating the tube wall and is characterized by evaporation at the vapor-liquid interface of the film.

Boiling Inside Vertical Tubes

Figure 2-9 [from Collier (1981)] depicts the flow patterns, heat transfer regions, and axial temperature variations for boiling in upward flow inside an electrically heated tube. Because of the hydrostatic head acting on the liquid feed at the bottom of the tube, the liquid enters subcooled. Therefore, it must first be heated by means of single-phase convection to the saturation temperature. The subcooled boiling regime may be activated before the liquid reaches this point ($x = 0$) if the wall temperature has first risen to the nucleation superheat at some point along the tube surface. The bubbles formed in the subcooled boiling regime condense as they enter the colder liquid core. When the bulk liquid reaches the saturation temperature corresponding to the local pressure at that point, saturated nucleate boiling begins. This regime is initially characterized by isolated bubbles in the flow. As these bubbles coalesce, bullet-shaped slugs of vapor are formed that occupy nearly the entire cross-section of the flow channel. Farther up the tube these slugs grow together to form a continuous vapor core in the center of the flow channel, with a climbing liquid film covering the tube wall. This regime is called annular flow. The two-phase velocity increases rapidly with vapor formation and is large compared to the velocity of the liquid at the entrance of the tube. Thus the convective heat transfer to the liquid is greatly increased and usually becomes the dominant heat transfer mode relative to the nucleate boiling heat transfer process. In fact, the formation of the slugs of vapor and the vapor core lead to a complete suppression of bubble growth at the tube wall.

Higher up the tube, the shear stress on the free interface of the slower-flowing liquid film causes waves to form, which can result in the entrainment of liquid droplets in the vapor core. Eventually, the liquid film is completely evaporated. This point is called the dryout point and is usually avoided in the design of heat transfer equipment because of fouling problems and the ensuing poor heat transfer rates in the mist flow regime that follows. If the tube were long, eventually the entrained liquid droplets

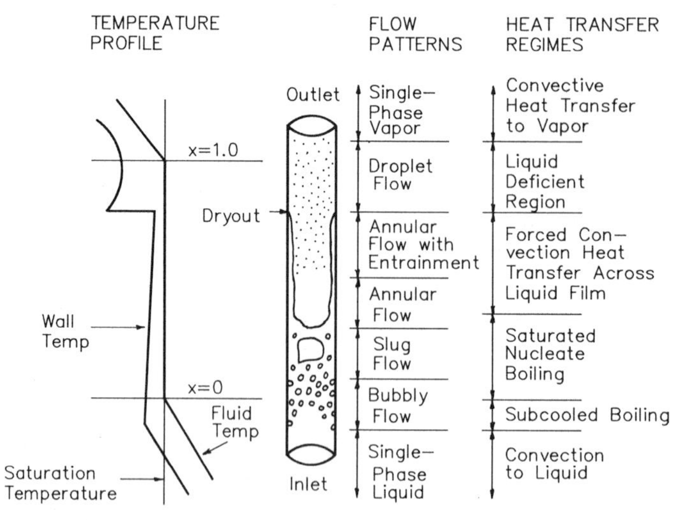

Figure 2-9 Boiling in upward flow inside a uniformly heated vertical tube.

would all be evaporated and only vapor would remain. At this point, the flow would finally become fully developed thermally and hydrodynamically.

Onset of Nucleate Boiling

The wall superheat required to initiate boiling in flow boiling is not the same as that for nucleate pool boiling because the thermal boundary layers created by the respective processes are quite different. To distinguish between the two phenomena, the initiation of boiling inside a tube is called the onset of nucleate boiling. Prediction of the point at which the onset of nucleate boiling occurs is important for an accurate thermal design because it divides the region of less effective single-phase convection from the much more effective regions of subcooled and saturated nucleate boiling. Therefore, if the onset of nucleate boiling occurs at a higher wall superheat than predicted, a larger percentage of the tube will be in the less effective single-phase convection regime, which can significantly reduce a heat exchanger's actual heat duty compared to its design value.

An expression for the superheat required for the onset of nucleate boiling can be obtained by utilizing the forced-convection temperature gradient rather than the uniformly superheated liquid condition assumed in deriving Eq. (2-1). The most widely quoted expression for the wall superheat required for the onset of nucleate boiling is that of Davis and Anderson (1966):

$$\Delta T_{\mathrm{ONB}} = \left(\frac{8\sigma T_{\mathrm{sat}}\, q}{\lambda_{\mathrm{L}}\, \rho_{\mathrm{v}}\, \Delta h_{\mathrm{v}}} \right)^{1/2} \tag{2-23}$$

where the physical properties are evaluated at the saturation conditions corresponding to the local pressure.

Heat Transfer Correlations for Boiling Inside Vertical Tubes

Single-phase forced convection. A correlation for the subcooled liquid flowing in the entrance of the tube is required to predict the heat transfer rates before the initiation of boiling. If the flow is laminar, care has to be used to take into account the thermal and hydrodynamic entrance length effects and the type of boundary condition at the wall. For example, with steam heating the wall is essentially isothermal, whereas for a single-phase heating medium the boundary condition may be closer to a uniform heat flux. The reader is referred to an appropriate text for these correlations.

For turbulent flow inside tubes, the entrance lengths required to reach fully developed flow are relatively short. Thus the correlation of Dittus and Boelter (1930) can be used:

$$\mathrm{Nu}_D = 0.023 \mathrm{Re}_D^{0.8} \mathrm{Pr}_L^{0.4} \tag{2-24}$$

where the fluid properties are evaluated at the bulk liquid temperature. This correlation is also applicable to the single-phase flows at the entrance of horizontal tubes (discussed below under boiling inside horizontal tubes).

Saturated boiling regime. The saturated boiling regime is modeled by superimposing the contribution of liquid-phase convection on that of nucleate boiling heat transfer. Many different expressions have been proposed to describe the interaction between these two processes. One widely used method is that of Chen (1966). This correlation was developed from data for organic liquids boiling at near atmospheric pressures and for water boiling at pressures in the range of 0.5 to 35 bar. The correlation predicts the local heat transfer coefficient along the tube. The general equation of the model for evaporating pure fluids is

$$\alpha = \alpha_L F + \alpha_{nb} S \tag{2-25}$$

where α_L is the single-phase convective heat transfer coefficient for the liquid flowing alone in the tube, α_{nb} is the nucleate pool boiling heat transfer coefficient for the specified wall superheat, F is the two-phase multiplier to correct for the augmentation in the liquid convection, and S is the nucleate boiling suppression factor, which accounts for the relative level of suppression of nucleate boiling caused by the convection process. The steps in the calculation of the local heat transfer coefficient α are as follows:

1. Calculate the liquid-only Reynolds number for liquid flowing alone in the tube:

$$\mathrm{Re}_L = \frac{\dot{m}_L D_i}{A \mu_L} \tag{2-26}$$

where the liquid mass flow rate is obtained from the local liquid quality $(1 - x)$ and the total mass flow rate of the two-phase fluid. The liquid viscosity is evaluated at the local saturation temperature.

2. Determine the liquid Prandtl number:

$$\mathrm{Pr}_L = \frac{c_{PL}\, \mu_L}{\lambda_L} \tag{2-27}$$

with all properties evaluated at the local saturation temperature.

3. Calculate the liquid-only convective heat transfer coefficient α_L using Eq. (2-24).
4. Determine the reciprocal of the Lockhart-Martinelli parameter:

$$\frac{1}{X_{tt}} = \left(\frac{\dot{m}_v}{\dot{m}_L}\right)^{0.9} \left(\frac{\rho_L}{\rho_v}\right)^{0.5} \left(\frac{\mu_v}{\mu_L}\right)^{0.1} \tag{2-28}$$

where the vapor mass flow rate is obtained from the total two-phase mass flow rate multiplied by the local vapor quality. The properties are evaluated at the saturation temperature.

5. Calculate the two-phase multiplier, which represents the ratio of the two-phase Reynolds number to the liquid-only Reynolds number, from one of the following two expressions:

$$F = 1.0 \quad \text{for} \frac{1}{X_{tt}} \leq 0.1 \tag{2-29}$$

$$F = 2.35\left(\frac{1}{X_{tt}} + 0.213\right)^{0.736} \quad \text{for } \frac{1}{X_{tt}} > 0.1 \tag{2-30}$$

6. Calculate the two-phase Reynolds number:

$$\mathrm{Re}_{TP} = \mathrm{Re}_L F^{1.25} \tag{2-31}$$

7. Determine the nucleate boiling suppression factor:

$$S = \frac{1}{1 + 2.53 \times 10^{-6}\mathrm{Re}_{TP}^{1.17}} \tag{2-32}$$

8. Calculate the nucleate pool boiling heat transfer coefficient with the correlation of Forster and Zuber (1955);

$$\alpha_{nb} = 0.00122\left(\frac{\lambda_L^{0.79} c_{pL}^{0.45} \rho_L^{0.49}}{\sigma^{0.5}\mu_L^{0.29}\Delta h_v^{0.24}\rho_v^{0.24}}\right)(T_w - T_{sat})^{0.24}(P_w - P_{sat})^{0.75} \tag{2-33}$$

   where P_w is the saturation pressure evaluated at the wall temperature T_w and P_{sat} is the local saturation pressure corresponding to T_{sat} (the pressures in this expression are in pascals). The fluid properties are determined at the local saturation temperature.
9. The local heat transfer coefficient can now be calculated by substitution of the respective values into Eq. (2-25).

The Chen (1966) model was developed from a data base with vapor qualities ranging from 0 to 0.7 and mass fluxes from 54 to 4070 kg/(m^2·s). For refrigerants such as R-11, R-12, and R-22, the Shah (1976) correlation will give better results. The reader is referred to Butterworth and Shock (1982) for an in-depth review of flow boiling heat transfer.

Maximum Heat Flux for Intube Boiling

The maximum heat flux for saturated boiling in vertical tubes operating as thermosyphons is limited by three different flow phenomena: film boiling, mist flow, and flow instabilities. The first refers to the departure from nucleate boiling to film boiling, the second to the dryout of the liquid film in annular flow, and the last to the periodic surging that can occur in thermosyphons. Palen et al. (1974) developed a simple, reduced property correlation based on data for a wide range of fluids for conditions representative of the above phenomena. It is given as

$$q_{t,max} = 23{,}660\left(\frac{D_i^2}{L}\right)P_c^{0.61}\left(\frac{P}{P_c}\right)^{0.25}\left(1 - \frac{P}{P_c}\right) \tag{2-34}$$

Figure 2-10 depicts a comparison of this expression with measured maximum heat

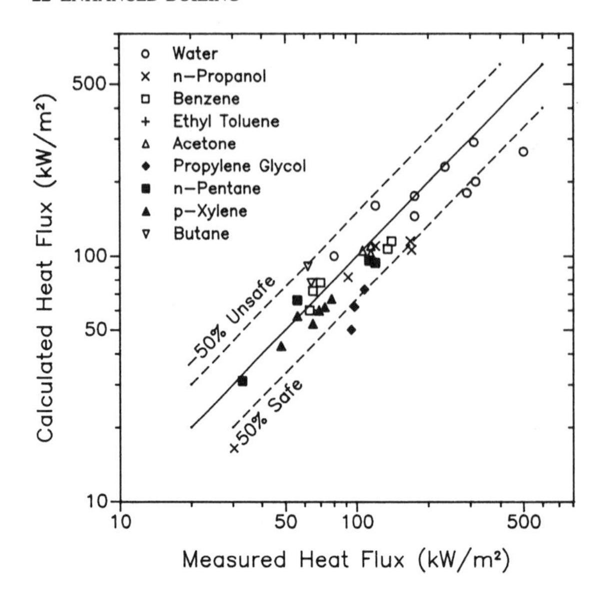

Figure 2-10 Comparison of maximum heat flux correlation with intube boiling data.

flux data. It should be pointed out that the correlation was developed from data for several different tube diameters and lengths. The outside diameters of the tubes studied ranged from 19.1 to 31.8 mm, and the tube lengths varied from 1.52 to 3.66 m.

Boiling Inside Horizontal Tubes

Boiling inside horizontal tubes differs from boiling inside vertical tubes because buoyancy and gravity act normal to the direction of flow. Therefore, the vapor formed tends to separate from the liquid phase and migrates toward the top of a horizontal tube. Figure 2-11 illustrates the various flow regimes that can occur in flow boiling in horizontal tubes. The bubbly flow regime is analogous to the nucleate boiling regime in vertical flow except that the individual bubbles rise toward the top of the flow channel. In slug and plug flows, the top of the tube is intermittently wetted by the

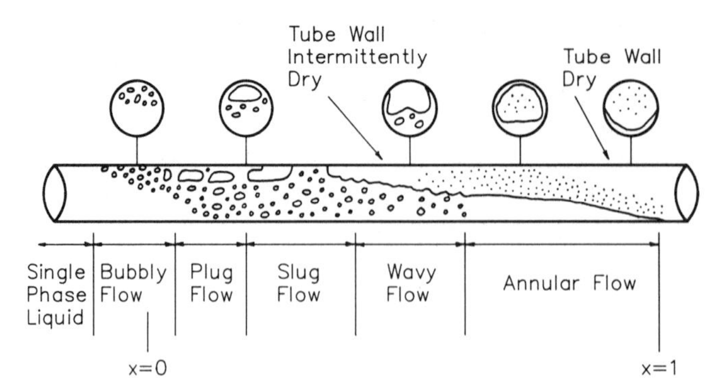

Figure 2-11 Flow regimes during boiling in horizontal tubes (flow is from left to right).

passing waves of liquid. Annular flow is similar to vertical flow except that the liquid film is thicker at the bottom of the channel than at the top. Annular flow with mist refers to the entrainment of liquid droplets into the vapor core under otherwise annular flow conditions. Stratified flow tends to occur at low liquid mass flow rates such that the vapor separates completely from the liquid and the channel wall is only partially wetted by the liquid. Wavy flow is similar to stratified flow but is distinguished by the waves created by the shear stresses acting on the liquid-vapor interface from the high-velocity vapor flow.

There are many different correlations available for predicting local heat transfer coefficients for boiling inside horizontal tubes. The Gunger and Winterton (1986) correlation is presented here. It was developed from a large data base (3,693 points) that covered both nucleate- and convective-dominated boiling regimes. The basic expression of their correlation is

$$\alpha = E\alpha_L + S\alpha_{nb} \tag{2-35}$$

where the liquid-only heat transfer coefficient is defined by the Dittus-Boelter correlation [Eq. (2-24)] and the new convection enhancement factor E is

$$E = 1 + 24{,}000\mathrm{Bo}^{1.16} + 1.37\left(\frac{1}{X_{tt}}\right)^{0.86} \tag{2-36}$$

The liquid Reynolds number is calculated from Eq. (2-26), and the Lockhart-Martinelli parameter is obtained from Eq. (2-28). The boiling number Bo is defined as

$$\mathrm{Bo} \equiv \frac{q}{G\Delta h_v} \tag{2-37}$$

The nucleate pool boiling heat transfer coefficient is calculated with the new expression of Cooper (1984):

$$\alpha_{nb} = 55P_r^{0.12}(-\log P_r)^{-0.55}M^{-0.5}q^{0.67} \tag{2-38}$$

where M is the molecular weight. A new boiling suppression factor S was given by Gunger and Winterton as

$$S = \left[1 + 0.00000115E^2\mathrm{Re}_L^{1.17}\right]^{-1} \tag{2-39}$$

At Froude numbers less than 0.05, several additional corrections to E and S were suggested.

The average heat transfer coefficient over the boiling region can be calculated by integrating the local heat transfer coefficient in Eq. (2-35) with respect to the vapor quality:

$$\overline{\alpha} = \frac{1}{x_{out} - x_{in}} \int_{x_{in}}^{x_{out}} \alpha\, dx \tag{2-40}$$

For quick approximations, the average heat transfer coefficient can be obtained with the Pierre (1964) correlation:

$$\overline{\alpha} = C\left(\frac{\lambda_L}{D_i}\right)\left[\mathrm{Re}_L^2 \frac{(x_{out} - x_{in})\Delta h_v}{L}\right]^n \tag{2-41}$$

where $C = 0.0009$ and $n = 0.5$ for exit vapor qualities less than 0.9 and $C = 0.0082$ and $n = 0.4$ for exit qualifies greater than 0.9 or vapor superheatings up to 6 K, according to Agrawal, Varma, and Lal (1986).

For further information about flow boiling inside horizontal tubes, the reader is referred to the review by Butterworth and Robertson (1977).

2-5 NOMENCLATURE

a_L	liquid thermal diftusivity (m^2/s)
A	heat transfer surface area (m^2
Bo	boiling number
c_{PL}	liquid specific heat [J/(kg·K)]
c_{PV}	vapor specific heat [J/(Kg·K)]
C	constant in Eq. (2-41)
d	bubble departure diameter (m)
D_b	tube bundle diameter (m)
D_i	tube inside diameter (m)
D_o	tube outside diameter (m)
E	convection enhancement factor
F	two-phase flow multiplier
F_b	bundle boiling correction factor
F_c	mixture boiling correction factor
F_P	pressure correction factor
g	gravitational acceleration (m/s^2)
G	total mass flux [kg/(m^2·s)]
Δh_v	latent heat of vaporization (J/kg)
$\Delta h_v'$	modified latent heat of vaporization (J/kg)
K	constant in Eq. (2-21)
L	tube length (m)
$\dot{m}_L$	local liquid mass flow rate (kg/s)
$\dot{m}_v$	local vapor mass flow rate (kg/s)
M	molecular weight
n	exponent in Eq. (2-41)
Nu_D	Nusselt number
P_c	critical pressure (N/m^2)
P_r	reduced pressure

P_{sat} saturation pressure (N/m^2)
P_w saturation pressure corresponding to wall temperature (N/m^2)
Pr_L liquid Prandtl number
$(dP/dT)_{sat}$ slope of vapor pressure curve [$N/(m^2 \cdot K)$]
q heat flux (W/m^2)
q_{max} peak nucleate heat flux (W/m^2)
$q_{b,max}$ maximum bundle heat flux (W/m^2)
$q_{t,max}$ maximum vertical intube heat flux (W/m^2)
q_{min} minimum bundle heat flux (W/m^2)
R radius of vapor nucleus (m)
Ra_D Rayleigh number
Re_D Reynolds number
Re_L liquid-only Reynolds number
Re_{TP} two-phase Reynolds number
S boiling suppression factor
T_b bulk liquid temperature (K)
T_f film temperature (K)
T_{sat} saturation temperature (K)
T_w wall temperature (K)
ΔT wall superheat (K)
ΔT_{ONB} superheat at onset of nucleate boiling (K)
ΔT_{sat} boiling nucleation superheat (K)
x vapor quality
x_{in} inlet vapor quality
x_{out} outlet vapor quality
X_{tt} Lockhart-Martinelli parameter
α heat transfer coefficient [$W/(m^2 \cdot K)$]
$\overline{\alpha}$ average convective boiling heat transfer coefficient [$W/(m^2 \cdot K)$]
α_b bundle boiling heat transfer coefficient [$W/(m^2 \cdot K)$]
α_{fb} film boiling heat transfer coefficient [$W/(m^2 \cdot K)$]
α_L liquid-only convective heat transfer coefficient [$W/(m^2 \cdot K)$]
α_{nb} nucleate boiling heat transfer coefficient [$W/(m^2 \cdot K)$]
α_{nc} natural convection heat transfer coefficient [$W/(m^2 \cdot K)$]
α_{rad} radiative heat transfer coefficient [$W/(m^2 \cdot K)$]
α_{sub} subcooled nucleate boiling heat transfer coefficient [$W/(m^2 \cdot K)$]
β contact angle
β_L coefficient of thermal expansion of liquid (K^{-1})
λ_L liquid thermal conductivity [$W/(m \cdot K)$]
λ_v vapor thermal conductivity [$W/(m \cdot K)$]
μ_L liquid dynamic viscosity ($N \cdot s/m^2$)
μ_v vapor dynamic viscosity ($N \cdot s/m^2$)
ν_L liquid kinematic viscosity (m^2/s)
ν_v vapor kinematic viscosity (m^2/s)
ρ_L liquid density (kg/m^3)
ρ_v vapor density (kg/m^3)

σ surface tension (N/m)
ϕ_b bundle geometric factor

REFERENCES

Agrawal, K. N., H. K. Varma, and S. Lal. 1986. Heat transfer during forced convection boiling of R-12 under swirl flow. *J. Heat Transfer* 108:567–73.

Bajorek, S. M. 1988. An experimental and theoretical investigation of multicomponent pool boiling on smooth and finned surfaces. Ph.D. diss., Michigan State University, East Lansing.

Berenson, P. 1960. *Transition boiling heat transfer from a horizontal surface* (M. I. T. Heat Transfer Laboratory report 17).

Brisbane, T. W. C., I. D. R. Grant, and P. B. Whalley. 1980. *A prediction method for kettle reboiler performance* (ASME paper 80-HT-42).

Bromley, A. L. 1950. Heat transfer in stable film boiling. *Chem. Eng. Progr.* 46:221–27.

Butterworth, D., and J. M. Robertson. 1977. Boiling and flow in horizontal tubes. In *Two-phase flow and heat transfer*, ed. D. Butterworth and G. F. Hewitt. London: Oxford Univ. Press, chap. 11.

Butterworth, D., and R. A. W. Shock. 1982. Flow boiling. *Proc. 7th Int. Heat Transfer Conf.* 1:11–30.

Chen, J. C. 1966. A correlation for boiling heat transfer to saturated fluids in convective flow. *Ind. Eng. Chem. Process Des. Dev.* 5(3):322–29.

Churchill, S. W., and H. H. S. Chu. 1975. Correlating equations for laminar and turbulent free convection from a horizontal cylinder. *Int. J. Heat Mass Transfer* 18:1049–53.

Collier, J. G. 1981. *Convective boiling and condensation*, 2d ed. New York: McGraw-Hill.

Cooper, M. G. 1984. Heat flow rates in saturated nucleate pool boiling—A wide-ranging examination using reduced properties. *Adv. Heat Transfer* 16:157–239.

Cornwell, K., N. W. Duffin, and R. B. Schuller. 1980. *An experimental study of the effects of fluid flow on boiling within a kettle reboiler tube bundle* (ASME paper 80-HT-45).

Davis, E. J., and G. H. Anderson. 1966. The incipience of nucleate boiling in forced convection flow. *AIChE J.* 12:774–80.

Dittus, F. W., and L. M. K. Boelter. 1930. *Univ. Calif. (Berkeley) Publ. Eng.* 2:443.

Forster, H. K., and N. Zuber. 1955. Dynamics of vapor bubble growth and boiling heat transfer. *AIChE J.* 1:531–35.

Gunger, K. E., and R. H. S. Winterton. 1986. A general correlaton for flow boiling in tubes and annuli. *Int. J. Heat Mass Transfer* 19:351–58.

Hui, T. O., and J. R. Thome. 1985. A study of binary mixture boiling: Boiling site density and subcooled heat transfer. *Int. J. Heat Mass Transfer* 28:919–28.

Kutateladze, S. S. 1948. On the transition to film boiling under natural convection. *Kotloturbostroenie* 3:10.

Mostinski, I. L. 1963. Applicaton of the rule of corresponding states for the calculation of heat transfer and critical heat flux. *Teploenergetika* 4:66; English abstract, *Br. Chem. Eng.* 8(8):580 (1963).

Palen, J. W. 1983. Shell-and-tube reboilers. *Heat Exch. Des. Handb.* 3:3.6.1-1–3.6.5-6.

———. C. C. Shih, A. Yarden, and J. Taborek. 1974. Performance limitations in a large-scale thermosyphon reboiler. *Proc. 5th Int. Heat Transfer Conf.* 5:204–208.

Palen, J. W., and W. M. Small. 1964. A new way to design kettle and internal reboilers. *Hydrocarbon Process.* 43(11):199–208.

Palen, J. W., A. Yarden, and J. Taborek. 1972. Characteristics of boiling outside large-scale multitube bundles. *Chem. Eng. Progr. Symp. Ser.* 68(118):50–61.

Pierre, B. 1964. Flow resistance with boiling refrigerants. *ASHRAE J.* 6:58–65, 73–77.

Rohsenow, W. M. 1973. Boiling. In *Handbook of heat transfer*, ed. W. M. Rohsenow and J. P. Hartnett. New York: McGraw-Hill, chap. 13.

Sardesai, R. G., J. W. Palen, and J. R. Thome. 1986. Nucleate pool boiling of hydrocarbon mixtures. Paper read at AIChE National Conference, November 2–7, Miami Beach (paper 127a).

Shah, M. M. 1976. A new correlation for heat transfer during boiling flow through pipes. *ASHRAE Trans.* 82:66–86.

Shakir, S., and J. R. Thome. 1986. Boiling nucleation of mixtures on smooth and enhanced boiling surfaces. *Proc. 8th Int. Heat Transfer Conf.* 4:2081–86.

Stephan, K., and M. Abdelsalam. 1980. Heat transfer correlations for natural convection boiling. *Int. J. Heat Mass Transfer* 23:73–87.

Thome, J. R., and S. Shakir. 1987. A new correlation for nucleate pool boiling of aqueous mixtures. *AIChE Symp. Ser.* 83(257):46–51.

———. and C. Mercier. 1982. Effect of composition on boiling incipient superheats in binary liquid mixtures. *Proc. 7th Int. Heat Transfer Conf.* 4:95–100.

Uhlig, E., and J. R. Thome. 1985. Boiling of acetone-water mixtures on smooth and enhanced surfaces. In *Advances in Enhanced Heat Transfer*, HTD Vol. 43, 49–56. New York: American Society of Mechanical Engineers.

Van Stralen, S. J. D., and R. Cole. 1979. *Boiling phenomena*. Washington, D.C.: Hemisphere, vols. 1 and 2.

Zuber, N. 1959. *Hydrodynamc aspects of boiling heat transfer* (AEC report AECU-4439, physics and mathematics).

CHAPTER

THREE

HISTORICAL DEVELOPMENT OF ENHANCED BOILING SURFACES

3-1 INTRODUCTION

This chapter discusses the historical development of enhanced boiling surfaces. Because there are literally thousands of publications treating various aspects of this technology's development, only the most important studies can be highlighted. The subject is divided into two parts. The first part reviews the development of externally enhanced boiling surfaces, with reference to those methods that apply to boiling on the outside of a tube or on the exterior of other surfaces. The second part briefly describes the advancements in enhancement geometries for boiling inside tubes. The heat transfer mechanisms controlling external and internal enhanced boiling processes are discussed separately in Chapters Six and Ten, respectively. The special fin geometries for boiling in plate-fin heat exchangers are described in Chapter Thirteen.

3-2 EARLY DEVELOPMENTS

The origins of enhanced boiling heat transfer can be traced back to the initial studies of nucleate pool boiling heat transfer on plain surfaces. Improved understanding of boiling phenomena on plain surfaces led to ideas about how to modify certain aspects of the process to increase the boiling heat transfer coefficient.

Two boiling parameters in particular have captured the most attention: boiling nucleation and boiling site density. Boiling nucleation is of paramount importance because its occurrence controls the wall superheat, or heat flux, at which the heat transfer process passes from the less effective natural convection mode to the much more effective nucleate pool boiling mode. Boiling site density, which is defined as the number of active boiling sites per unit area on the heated surface, was also recognized as significant because it was easily observed at low heat fluxes that the increase in the number of active boiling sites with increasing heat flux coincided with the rise in the boiling heat transfer coefficient. Thus most early enhanced boiling surfaces attempted to promote boiling nucleation at lower wall superheats to increase the number of active boiling sites on the surface. The use of fins to increase the heat transfer area per unit length of tubing, an already common practice for single-phase heat transfer, was another approach taken to augment boiling heat transfer per unit length of tubing.

Apparently the earliest study on augmentation of nucleate pool boiling was carried out by Jacob and Fritz (1931). They studied the effect of surface preparation on water boiling at atmospheric pressure on several specially prepared, horizontal copper plates. They also investigated the effect of prolonged boiling on the performance of these surfaces. One of their test surfaces was prepared by sandblasting, and another had a crosshatched pattern of grooves 0.172 mm wide and 0.152 mm deep at spacings of 0.477 mm. The control surface was a smooth chrome-plated surface. Figure 3-1 depicts some of their results. From their boiling curves, the following conclusions were made:

1. Sandblasting a surface increased the boiling heat transfer coefficient up to about 25% at a fixed heat flux; at a fixed wall superheat, performance could be increased more than fourfold.

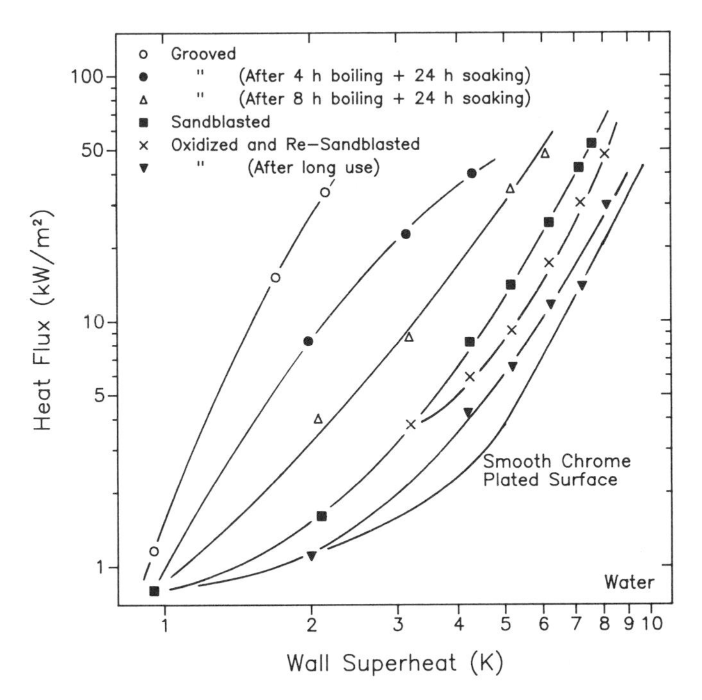

Figure 3-1 Effect of surface preparation and time on boiling.

2. Machining grooves into the heated surface yielded more augmentation, much larger than the percentage increase in the wetted surface area of the grooved plate, by providing better locations for nucleation sites.
3. Boiling augmentation diminished quite rapidly with time, tending toward the smooth chrome plate's boiling curve.

Thus this early study demonstrated that notable enhancement could be obtained by using common manufacturing methods. The prospects for long-term use in practical applications did not look promising, however.

In another study carried out several years later, Sauer (1935) observed similar results of enhancement and subsequent deterioration for a tube with 0.4-mm square grooves cut at 1.6-mm spacings. These two studies apparently dampened interest in developing boiling enhancements until the 1950s because Webb, Bergles, and Junkhan (1983) list only a few enhanced boiling heat transfer patents that were issued during this period.

Boiling Nucleation and Boiling Site Density Studies

Boiling nucleation theory. It was not until the 1950s that advances in boiling nucleation began to be made that could be applied to the development of enhanced boiling surfaces. Corty and Foust (1955) and Bankoff (1956, 1957, 1958) postulated that boiling initiated from preexisting vapor embryos trapped in pits and cracks in the heated wall. Bankoff (1957) also showed analytically that the free energy, or work of formation, required to create a vapor nucleus at a planar solid surface in a liquid was less than that required for homogeneous nucleation in the liquid phase alone. On the basis of his physical model, depicted in Fig. 3-2, Bankoff derived the following expression for the reduction in the free energy required to form a vapor nucleus at a solid wall relative to homogeneous nucleation in the liquid phase:

$$\phi = \frac{2 + 2\cos\beta + \cos\beta\sin^2\beta}{4} \tag{3-1}$$

where the molar free energy of bubble formation for a pure fluid is given as

$$\Delta\tilde{g} = \frac{16}{3}\frac{\pi\sigma^3\tilde{v}_g^2}{\Delta T_{sat}^2(\Delta\tilde{h}_v/T_{sat})^2}\phi \tag{3-2}$$

The contact angle in the expression is defined as the angle through the liquid between

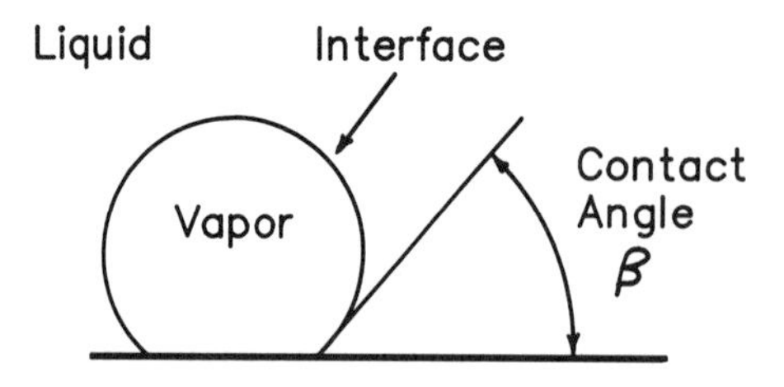

Figure 3-2 Nucleation of vapor nuclei at a solid surface.

the liquid-vapor interface and the solid surface at the point of contact. Therefore, this pioneering work showed that boiling nucleation for ordinary fluids, with contact angles typically ranging between 25° and 90°, would preferentially occur at the heated surface rather than in the superheated liquid bulk. In addition, this theory demonstrated that the contact angle was an important physical parameter in the heterogeneous nucleation process.

Vapor trapping mechanism. The process of trapping vapor or gas during the spreading of a liquid film over a surface containing scratches and cavities was also considered by Bankoff (1958), who assumed preexisting vapor or gas nuclei trapped in the surface to be the origin of active boiling sites, perhaps on the basis of deactivation boiling studies carried out earlier by Harvey et al. (1944, 1945). As illustrated in Fig. 3-3, Bankoff hypothesized that there was a minimum contact angle required to trap vapor or gas in a groove. Thus one condition for vapor or gas to be trapped in the groove is that the contact angle β must be larger than the wedge angle θ.

Clark, Strenge, and Westwater (1959) confirmed this hypothesis by obtaining high-speed cine films of active boiling sites on polished surfaces with a microscope. They observed bubbles emerging from cavities and scratches in the heated surface. The cavity mouths ranged from 10 to 100 μm in diameter, and the scratches were on the order of 10 μm wide. Cornwell (1975, 1977) again confirmed this conclusion experimentally by obtaining scanning electron microscope photographs of locations identified as boiling sites.

It should be pointed out, however, that the Bankoff vapor trapping model and the more complete theory of Lorentz, Mikic, and Rohsenow (1974) are only valid when a thin, slow-moving, liquid film is advancing across the surface. In practice, there are many exceptions to this situation. For instance, the trapping of vapor when a boiling site deactivates represents a completely different physical process.

For a real fluid and surface, the contact angle is not constant but varies with dynamic motion along the surface and with flow around the edge of the groove. In fact, the contact angles usually quoted in boiling nucleation studies are "macroscopic" values

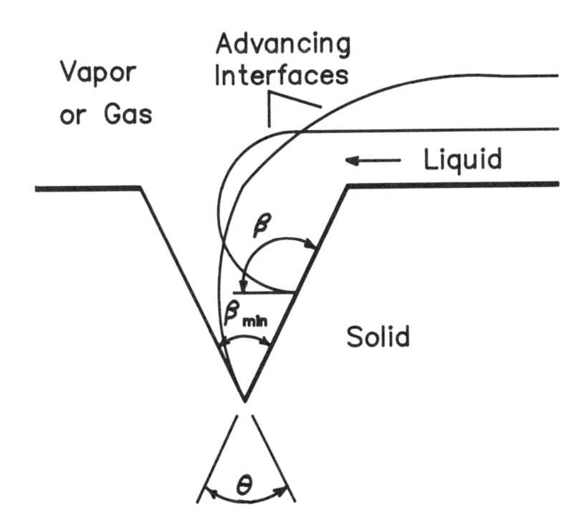

Figure 3-3 Advance of a liquid film over a gas-filled groove.

obtained from a magnified photograph or with a goniometer. These values are not really applicable at the microscopic level because they are determined relative to a hypothetical planar surface at the wall rather than the microscopic topography existing at the point of the triple interface, where the surface is anything but planar. As shown in Fig. 3-4, it is unlikely that there is any relationship between macroscopic and microscopic contact angles. In addition, at the microscopic level Chappius (1982) described a thin liquid "foot" that exists at the front of a liquid film spreading on a surface, as illustrated in Fig. 3-5.

In summary, there is still much to be learned about vapor trapping on real surfaces. This is especially true for the complex macroscopic and microscopic geometries of enhanced boiling surfaces. Liquid trapping on these surfaces can be hypothesized to be dependent on the radius of the advancing liquid meniscus as depicted in Fig. 3-6, which is in turn a function of the geometry of the passageway, the capillary forces, the buoyancy of the vapor, and the triple interface (solid-liquid-vapor).

Boiling nucleation in a thermal boundary layer. Boiling nucleation occurring at a heated wall with the bulk liquid saturated rather than uniformly superheated is affected by the presence of the thermal boundary layer. Hsu (1962), using a model proposed by Hsu and Graham (1961), developed a nucleation theory for a vapor nucleus resting at the mouth of a cavity and situated in the thermal boundary layer forming in the liquid

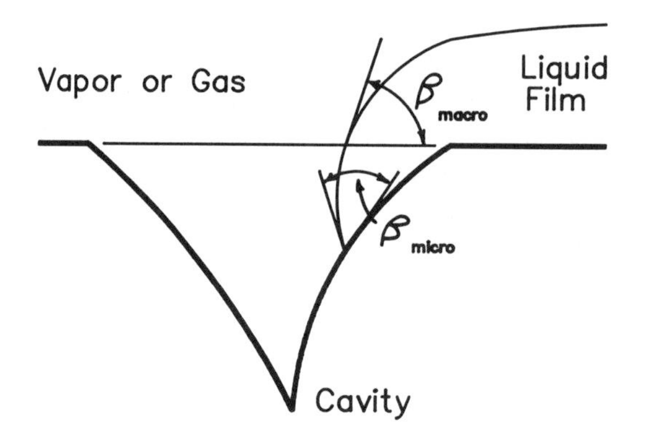

Figure 3-4 Microscopic and macroscopic contact angles.

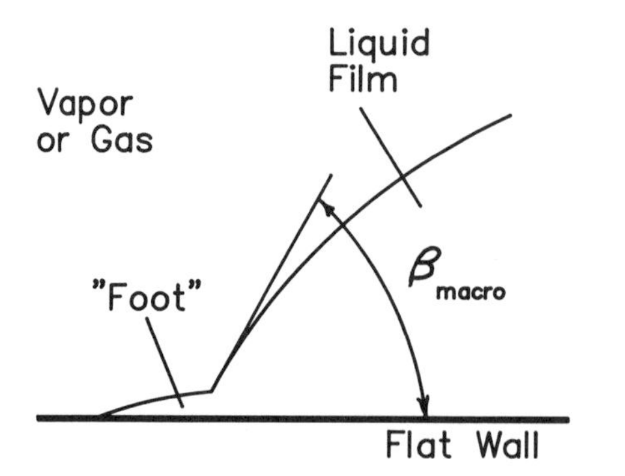

Figure 3-5 Liquid "foot" of a spreading liquid film.

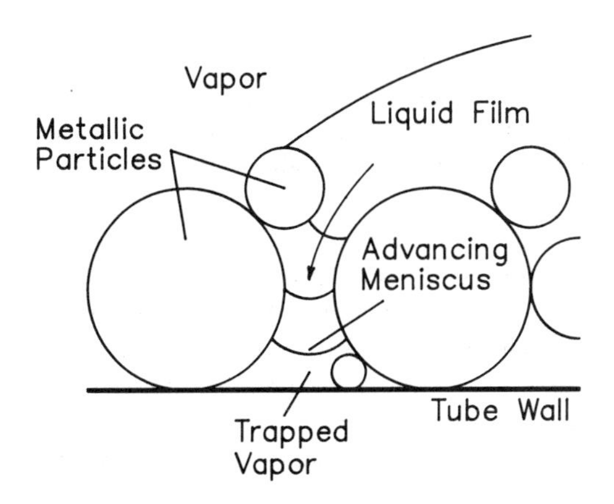

Figure 3-6 Advancing liquid meniscus over an idealized porous surface.

adjacent a wall on application of a step increase in the wall temperature. As the thermal boundary layer develops, as depicted in Fig. 3-7 for linear temperature profiles, the superheat in the liquid boundary layer at the top of the vapor nucleus may at some point become greater than the bubble nucleation superheat given by Eq. (2-1):

$$\Delta T_{\text{sat}} = \frac{2\sigma}{R(dP/dT)_{\text{sat}}} \tag{3-3}$$

At this point the vapor nucleus can begin to grow, and an active boiling site is established. Very small and very large vapor nuclei (when in thermal isolation from other active boiling sites) do not activate because the local superheating at their respective tips do not attain the necessary nucleation superheat given by Eq. (3-3). Therefore, Hsu (1962) demonstrated that it is not sufficient only to trap vapor or gas in a cavity to provide an embryo for bubble growth in a plain or enhanced boiling surface but also R must be within a prescribed range that depends on the local temperature gradient in the liquid.

Although Hsu's (1962) theory is helpful in understanding the effect of the transient liquid temperature profile on bubble nucleation, several assumptions limit its effective-

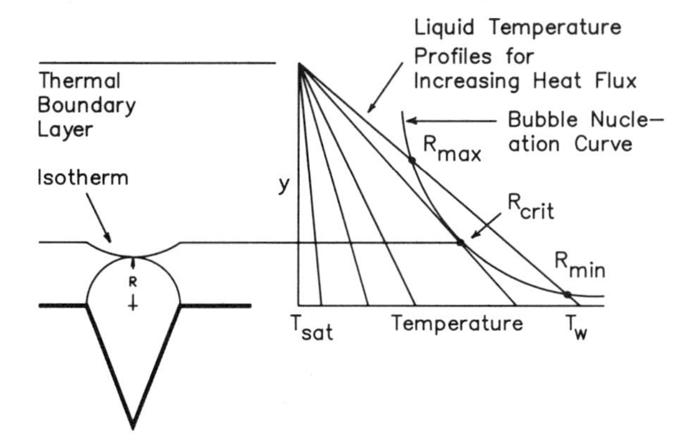

Figure 3-7 Boiling nucleation in a thermal boundary layer.

ness. For instance, the preexisting vapor nucleus assumed at the mouth of a cavity would be unstable in the initial, uniformly saturated liquid and also in the thermal boundary layer up until its "nucleation." Instead, vapor nuclei that are stable even when the liquid is uniformly saturated (or subcooled) must be available. Cavities that can contain such vapor nuclei are described below.

Reentrant cavities. Griffith and Wallis (1960) demonstrated that the geometry of a cavity containing trapped vapor is important to the boiling nucleation process. First of all, the diameter of the mouth of the cavity defines the nucleation radius and hence the wall superheat at which the cavity will become activated. Second, the stability of a trapped vapor nucleus is determined by the cavity's shape. These two points are illustrated in Fig. 3-8, where the liquid-vapor interface, its radius of curvature, and the corresponding vapor volume are shown for a liquid with a contact angle of 90°. Note that when the interface becomes concave inside the cavity the nucleation superheat becomes negative, and hence the vapor nucleus can withstand liquid subcooling. Griffith and Wallis (1960) therefore surmised that a reentrant cavity as shown in Fig. 3-8 would be more stable than a conical cavity or groove while also being an excellent geometry for trapping vapor or gas. Their experiments with cavities of various geometries also substantiated this, although their measured activation superheats did not always agree closely with those calculated from Eq. (3-3).

Shortly after this work, Benjamin and Westwater (1961) reported the construction of an artificial reentrant cavity with a mouth diameter of 0.004 inches (102 μm). Benjamin (1960) described the site to be stable, remaining active down to low wall superheats for boiling mixtures of water and ethylene glycol. Therefore, these two independent studies encouraged investigators to try to produce reentrant cavities in their enhancements to act as stable, easily activated boiling sites.

Modification of the boiling nucleation process to initiate boiling at lower wall superheats, however, apparently predated the above research. Chemists have used boiling balls or a glass rod for perhaps centuries to promote boiling in glass and Pyrex beakers. The author at various times has resorted to "alchemy" to try to understand what makes boiling balls work. Heterogeneous nucleation theory is based on the idea of trapping vapor or noncondensables (or both) in cavities and then surpassing the wall superheat required to maintain thermal equilibrium between the vapor nucleus and its surroundings to make the nucleus grow to become an active boiling site. Placing a

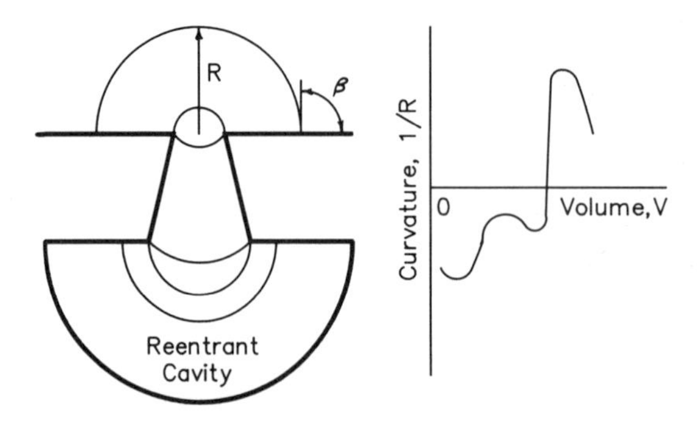

Figure 3-8 Liquid-vapor interface emerging from a reentrant cavity.

smooth glass ball at the bottom of an equally smooth beaker, however, or touching the bottom of the beaker with the edge of a glass rod as shown in Fig. 3-9 makes boiling initiate at the point of contact for conditions at which boiling was not originally occurring. When the glass ball or rod is removed from the surface, boiling ceases.

In a similar experiment with an imbedded thermocouple in the wall, the author placed a highly polished stainless steel ball bearing in a container of double-distilled and degassed water on top of a plain, horizontal brass disk heated electrically from below with a Nichrome resistance heater. The ball bearing caused boiling to occur underneath itself as it rolled around the heated surface, which was otherwise in the quiescent natural convection mode of heat transfer. Boiling occurred only underneath the ball bearing and ceased immediately after its passage of that point on the surface. Therefore, the ball bearing must have briefly modified the local boiling nucleation criterion as it passed over the surface.

The author next attached an array of the same ball bearings to the surface with a thin uniform coating of soft solder in a similar experiment to measure the nucleation wall superheats. For saturated double-distilled and well-boiled water at 1.0 bar, boiling initiated at wall superheats so small (about 0.1 K) that they were not measurable. Therefore, either very large vapor nuclei (on the order of 0.33 mm according to Eq. [3-3] with a wall superheat of 0.1 K) were trapped between the ball bearing and the flat disk or some other previously unidentified phenomenon must have promoted the inception of boiling. Because boiling was also initiated under similar circumstances (e.g., when a ball bearing was dropped on the heated solder-coated disk), the vapor trapping mechanism of boiling nucleation theory does not seem to explain these curious results. Therefore, the low wall superheats required to activate nucleate pool boiling on enhanced boiling surfaces are not only the result of their surfaces' vapor trapping characteristics and favorable cavity geometries.

It is speculated that the liquid trapped in the narrow channel between the ball bearing

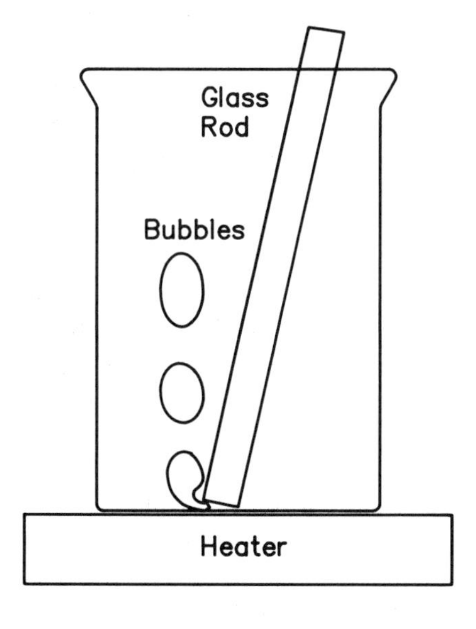

Figure 3-9 Boiling activation under a glass rod.

and the flat plate (or boiling ball and Pyrex plate) becomes uniformly superheated to the wall temperature because the liquid layer is extremely thin and also because the wall tends to be hotter as a result of the poorer heat transfer. Thus a small inactive nucleus may be able to activate. Consequently, it appears that a boiling enhancement not only serves to trap large vapor nuclei in reentrant cavities and to make them more resistant to the effects of subcooling but also provides a better thermal environment for their eventual activation.

Boiling site density. Kurihari and Myers (1960) studied the relationship between the boiling site density and the boiling heat transfer coefficient for water and several organic fluids boiling on copper surfaces roughened with emery paper. Using successive grades of grit size, they also investigated the effect of surface roughness on the boiling site density. They observed that the boiling heat transfer coefficient increases as the surface roughness increased and that the boiling site density increased concurrently. Their studies showed that the boiling heat transfer coefficient was directly proportional to the boiling site density:

$$\alpha \propto (N/A)^{0.43} \tag{3-4}$$

Therefore, they concluded that increasing the number of active boiling sites resulted in an improvement in heat transfer performance.

Other Studies of Boiling on Roughened Surfaces

Several other early studies on surface roughness influenced the development of enhanced boiling surfaces. Corty and Foust (1955) were apparently the first to investigate extensively the effect of surface roughness on the nucleate pool boiling curve. Using different grades of emery paper, they produced nickel surfaces with scratches ranging from 0.254 to 25.4 μm across by 0.05 to 0.635 μm deep. Figure 3-10 depicts the variation that they observed in the wall superheat as a function of the root-mean-square surface roughness for a constant boiling heat transfer coefficient. As demonstrated by the asymptotic trend in the data points with increasing roughness, there is a point at which higher roughness has little effect on the boiling process. Therefore, using surface roughness alone as an enhancement appears to have its limitations.

Berenson (1960, 1962) studied the effect of surface finish on the boiling of *n*-pentane on a copper plate. He used both a mechanical roughening technique (different grades of emery paper) and chemical treatment (to cause oxidation of the surface). Oxidation of a given surface was shown to translate the boiling curve to the right, reducing boiling performance by about 15%. His pool boiling curves for different surface finishes are shown in Fig. 3-11. He observed that:

1. the nucleate pool boiling curve shifts to the left with increasing roughness;
2. the slope of the boiling curve is higher for the three roughened surfaces than for the mirror-finished surface; and
3. the peak nucleate heat flux is hardly affected by the variation in surface roughness.

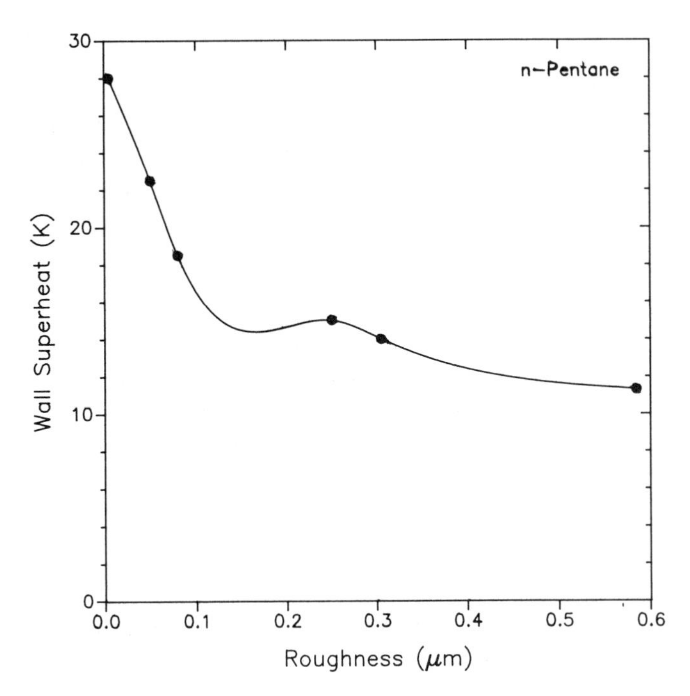

Figure 3-10 Variation in wall superheat with surface roughness [to maintain α = 5,670 W/(m^2·K)].

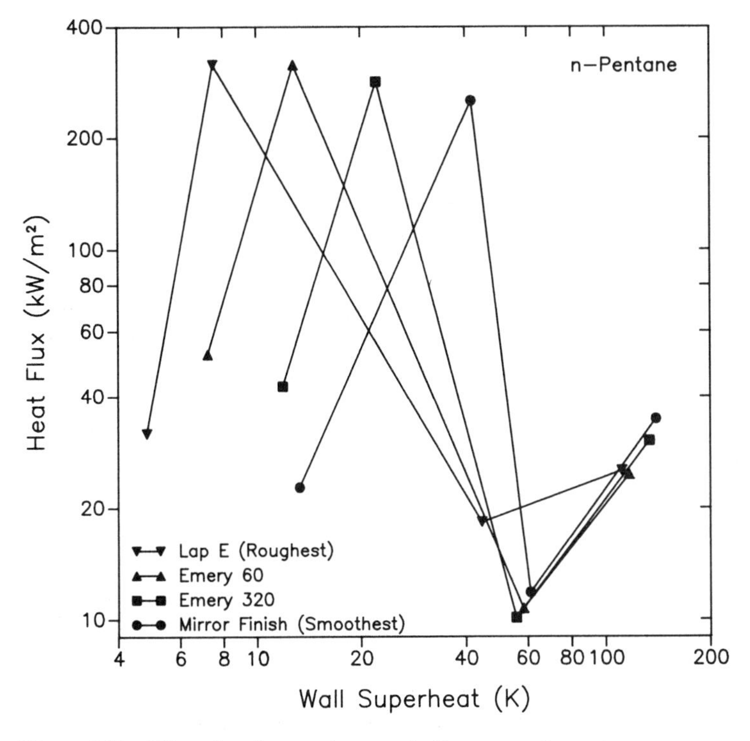

Figure 3-11 Effect of surface roughness on boiling curve of *n*-pentane.

The above studies, and many that followed, firmly established that nucleate boiling heat transfer is dependent on the microgeometry of the boiling surface. This realization encouraged the development of special surfaces whose improved performance was stable over time, particularly for applications of commercial interest (e.g., evaporators in the refrigeration and air separation industries).

Early Studies on Externally Finned Tubes

According to the extensive patent literature listing of Webb, Bergles, and Junkhan (1983), the first patent for an extended surface for heat transfer augmentation was obtained by Still (1929) for forced convection to air on the outside of a tube. Stone and Tilley (1938) followed with another patent for the same application. Lea (1941) then patented the first externally finned tube for condensation. By then externally finned tubes had already been applied to boiling since Jones (1941) described tests of an industrial-size, finned tube evaporator bundle with tubes having 16 fins per inch (fpi), or 632 fins per meter, for evaporating R-11. Webber (1960) reported his successful operational experiences with finned tube reboilers in a refinery light ends unit. These studies also played an important role in the historical development of enhanced boiling surfaces by demonstrating their viability in industrial practice.

During the late 1940s and early 1950s extensive research on boiling on the outside of integral low-finned tubes was carried out at the University of Michigan by Katz and co-workers; Zieman and Katz (1947), Robinson and Katz (1951), Myers and Katz (1952, 1953), and Katz, Myers, and Young (1955). In the first study, nucleate pool boiling curves were obtained for hexane and isobutane boiling on a single horizontal tube with 14.5 fpi (572 fins per meter). In the subsequent studies, four horizontal low-finned tubes in a vertical array were tested. These copper tubes had 19.5 fpi (770 fins per meter) with an 18.8-mm diameter over the fins and a base diameter of 16.0 mm, producing a wetted surface area on the finned tube 2.7 times that of a comparable plain tube with the same diameter as at the tip of the fins.

Figure 3-12 depicts the Myers and Katz (1953) finned tube and plain tube boiling curves (represented by best line fits through their data points) for the bottom tube in the array. The finned tube heat flux was calculated from the outside circumference over the fins and the heated length of the tube, not the total outside wetted area. (For enhancements with complex geometries, the outside nominal surface area is used to calculate the heat flux because the actual wetted surface area is not known. Thus, to be consistent, the same area should also be used for finned surfaces when comparing performances.) The finned tube gave substantial improvement compared to the plain tube performances for all the fluids, evaluated at either the same wall superheat or the same heat flux. Because the fluids in Fig. 3-12 cover a wide class of fluids (organics, an inorganic, and a common refrigerant), these results can be expected to be a representative sample of single-finned tube boiling performance. If the finned tube boiling curves are determined instead from the total wetted surface area (which would translate the curves downward), the boiling heat transfer coefficients would still be equal to or greater than those for the plain tube at the same heat flux. Therefore, it was demonstrated that the fins not only increased the surface area but also modified the boiling process

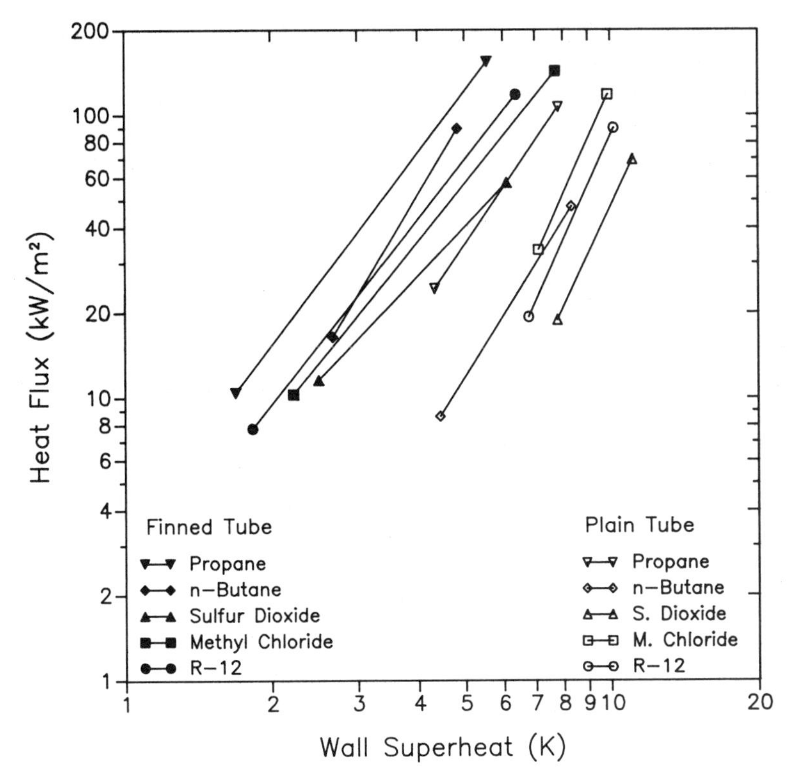

Figure 3-12 Comparison of finned tube and plain tube boiling curves at the following pressures: propane, 5.1 to 6.7 bar; *n*-butane, 1.6 to 1.9 bar; sulfur dioxide, 2.8 bar; methyl chloride, 3.7 bar; and R-12, 4.2 bar.

to augment performance. This work was crucial to the subsequent search for an optimumal fin density and fin profile for low-finned tubes for boiling.

3-3 NOVEL ENHANCEMENT TECHNIQUES

This section describes the historical development of special geometries to enhance boiling performance. As will be seen, the innovations diverged rapidly from the developments from the early studies on surface roughness in an effort to attain long-term stability in enhanced boiling performance.

Classification of Enhancement Techniques

Eleven classifications to categorize enhancement techniques for nucleate pool boiling were proposed by Bergles and Webb (1978–1980). These are:

1. treated surfaces, which refer to surfaces coated, plated, or covered with another material;

2. rough surfaces, which refer to an abrasive treatment of the surface such as rubbing with emery paper;
3. extended surfaces, which refer to the use of fins of any shape and form, including deformed circumferential fins and the like;
4. surface tension devices, which involve wicks and slots in the wall to transport liquid to the evaporation zone;
5. additives, which refer to chemicals or other fluids added to the liquid;
6. mechanical aids, which refer to rotation, scraping, or wiping of the heated surface;
7. surface vibrations, in which the heated surface is vibrated;
8. fluid vibrations, in which the fluid is vibrated by ultrasound or oscillations in pressure;
9. electrostatic fields, which refers to the application of alternating or direct current fields to the boiling fluid;
10. injection, which refers to the introduction of a gas to the evaporating fluid through a porous wall; and
11. compound enhancement, which refers to the use of two or more enhancements such as a coating on an extended surface.

Techniques 1 through 4 are said to be passive techniques and 5 through 10 are defined as active techniques (the latter are so called because they utilize an external agent or force to attain augmentation). The scope of the present review is limited to the passive types because they have the most practical importance. Of these, treated surfaces and extended surfaces are the most commercially viable enhancement techniques.

Treated Surfaces

Many variations of treated surfaces have been developed. Most of these can be subdivided into the following categories: porous metallic coatings, nonwetting coatings, and attached promoters (such as screens and wires).

Porous coatings. One of the first surfaces developed specifically to enhance nucleate pool boiling was the porous sintered metallic coating of Milton (1968) developed at Union Carbide, which originally filed for patent in 1956. This porous metallic boiling surface, now well known by its trade name High Flux after further improvements by Milton (1970, 1971), is shown in Fig. 3-13 (from Union Carbide [1979]). Milton and Gottzmann (1972) presented a brief history of this surface's development, which began back in 1947, noting that the primary application in mind was for the main condenser in air separation plants, with nitrogen condensing on one side of the tube while oxygen boiled on the other.

Initially, Union Carbide tested plain surfaces, noting that the boiling heat transfer coefficients varied for different smooth copper surfaces; the higher values were for surfaces with a larger number of active boiling sites. Examination with a low-power microscope revealed that the bubbles appeared to originate from cavities in the surface by nucleation of the vapor left behind by the previously departed bubble. This work led to the study of mechanically roughened surfaces, which were observed to initiate

Figure 3-13 Photograph of High Flux coating (courtesy Union Carbide). Top view of surface (20×).

boiling at lower wall superheats and to provide many more boiling sites while also increasing the boiling heat transfer coefficient to two to three times that for the smooth surfaces. Then, in a further effort to augment heat transfer by increasing the number of boiling nucleation sites and to keep them in intimate thermal contact with the heated wall, the surface was coated with a thin porous layer of nickel. This coating produced a large number of active boiling sites and gave a 20-fold improvement in the boiling heat transfer coefficient for liquid oxygen at atmospheric pressure, as shown in Fig. 3-14. By using copper instead of nickel, an additional improvement was achieved. This enhancement was commercialized and the coating became known by its trade name High Flux (now belonging to UOP).

Figure 3-15 (from Gottzmann, O'Neill, and Minton [1973]) depicts some nucleate pool boiling curves of High Flux for four typical fluids compared to those for smooth surfaces, all at 1.01 bar. Performance at a given heat flux is about ten times or more that for a plain tube in nucleate pool boiling. Figure 3-16 (from Starner and Cromis [1977]) shows an independently obtained comparison between a copper High Flux tube (nominal outside diameter, 19.0 mm) and a copper, integral low-finned tube with 26 fpi (1,026 fins per meter) of the same diameter for nucleate pool boiling of refrigerant-12. Here, the heat duty per unit length of tube is plotted against the wall superheat. The High Flux boiling performance was found to be six times that of the finned tubing when compared at the same heat duty. Thus these studies and others in the refrigeration industry proved that enhanced boiling tubes could greatly improve evaporator performance.

There were several other milestones related to the High Flux tube. First, the porous metallic surface was found to perform just as well under flow boiling conditions as for nucleate pool boiling. Second, the High Flux tube proved that boiling enhancement could be stable over a long period of time in actual operation in evaporators and reboilers, as illustrated in Fig. 3-17 (from Gottzmann, O'Neill, and Minton [1973]) for a cascade refrigeration condenser boiling propylene at 1.4 bar on its shell side and condensing ethylene at 20 bar on its tube side. This prototype heat exchanger was installed as a small satellite unit on an operating ethylene column in December 1967. Test data showed that its performance had not deteriorated during 4½ years of operation

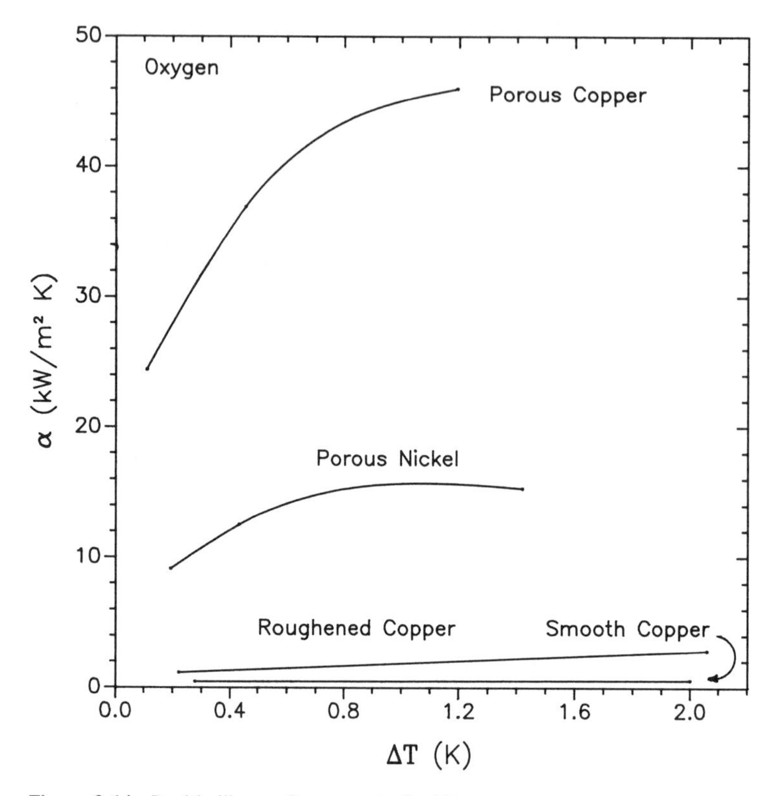

Figure 3-14 Pool boiling performances in liquid oxygen.

(at the time of the publication of the data in 1972). The only cleaning to which the High Flux test unit was subjected was a hot methane thaw. Because of an old reciprocating compressor in the cascade refrigeration system, the oil content in the evaporating propylene at times was reported to be as high as 9%, although no fouling occurred. Overall thermal performance was observed to be about five times that of a conventional plain tube unit.

Other metallic porous coatings have since been developed. Some of these (in chronological order) are summarized as follows:

1. Marto and Rohsenow (1965, 1966) produced porous welds and a sprayed stainless steel coating on plain surfaces for augmenting the boiling of liquid sodium.
2. Almgren and Smith (1969) developed a flame spray process to augment the boiling of liquid nitrogen (a plain surface was first sandblasted and then a 0.13-mm thick porous layer was applied to the surface from two separate guns to deposit copper and zinc simultaneously).
3. Czikk et al. (1970) described the application of the High Flux coating to the outside of a corrugated tube for evaporating refrigerant-11.

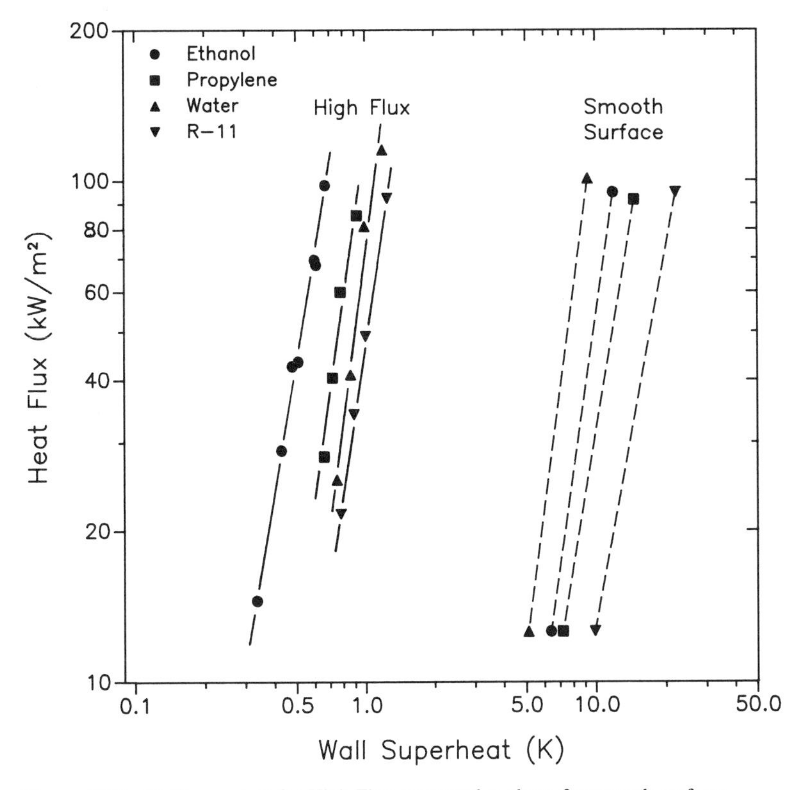

Figure 3-15 Boiling curves for High Flux compared to those for smooth surfaces.

4. Oktay and Schmeckenbecher (1972) developed a method to form dendritic heat sinks on semiconductor chips with the use of nickel-plating process, which produced a high density of dendrites about 1.0 mm high. Oktay (1982) described the thermal performance of this surface for boiling a dielectric coolant.
5. Inoue (1974) developed a new method for producing a porous sintered surface by means of electrical sintering rather than baking in an furnace.
6. Grant (1974, 1977) improved Union Carbide's manufacturing method of the High Flux tube.
7. Dyundin et al. (1975) used both an electric arc and a flame spraying technique to produce porous surfaces of various thicknesses to enhance the boiling of R-11, R-12, and R-22.
8. Dahl and Erb (1976) developed a method for flame spraying aluminum powder onto a surface to produce a porous metallic layer.
9. Albertson (1977) produced a porous metallic coating by using a high–current density electroplating technique to form a large number of dendrites on a surface. These dendrites were then compacted for further improvement of performance.
10. Janowski, Shum, and Bradley (1978) developed a porous surface called the Korotex II tube, which was commercially available for a brief period. This enhanced

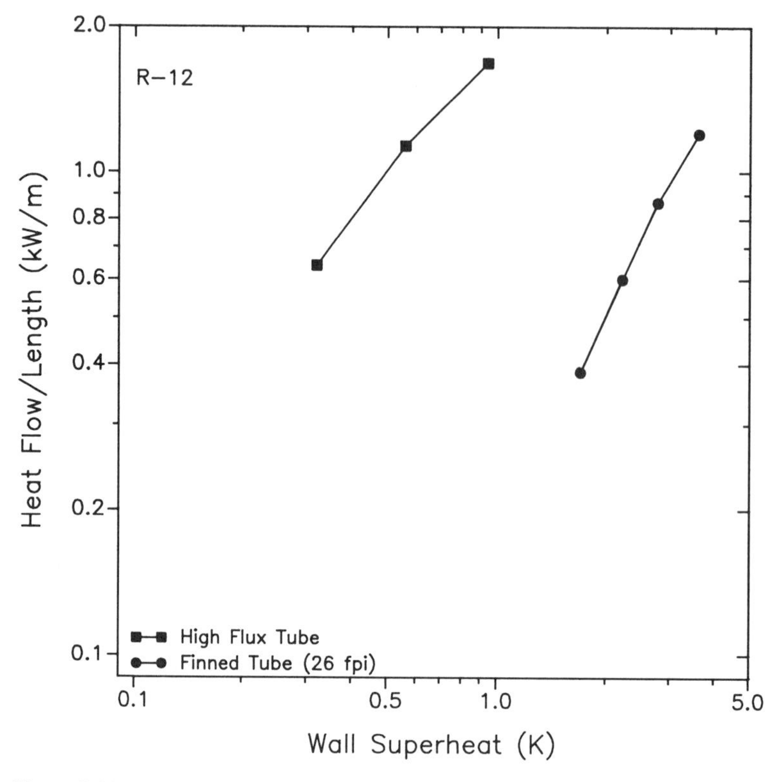

Figure 3-16 Comparison of boiling on High Flux and finned tubes.

boiling tube was manufactured by wrapping plain, drawn tubing with an open-cell polyurethane foam and then electroplating with copper. Pyrolysis was used to remove the foam, leaving behind a porous layer with a skeletal-type structure. This method was apparently abandoned because of its high production cost.

11. Fujii, Nishiyama, and Yamanaka (1979) produced porous copper coatings by pressing spherical copper particles firmly to a plain surface and then plating with copper to join the particles metallurgically to the surface. With this technique, they were able to vary the coating by using different powder sizes and layer thicknesses. Fujii (1984) describes more recent work to optimize the performance of this surface.
12. Shum (1980) invented an enhanced boiling surface for refrigerants by producing a porous coating on an integral low-finned tube with the use of an electroplating process whose bath contained graphite powder. The graphite particles were attracted to the surface by an electric field and then plated over to produce a fine porous layer on the sides of the fins.
13. Taborek (1980) developed a method for flame spraying copper wire onto a surface.
14. Trepp and Hoffman (1980) coated tubes with porous layers by plasma spraying of copper and also by electrogalvanizing the surface.

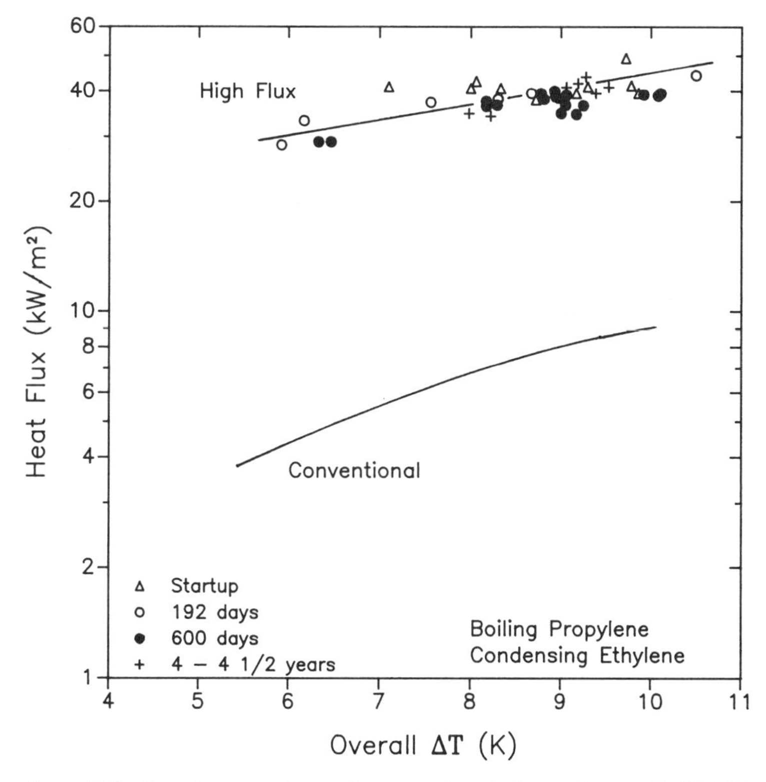

Figure 3-17 Long-term operating performance of a reboiler-condenser with High Flux tubing.

15. Venart, Sousa, and Jung (1986) flame sprayed carbon steel onto surfaces.

In summary, porous coatings can be produced by means of many different processes in metals of industrial interest, such as copper, aluminum, and steel. The thickness of the porous layer, its porosity, and its particle size are the primary production control variables.

Nonwetting coatings. A completely different approach to enhancement also originated from the work on boiling nucleation described above. After Bankoff (1957) analytically showed that the free energy or work of formation of a vapor nucleus was less for fluid-surface combinations with smaller contact angles (see Eqs. [3-1] and [3-2]), Griffith and Wallis (1960) hypothesized that coating the inside of cavities with a nonwetting substance would improve the stability of the cavity for boiling nucleation. The first effect of a nonwetting substance would be to reduce the contact angle between the liquid-vapor interface and the metal wall. This, however, would tend to decrease the volume of vapor trapped, as can be deduced from Fig. 3-3. Yet the smaller contact angle would result in a larger radius of curvature of the liquid-vapor interface of the trapped vapor nucleus, which makes the nucleus stable down to a lower superheat according to Eq. (3-3). Griffith and Wallis then tested their ideas by producing artificial cavities by forcing phonograph needles into a flat metallic surface, forming hemispherical cavities about 0.080 mm in diameter. Some cavities were coated with paraffin; these proved to be more stable as boiling sites than uncoated cavities.

Young and Hummel (1954) later produced a nonwetting surface by spraying Teflon onto a smooth stainless steel surface. This technique produced many small spots less than 0.25 mm in diameter. They made a second type of surface by pitting the stainless steel surface and then coating only the pits with Teflon. The boiling performances for water on these surfaces, together with those of the original smooth surface and a pitted surface without Teflon, are shown in Fig. 3-18. At a given wall superheat, the surface with Teflon-coated pits increased the heat transfer coefficient by about a factor of 10 compared to the same surface without Teflon in its pits. Hummel (1965) received a patent for this enhanced boiling surface. Apparently, the Teflon spots tended to peel off the surface with prolonged use, however (Anonymous [1970]).

Marto, Moulson, and Maynard (1968) studied special boiling surfaces for boiling liquid nitrogen, which is a nonwetting fluid with a contact angle believed to be only about 5°. They observed that a polished surface with an array of pressed cavities about 0.38 mm in diameter and 0.76 mm in depth was only somewhat successful in

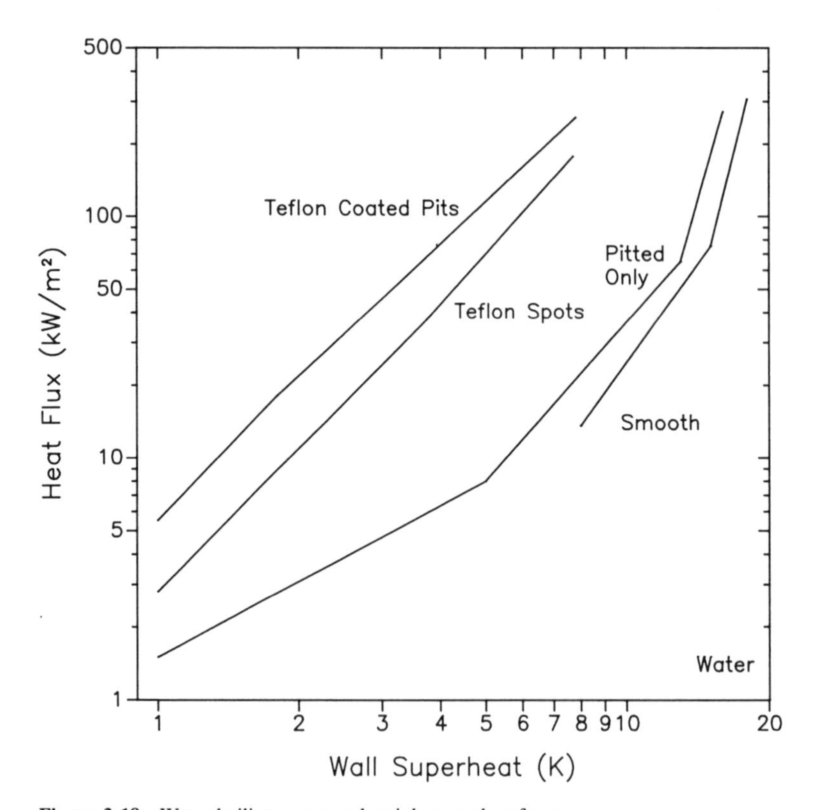

Figure 3-18 Water boiling on treated stainless steel surfaces.

augmenting the boiling performance of nitrogen. Because the boiling curves for their plain, polished surface and for their pitted surface were steep but close together, significant enhancement was found when comparing them at the same wall superheat and moderate enhancement when comparing them at the same heat flux. The cavity diameters were large for liquid nitrogen and probably not effective for trapping nitrogen vapor. For example, Thome (1978) tried these types of cavities and found them to be unstable and difficult to activate. He found instead that microdrilled cavities 0.021 to 0.050 mm in diameter and cylindrical in shape were much more stable as boiling sites and easier to activate.

Another nonwetting study was carried out by Vachon et al. (1969) for water boiling on Teflon-coated surfaces. They tested layers from 7.5 to 35 μm thick. Only the thinnest layer improved heat transfer, but by much less than that observed by Young and Hummel (1965) for Teflon spots. The insulating effect of the coating reduced the boiling performance of the thicker layers, becoming the dominant thermal resistance. Also, Hummel (1965) and Gaertner (1967) cautioned against utilizing a continuous coating rather than spots lest the entire nonwetted surface function in the film boiling regime.

Schade and Park (1979) investigated the boiling enhancement capabilities of a plasma-deposited polymer coating for evaporating refrigerant-113. Kotchaphakdee and Williams (1969) had already shown that adding trace additives of polymers to water augmented nucleate pool boiling heat transfer. Notable boiling enhancement with a splattered layer of polymer was also obtained.

Thome (1978) observed that boiling performance deteriorated substantially when a thin film of ice (frozen water vapor and carbon dioxide) coated a polished heated surface during the boiling of liquid nitrogen and liquid argon. The coating was formed by white flakes of ice observed floating in the liquid when a cold trap was not used to "dry" gases of 99.996% purity! Thus a nonwetting surface in combination with a nonwetting liquid produced poor results, a conclusion that was also reached by Marto and Lepere (1981) for refrigerants.

Apparently no commercial use of nonwetting coatings or spots to enhance boiling heat transfer has resulted from the above work. Concerns about the long-term durability of spotted surfaces, the effect of even a little fouling on the surface's wetting characteristics, and the like have dulled interest in this enhancement technique.

Attached promoters. Various materials have been attached to the surfaces of plates and wound on tubes in an effort to augment nucleate pool boiling heat transfer. These promoters can be categorized as follows:

- wires, both wetting (metallic) and nonwetting (nylon)
- meshes and screens, again both wetting and nonwetting
- spheres, hemispheres, and spines
- specially designed cover plates and shrouds
- various wicking materials for heat pipes

Webb (1970) wound metallic wires and nonmetallic cords between the fins of integral low-finned tubes to augment their performance for evaporating refrigerants.

Schmittle and Starner (1978) further simplified this procedure by wrapping a nonwetting cord (nylon) around plain, smooth, drawn tubing. To create additional nucleation sites and to reduce the cross-sectional area in the channel underneath the T-shaped fins of the Gewa-T enhanced boiling surface of Wieland-Werke (personal communication, 1985), Marto, Wanniarachchi, and Pulido (1985) wrapped two to five metallic wires 0.1 mm in diameter around the channel. The resulting boiling performances for refrigerant-113 at 1.01 bar are shown in Fig. 3-19. Substantial improvement was noted at low heat fluxes compared to the plain Gewa-T tube.

Meshes and screens have been tested by quite a few investigators. Abhat and Seban (1974), for instance, wound copper mesh around a cylinder and tested this configuration in water. Hasegawa, Echigo, and Irie (1975) tested various woven meshes of one or two layers for boiling water on a heated plate. Asakavicius et al. (1979) tested surfaces with eight to twelve layers of copper screen for water, refrigerant-113, and ethanol. Augmentations typically were not large, only about 100% at low heat fluxes and less or zero at higher fluxes. Nishikawa, Ito, and Tanaka (1983) demonstrated that the degree of contact of the screen with the heated wall is of paramount importance. They observed that good contact (e.g., metallurgical contact by plating over the attached promoter or sintering it to the surface) produced high augmentation, whereas poor

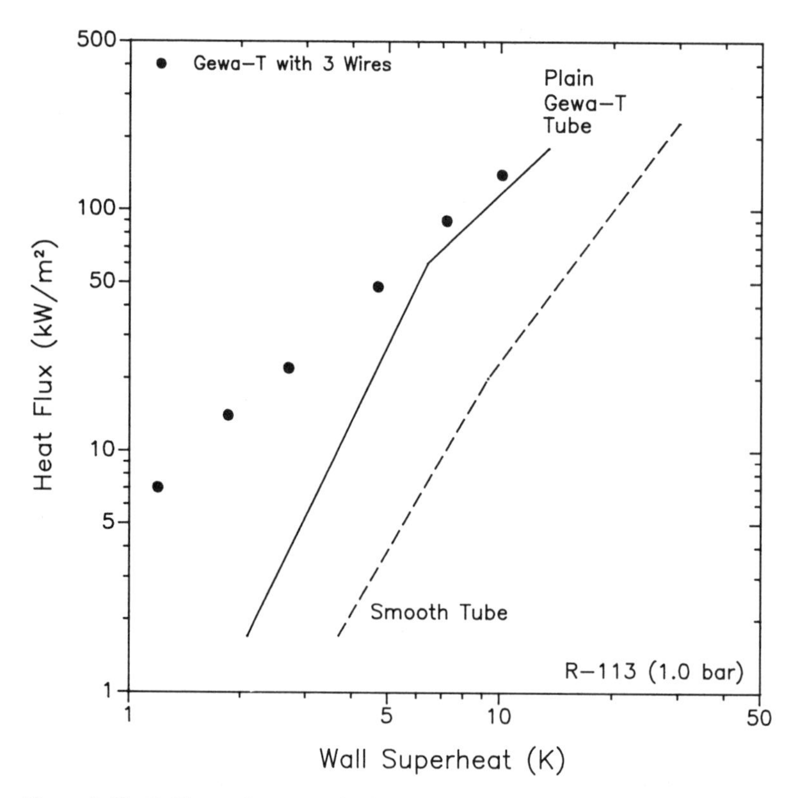

Figure 3-19 Boiling performance of a Gewa-T tube wth three wires wrapped in its channel.

contact (e.g., mechanical contact with or without pressure) gave much lower levels of augmentation.

Ma et al. (1986) tested surfaces with triangular and rectangular grooves covered by metallic screens and porous plates of differing geometries. The screens (and porous plates) were sintered to the surfaces to establish good contact. The boiling performances for the screens with a porous cover plate on top gave the best results. For the best configuration, boiling performance was about 10 times that of a smooth surface when compared at the same moderate heat flux level. At higher heat fluxes, performance deteriorated toward that of the plain surface tested.

Spheres and particles metallurgically fixed to a surface by plating, soldering, or sintering were described above under the porous enhancement category. Ferrel and Alleavitch (1970), however, tested spherical monel particles simply piled on a heated plate. This yielded better enhancement in water than that obtained by Abhat and Seban (1974) for a copper mesh but was still not as good as promoters metallurgically attached to the surface. Shih and Westwater (1974) also reported testing spheres, hemispheres, and disks as high-performance fins for boiling.

A unique cover plate was developed by Ragi (1972) by stamping a 0.08-mm thick sheet of metallic foil to produce an embossment with a large density of four-sided pyramid-shaped projections. This cover plate was then punctured at the peaks of the pyramids and brazed at its base to the heated wall, effectively forming a reentrant cavity under the foil at each pyramid. Boiling performance for water was reported to be about 10 times that of a smooth surface at the same heat flux.

The effects of placing a shroud over the Gewa-T surface have been studied by Marto and Hernandez (1983) and Marto, Wanniarachchi, and Pulido (1985) for boiling refrigerant-113. The aluminum shrouds were thin-walled cylinders with longitudinal openings along the bottom and top. These investigators found that a shroud whose top opening had an apex angle of 60° and whose bottom opening was 8.5° gave the greatest improvement in the Gewa-T boiling performance; approximately a factor of 2 greater than that of an unshrouded Gewa-T tube.

Arshad and Thome (1983) soldered microdrilled cover plates to grooved surfaces to form two different boiling enhancements. The cover plates were made with 0.005-mm thick copper shimming stock, and each had hundreds of microdrilled holes. A thin layer of soft solder was applied before cutting the grooves so that the cover plates could be "baked" on in an oven. Figure 3-20 compares their boiling curves in water at 1.0 bar to the data of Nakayama et al. (1979) for different Thermoexcel-E enhanced boiling surfaces (described below) and several plain surfaces. The rectangular grooved surface with 0.25-mm diameter holes in its cover plate had the better performance of the two.

Various wicking materials have been studied for their ability to enhance the boiling process inside heat pipes. The desired characteristics of a wick in a heat pipe are not, however, the same as for enhanced boiling surfaces used in shell-and-tube heat exchangers. Besides its evaporation characteristics, the capillary and pressure drop characteristics are important to establish the wick's pumping ability, which in turn is related to dryout of the wick and its maximum operational heat flux. Corman and McLaughlin (1976) discussed the desirable characteristics of wicks in detail and presented some boiling test data for heat transfer and maximum heat flux for wicks

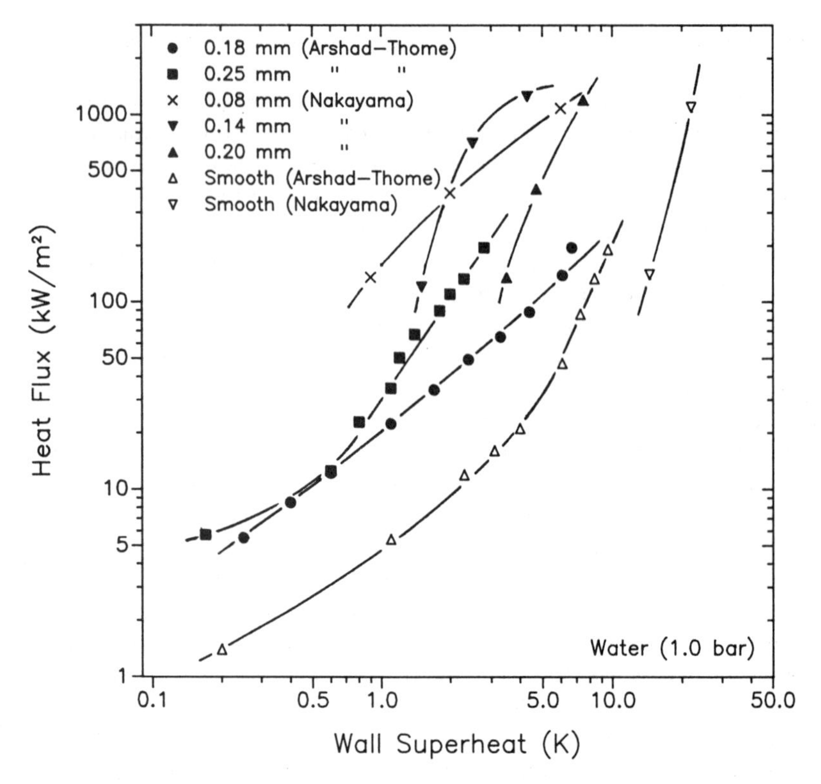

Figure 3-20 Boiling curves of enhanced boiling surfaces in water at 1.0 bar.

constructed with nickel and copper felt metals, a nickel screen, and several finned surfaces. Evaporative heat transfer performances during nucleate pool boiling for a 20% density nickel felt metal wick, for instance, were several times better than those for a plain surface depending on the type of liquid feed (liquid feed from one end of the wick; liquid feed from the entire periphery of the wick; or a flooded wick, that is one submerged under liquid).

In summary, treated surfaces have in some cases been excellent for augmenting nucleate pool boiling. Sintering, brazing, or plating of metallic promoters to the wall is necessary to obtain good thermal contact and thus substantial improvement.

Integral Extended Surfaces

Extensive development work has been done on external, integral extended surfaces. The first type to be used in boiling was almost certainly the low-finned tube. Because these were applied to many processes other than boiling, however, it is difficult to pinpoint the origin of their use for evaporation. Referring to Webb, Bergles, and Junkhan (1983), the first externally finned tube patented specifically for use with liquids (condensation) was developed by Lea (1941).

Katz and co-workers (see the discussion of early studies, above, for their data) in the late 1940s and mid-1950s studied the boiling performance of single horizontal finned tubes and the effects of stacking these finned tubes one over the other to model boiling in a tube bundle. The rising vapor from the lower tubes was discovered to increase the heat transfer coefficient of the upper tubes, especially at low wall superheatings. This valuable work also demonstrated that the boiling heat transfer coefficients for a finned tube per unit of wetted surface area were equal to or greater than those for a plain tube operating under the same conditions. This result was important to practicing engineers because they could now design an evaporator using the total wetted surface area and a plain tube nucleate pool boiling heat transfer coefficient and still attain a conservative, reliable design. Thus the stage was set for the development and use of still more effective boiling surfaces.

Integral low-finned tubes. External boiling surfaces are primarily utilized in the form of horizontal bundles of tubes. In this configuration, helical circumferential-finned tubes allow the two-phase fluid to rise vertically between adjacent fins and on up to the next row of tubes. These tubes usually have a fin height of about 1⁄16 inch (1.6 mm), and it is normal practice to reduce the inside tube diameter during the finning process so that the diameter over the fins is equal to or slightly less than the plain unfinned ends. These finned tubes can thus be fed through the holes in tube sheets and baffles and therefore can also replace an existing plain tube bundle.

A wide variety of integral low-finned tubes are available. Initially, only ductile metals could be finned. Manufacturing methods have now been developed to fin even the hardest materials. For instance, Mascone (1986) reports that High Performance Tube, Inc. developed a new method for extruding integral low-finned tubing under pressure for metals such as zirconium and Hastelloy C22. Typical fin densities used for boiling vary from 11 to 30 fpi (434 to 1,184 fins per meter), with about 19 to 26 fpi (750 to 1026 fins per meter) being perhaps the most common, although not necessarily the optimum, fin density (there are also many medium- and high-finned tubes available commercially, but so far little interest has been shown in using them for boiling, probably because they cannot be drawn though a tube sheet). Integral low-finned tubes are described further in Chapter Four; fundamentals of their boiling characteristics are discussed in Chapter Six.

Development of special extended surface geometries. Haley and Westwater (1966) conducted one of the first studies on ways to improve the boiling performance of extended surfaces. They sought to optimize the shape of a single spine for pool boiling by minimizing its volume. They considered the situation in which the heated wall at the base of the fin was sufficiently hot to be in the film boiling regime. Because of the large variation in boiling heat transfer coefficient with wall superheat over the entire range of the pool boiling curve, they found that a turnip-shaped fin was the most effective (where the stem is joined to the wall). This is not the optimum shape when only nucleate pool boiling occurs on the fin, however, or when fins are in close proximity to one another.

The evolution of special extended surface geometries centered primarily on ways to alter low-finned tubes rather than the shape of just one fin. These techniques attempted to produce large densities of reentrant cavities to promote nucleate boiling at low wall superheats and to attain a high boiling site density. Because the tubes must slide through tube sheets and baffles when shell-and-tube heat exchangers are being constructed, an important restriction was that the diameter of the enhanced section had to be equal to or less than that of the plain ends of the original tube.

The following is a chronological summary of various extended surfaces developed around the world. Sketches of some of these (from Webb [1981]) are given in Fig. 3-21 (others are shown in Chapter Four).

1. Ware (1967) developed a knurling process to alter low-finned tubes to produce knob-shaped fins. This geometry, however, did not create reentrant cavities.
2. Kun and Czikk (1969) made a surface with grooves cut in two directions to produce a multitude of reentrant cavities and channels.
3. Szumigala (1971) patented a knurling process that cut the tops of helical circumferential fins into a Y-shape and bent them over to form reentrant channels.
4. Webb (1972) developed an effective extended surface geometry by rolling low-finned tubes to bend the fins over one another and thus to create a continuous reentrant cavity with a controlled size of the opening.
5. Fujie et al. (1977) apparently took Webb's idea a step further and rolled the saw-toothed Hitachi Thermoexcel-C tube that was developed 2 years earlier for condensation, bending it over to form circumferential channels that were con-

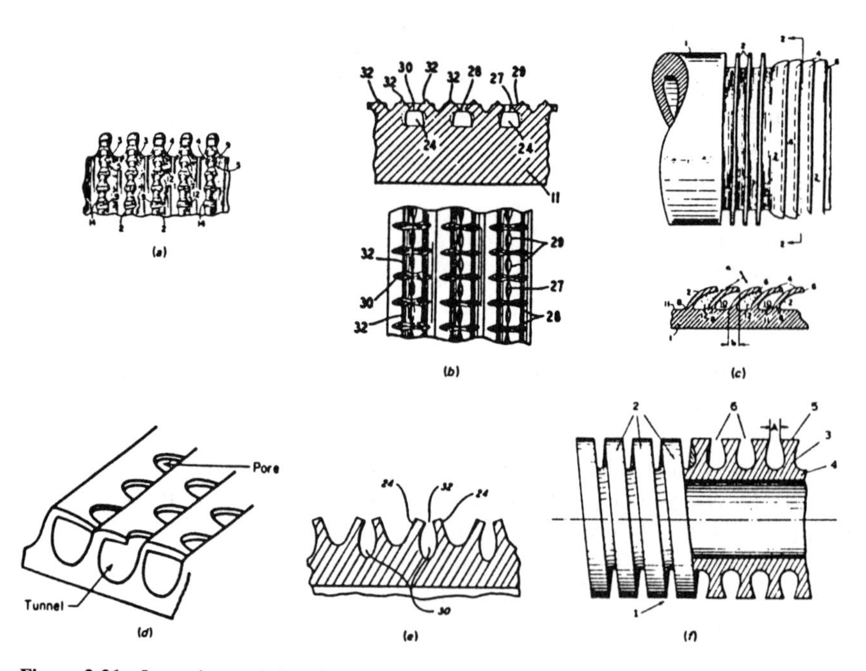

Figure 3-21 Integral extended surface geometries developed for nucleate pool boiling (From "Evolution of Enhanced Surface Geometries for Nucleate Boiling" by Ralph L. Webb, *Heat Transfer Engineering*, 1981, Vol. 2, Nos. 3–4, pp. 46–69. Copyright 1981 by Hemisphere Publishng Corporation. Reprinted by permission of publisher and author.) (*a*) Ware (1967), (*b*) Szumigala (1971), (*c*) Webb (1972), (*d*) Fujie et al. (1977), (*e*) Brothers and Kallfelz (1979), and (*f*) Saier, Kastner, and Klockler (1979).

nected to the outside by triangular holes. This tube is known commercially as the Thermoexcel-E (see Yukiteru and Kiyoshi [1986] for a history of the tube's development and manufacture).

6. Brothers and Kallfelz (1979) produced helical circumferential fins with an alternating size of root diameters between adjacent fins. These fins were then bent toward one another to produce alternating channels, one reentrant and the other open. This configuration was somewhat similar to the earlier designs by Szumigala (1971) and Webb (1972).
7. Saier, Kastner, and Klockler (1979) formed V-shaped fins by cutting the tips of circumferential fins and splitting them to form both reentrant channels and external open channels. They also formed T-shaped fins by rolling the Y-shaped fins to flatten their tops, giving the tube that is commercially known as the Gewa-T.
8. Fujikake (1980) patented a new type of cross-grooved surface similar to that developed by Kun and Czikk (1969). They cross-cut a low-finned tube having 40 fpi (1,579 fins per meter) and then compacted the spines formed, producing reentrant channels with a grid pattern. This tube is commercially known as the ECR-40 of Furukawa Electric.
9. Hitachi Cable (1984) developed a new version of their tube, the Thermoexcel-HE, which has fine projections formed just beneath the triangular holes to improve the surface's resistance to subcooling and to increase its boiling performance.
10. Wolverine Tube (1985) introduced another tube with a grid pattern of reentrant channels that is somewhat similar to the ECR-40, but they compacted the spines to form a narrow channel in one direction and a larger channel in the other. The geometry of the channels and their spacings were optimized for boiling refrigerants. This tube is known commercially as the Turbo-B.
11. Wieland-Werke (personal communication, 1985) developed the Gewa-TX tube, a new version of their earlier tube. It has periodic notches in the channels formed under the T-shaped fins to augment the convective two-phase heat transfer process inside the channels.

The above examples of integrally enhanced boiling tubes represent quite a wide range of the geometries produced with metalworking processes. The key feature common to all of them (with the exception of the first one) is the formation of one or more reentrant grooves. The geometry of the groove (or grooves) and deformed fin together with the size and shape of the opening to the outside pool of liquid are the crucial parameters controlling their boiling performances. Boiling performances for many of these tubes are presented in Chapter Seven. Mixture boiling data (where available) are described in Chapter Nine.

3-4 DEVELOPMENT OF ENHANCED SURFACES FOR INTUBE BOILING

Enhancements developed for intube boiling are typically different from external ones because convection tends to dominate the evaporation process and because pressure drop is an important consideration. Flow boiling in tubes is a strong function of many additional parameters, such as the flow pattern, the mass flux, the local vapor quality, and the orientation of the tube (horizontal and vertical).

Historical Developments

For boiling inside a tube, the heat transfer process is controlled by the contributions of convection and nucleate boiling. Most intube boiling enhancements have sought to modify the convection process. The reasoning was that this would not only augment the contribution of liquid-phase convection to boiling but would also increase heat transfer to subcooled liquids entering the tube and augment boiling at large exit vapor qualities or superheating dry vapor. Therefore, the main objective of these development programs has been to promote turbulence to increase convective heat transfer while minimizing the associated increase in pressure drop through the tube.

Reviewing the list of patents in Webb, Bergles, and Junkhan (1983) to reconstruct the historical development of enhanced boiling techniques for intube boiling, it can be seen that the first patent for intube heat transfer augmentation found by their search was one by Kemnal (1930) for the promotion of swirl flow for forced convection to air. The first intube boiling patent was a displacement enhancement device developed by Phillip (1933). Later, Bailey (1942) patented a roughened surface for intube boiling. The major step forward appears to be the Rogers (1949) patent, a method for manufacturing internally finned tubes, because most of those commercially available today are of that variety. Another significant advance was the metallic porous layer (High Flux) patented by Milton (1968), which could be applied to either the inside or outside wall of a tube. This coating augments the nucleate boiling contribution to flow boiling rather than the liquid convection contribution.

Integral enhancement techniques for intube boiling consist of manufacturing tubing with internal longitudinal fins or internal spiral or helical fins, corrugating or fluting the tube, or roughening its inside surface. The inside diameter of the tube, the tube wall material, the number of fins, the fin shape and dimensions, and the helix angle of the fins relative to the axis of the tube have been found to be the primary factors controlling enhancement.

Inserts placed inside tubes can be either loosely or snugly fit. For instance, one widely used enhancement is the aluminum star-shaped insert that is placed inside a tube before mechanically stretching the tube to reduce its diameter to produce a tight fit on the insert. Other inserts used or studied include spirally twisted metallic tapes, coiled wire inserts, inlet vortex generators, brushes, and mesh inserts.

Doubly enhanced tubes for boiling inside tubes with condensation (or single-phase convection) on the outside have also been developed. For instance, if a boiling enhancement is applied to the inside of tubes in a steam-heated vertical thermosyphon reboiler, it may be economically advantageous to use longitudinal fluted fins on the outside to enhance the condensation process. Tubes with integral internal boiling enhancements can be corrugated or externally finned to augment the shell-side coefficient for horizontal tube bundles.

The metalworking processes used to make internal integral extended surfaces are restricted by the narrow channel size of the tube and the long unsupported length of the tooling. Most internally enhanced tubes therefore have much simpler geometries than those used externally. Tubes with coatings or inserts are not affected by this restriction, but they may suffer from poor thermal contact if they are not metallurgically joined to the inner tube wall, which was shown to be a crucial factor in the use of metallic screens on externally enhanced boiling surfaces by Nishikawa, Ito, and Tanaka (1983) (see the discussion of treated surfaces, above).

Intube Boiling Augmentation

Boiling performances of internal boiling enhancements are not easy to compare because most of the data, even when obtained for the same fluid and at the same pressure, have not been gathered under the same test conditions (inlet subcooling, flow rate, tube diameter and length, exit quality, and so forth). The most valid comparison from a design standpoint is one conducted under the actual operating conditions of the evaporator to be designed. For instance, an enhanced boiling tube found to be superior by its manufacturer for evaporation at low mass fluxes may not be as effective when applied at larger mass fluxes. Nevertheless, to round out the historical development of these enhancements, several performance comparisons are shown below (this subject is treated in depth in Chapter Ten, Sec. 10-1).

Figure 3-22 (from Kubanek and Miletti [1979]) depicts a comparison of the average boiling heat transfer coefficients of different horizontal tubes for a fixed change in vapor quality (from inlet to outlet) of 0.7 at a nominal saturation pressure of 5.8 bar (4.4°C) for boiling R-22. The tubes represented are three internal spirally finned tubes (Forge-Fin tubes, Noranda Metal Industries), one tube with a twisted star-shaped insert, and a plain tube. Table 3-1describes the tube geometries tested. The tube with the star-shaped insert (24C) had the best heat transfer performance at low mass fluxes, providing an average heat transfer coefficient about five times that of the plain tube and several times that of each of the finned tubes. At the highest mass flux studied, the internally finned tube (25) performed best, giving an average heat transfer performance about three to four times that of the plain tube and about twice that of the tube with the insert. These increases were significantly larger than the corresponding area ratios in Table 3-1; thus the boiling process itself was augmented.

The pressure drop data for these tube test sections showed that the finned tubes had increases in pressure drop per unit length relative to the plain tube of 10% to 290%; the tube with the star-shaped insert had much greater increases of 320% to 1,500%. This study probably had a large influence on the development of intube boiling enhancements, principally by demonstrating that internally finned tubes could provide as good or better heat transfer performance than the widely used star inserts but at much smaller pressure drop penalties.

Figure 3-23 shows intube boiling data for oxygen obtained by Czikk, O'Neill, and Gottzmann (1981) for an 18.7-mm internal diameter High Flux tube with vertical upward flow, apparently at 1.01 bar. The local boiling heat transfer coefficients are plotted with the exit vapor quality indicated in parentheses. For reference, plain tube curves were calculated from the Chen (1966) correlation described in Chapter Two for several different local vapor qualities and mass fluxes. In addition, the nucleate pool

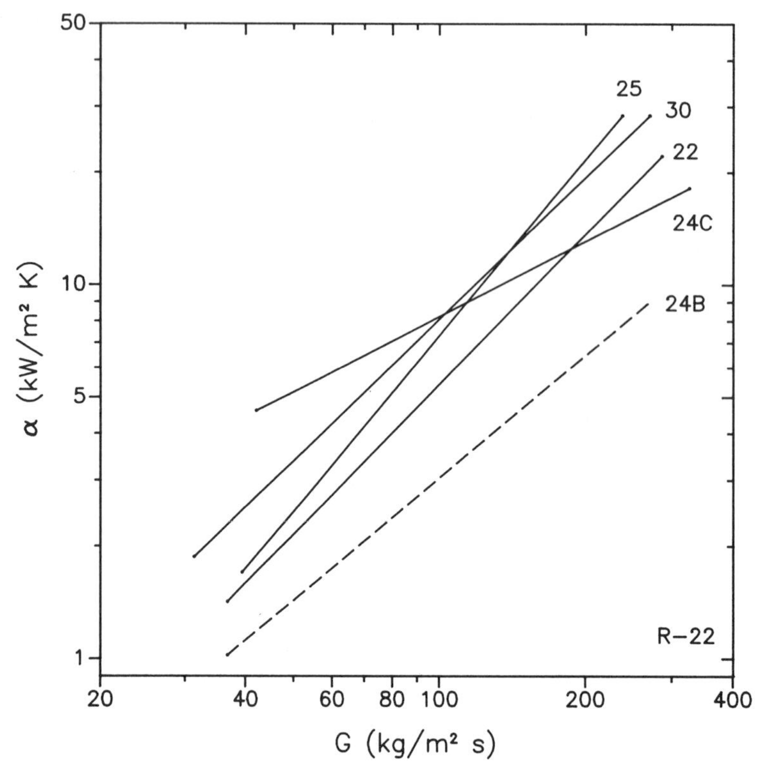

Figure 3-22 Average boiling heat transfer coefficient for evaporation inside enhanced tubes (see Table 3-1).

Table 3-1 Specifications of tubes in Fig. 3-22

Tube type and number	Number of fins	Tube internal diameter, mm	Tube hydraulic diameter, mm	Fin height, mm	Fin pitch, mm	Wetted area per unit length, $mm^2/m \times 10^{-3}$	Nominal area per unit length, $mm^2/m \times 10^{-3}$	Area ratio	Heated length, m
Plain, 24B		14.4	14.4			45.3	45.3	1.00	0.80
Insert, 24C	5	14.4	4.09		610	90.8	45.2	2.00	0.80
Finned, 22	32	14.7	7.57	0.635	305	87.0	46.2	1.88	0.80
Finned, 25	32	14.7	7.57	0.635	152	87.2	46.2	1.89	0.80
Finned, 30	30	11.9	6.30	0.508	102	68.0	37.4	1.82	0.80

boiling design curve for oxygen at 1.01 bar given in Antonelli and O'Neill (1981) for the High Flux tube is shown.

From the above comparison, the porous enhancement is seen to increase the intube boiling heat transfer coefficient for oxygen by a factor of 10 or more relative to flow boiling in a plain tube evaluated at the same heat flux. In addition, the flow boiling heat transfer coefficient for vertical upward flow inside the High Flux tube is insensitive

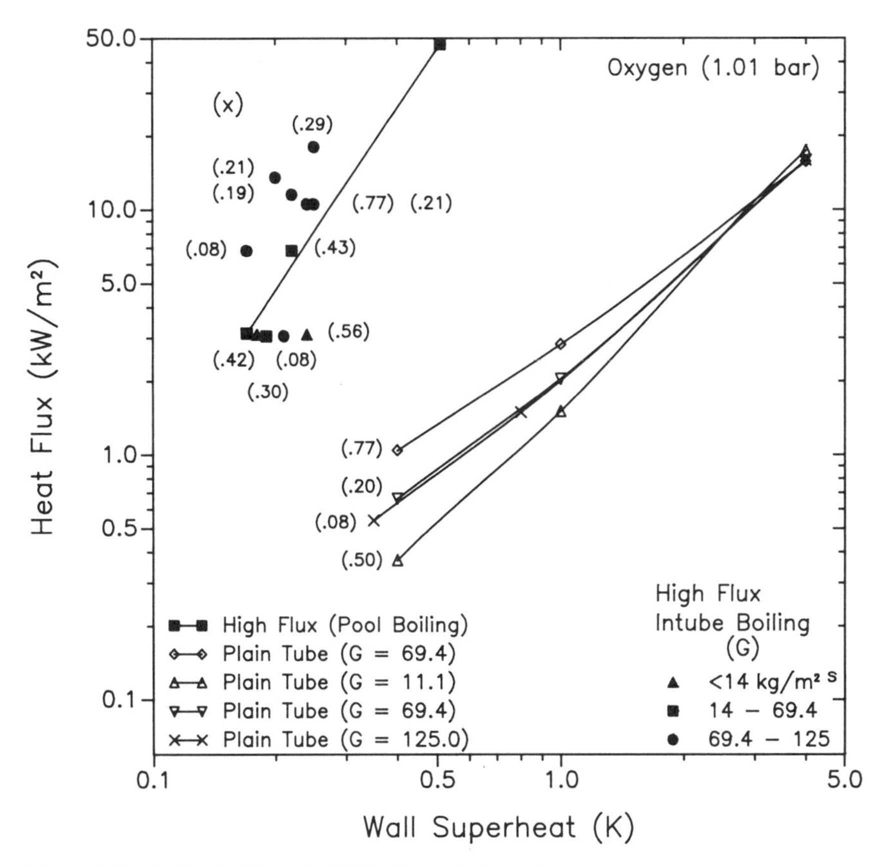

Figure 3-23 Boiling inside vertical High Flux and plain tubes.

to the exit vapor quality, which ranged from 0.08 to 0.77, and the mass flux. In contrast, the flow boiling heat transfer coefficient for the plain tube is a function of both vapor quality and mass flux.

Agrawal, Varma, and Lal (1986) recently studied the effect of twist ratio for 0.5-mm thick stainless steel twisted-tape inserts fitted snugly inside a 10.0-mm internal diameter tube. Refrigerant-12 was tested in the 2.1-m horizontal test section at a constant exit pressure of 1.38 bar. Four twist ratios were studied (the twist ratio is defined as the ratio of half the pitch of the helix of the tape to the inside diameter of the tube). Figure 3-24 depicts the variation in the local boiling heat transfer coefficient with the local vapor quality for the four different twisted tapes and the plain tube with the heat flux and the mass flux held constant. The local boiling heat transfer coefficient varied with vapor quality, but the trend was different for each twist ratio. The tape with the lowest twist ratio (i.e., the tape with the highest number of twists per unit length) gave the best heat transfer performance, being on average about 1.8 times better than that of the plain tube for these particular conditions. The two tapes with the highest twist ratios only out-performed the plain stainless steel tube at vapor qualities higher than 0.45, and then not by more than 40%.

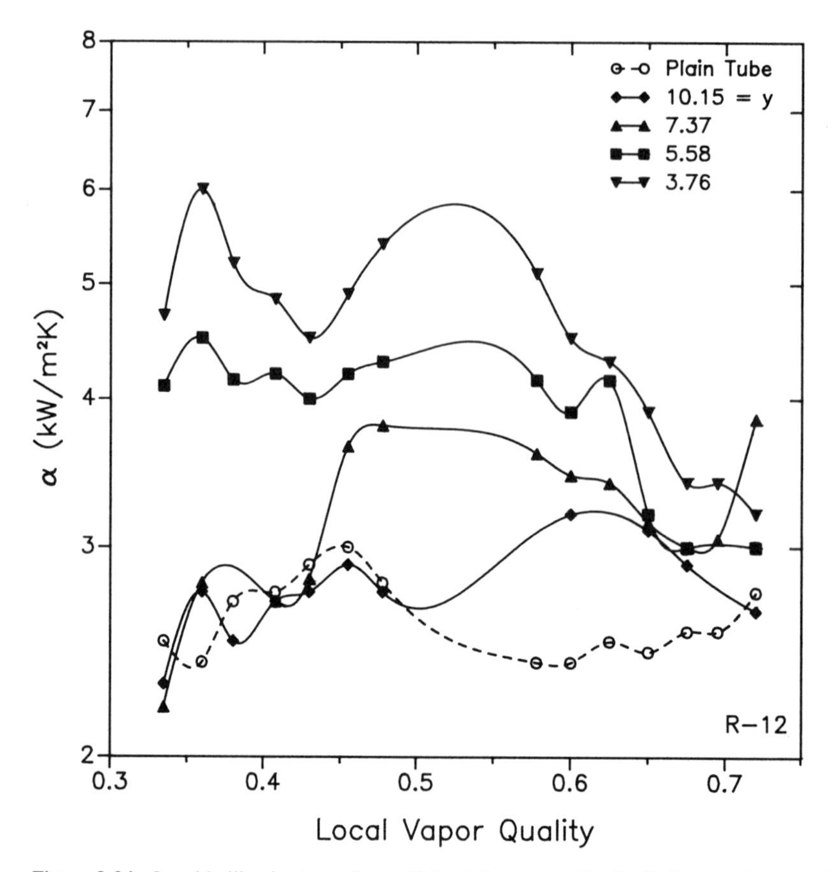

Figure 3-24 Local boiling heat transfer coefficients for evaporation inside horizontal tubes with different twisted tapes.

Although enhanced intube boiling research has emphasized maximizing the boiling heat transfer coefficient (or heat duty per unit length of tubing), some studies have sought instead to augment the critical heat flux, primarily for the cooling of fossil-fuel and nuclear steam generators. A historical survey of this subject is beyond the scope of the present review. Readers desiring further information are referred to Bergles (1976) and Collier (1981).

3-5 NOMENCLATURE

$\Delta\tilde{g}$	molar free energy of formation (J/mole)
G	mass flux [kg/(m^2·s)]
$\Delta\tilde{h}_v$	molar latent heat of vaporization (J/mole)
N/A	boiling site density (site/m^2)
$(dP/dT)_{sat}$	slope of vapor pressure curve (N/m^2)

R	radius of curvature of liquid-vapor interface (m)
R_{crit}	critical radius of curvature at cavity mouth (m)
R_{max}	maximum radius of curvature (m)
R_{min}	minimum radius of curvature (m)
T_{sat}	saturation temperature (K)
T_w	wall temperature (K)
ΔT	wall superheat (K)
ΔT_{sat}	boiling nucleation superheat (K)
$\tilde{v}_g$	molar density of vapor (mole/m^3)
V	vapor volume (m^3)
y	height from surface (m)
α	boiling heat transfer coefficient [W/(m^2·K)]
β	contact angle
β_{macro}	contact angle observed macroscopically
β_{micro}	contact angle observed microscopically
β_{min}	minimum contact angle
θ	wedge angle of groove or conical cavity
σ	surface tension (N/m)
ϕ	heterogeneous nucleation correction factor

REFERENCES

Abhat, A., and R. A. Seban. 1974. Boiling and evaporation from heat pipe wicks with water and acetone. *J. Heat Transfer* 96:331–37.

Agrawal, K. N., H. K. Varma, and S. Lal. 1986. Heat transfer during forced convective boiling of R-12 under swirl flow. *J. Heat Transfer* 108:567–73.

Albertson, C. E. 1977. Boiling heat transfer surface and method. U. S. Patent 4,018,264, April 19.

Almgren, D. W., and J. L. Smith, Jr. 1969. The inception of nucleate boiling with liquid nitrogen. *J. Eng. Ind. Trans. ASME Ser. B* 91:1210–16.

Anonymous. 1970. New heat exchanger tubes make their commercial debut. *Chem. Eng.* March, pp. 60–62.

Antonelli, R., and P. S. O'Neill. 1981. Design and application considerations for heat exchanger with enhanced boiling surfaces. Paper read at International Conference on Advances in Heat Exchangers, September, 1981, Dubrovik, Yugoslavia.

Arshad, J., and J. R. Thome. 1983. Enhanced boiling surfaces: Heat transfer mechanism and mixture boiling. *Proc. ASME-JSME Therm. Eng. Joint Conf.* 1:191–97.

Asakavicius, J. P., A. A. Zukauskav, V. A. Gaigalis, and V. K. Eva. 1979. Heat transfer from Freon-113, ethyl alcohol, and water with screen wicks. *Heat Transfer Sov. Res.* 11(1):92–100.

Bailey, E. G. 1942. Liquid vaporizing tube. U. S. Patent 2,279,548, April 14.

Bankoff, S. G. 1956. Ebullition from solid surfaces in absence of pre-existing gaseous phase. *Proc. Heat Transfer Fluid Mech. Inst.*, Stanford University.

———. 1957. Ebullition from solid surfaces in the presence of a pre-existng gaseous phase. *Trans. ASME* 79:735.

———. 1958. Entrapment of gas in the spreading of a liquid over a rough surface. *AIChE J.* 4(1):24–26.

Benjamin, J. E. 1960. Bubble growth in nucleate boiling of a binary mixture. Ph.D. diss., University of Illinois.

———. and J. W. Westwater. 1961. Bubble growth in nucleate boiling in a binary mixture. *Int. Dev. Heat Transfer* 212–218.

Berenson, P. J. 1960. Transition boiling heat transfer from a horizontal surface. Sc.D. diss., Massachusetts Institute of Technology.

———. 1962. Experiments on pool boiling heat transfer. *Int. J. Heat Mass Transfer* 5:985–99.

Bergles, A. E. 1976. Survey of augmentation of two-phase heat transfer. *ASHRAE Trans.* 82(part 1): 881–90.

———. and R. L. Webb. 1978–1980. Bibliography on augmentation of convective heat and mass transfer. *Previews Heat Mass Transfer* 4(2):61–73 (1978), 4(4):89–106 (1978), 5(2):83–103 (1979), 5(3):83–102 (1979), 6(1):89–106 (1980), 6(3):292–314 (1980).

Brothers, W. S., and A. J. Kallfelz. 1979. Heat transfer surface and method of manufacture. U. S. Patent 4,159,739, July 3.

Chappius, J. 1982. Contact angles. *Multiphase science and technology*, ed. G. F. Hewitt, J. M. Delhaye, and N. Zuber, 387–505.

Chen, J. C. 1966. A correlation for boiling heat transfer to saturated fluids in convective flow. *Ind. Eng. Chem. Process Des. Dev.* 5(3):322–29.

Clark, H. B., P. H. Strenge, and J. W. Westwater. 1959. Active sites for nucleate boiling. *Chem. Eng. Prog. Symp. Ser.* 55(29):103–10.

Collier, J. G. 1981. *Convective boiling and condensation.* 2d ed., 363–79. New York: McGraw-Hill.

Corman, J. C., and M. H. McLaughlin. 1976. Boiling augmentation with structured surfaces. *ASHRAE Trans.* 82(part 1):906–18.

Cornwell, K. 1975. Paper read at AIChE Annual Meeting, November, Los Angeles (paper 114c).

———. 1977. *Lett. Heat Mass Transfer* 4(1):63.

Corty, C., and A. Foust. 1955. Surface variables in nucleate boiling. *Chem. Eng. Progr. Symp. Ser.* 51(17):1–12.

Czikk, A. M., C. F. Gottzmann, E. G. Ragi, J. G. Withers, and E. P. Habdas. 1970. Performance of advanced heat transfer tubes in refrigerant-flooded liquid coolers. *ASHRAE Trans.* 76(part 1):96–109.

Czikk, A. M., P. S. O'Neill, and C. F. Gottzmann. 1981. Nucleate boiling from porous metal films: Effect of primary variables. In *Advances in Enhanced Heat Transfer*, HTD Vol. 18, 109–122. New York: American Society of Mechanical Engineers.

Dahl, M. M., and L. D. Erb. 1976. Liquid heat exchanger interface and method. U. S. Patent 3,990,862, November 9.

Dyundin, V. A., G. N. Danilova, A. V. Borishanskaya, V. N. Krotkov, V. A. Gogolin, V. A. Vakhalin, and G. A. Protasov. 1975. Enhancement of heat transfer with boiling refrigerants on coated surfaces. *Khim. Neft. Mashinostr.* 9:22–23.

Ferrel, J. K., and J. Alleavitch. 1970. Vaporization heat transfer in capillary wick structures. *Chem. Eng. Prog. Symp. Ser.* 66(102):82–91.

Fujie, K., W. Nakayama, H. Kuwahara, and K. Kakizaki. 1977. Heat transfer wall for boiling liquids. U. S. Patent 4,060,125, November 29.

Fujii, M. 1984. Nucleate pool boiling heat transfer from a porous heating surface (optimum particle diameter). *Heat Transfer Jap. Res.* 13(1):76–91.

Fujii, M., E. Nishiyama, and G. Yamanaka. 1979. Nucleate pool boiling heat transfer from micro-porous heating surface. In *Advances in Enhanced Heat Transfer*, HTD Vol. 18, 45–51. New York: American Society of Mechanical Engineers.

Fujikake, J. 1980. Heat transfer tube for use in boiling-type heat exchangers and method of producing the same. U. S. Patent 4,216,826, August 12.

Gaertner, R. F. 1967. Methods and means for increasing the heat transfer coefficient between a wall and a boiling liquid. U. S. Patent 3,301,314, January 31.

Gottzmann, C. F., P. S. O'Neill, and P. E. Minton. 1973. High efficiency heat exchangers. *Chem. Eng. Prog.* 69(7):69–75.

Grant, A. C. 1974. Porous metallic layer formation. U. S. Patent 3,821,018, June 28.

———. 1977. Porous metallic layer and formation. U. S. Patent 4,064,914, December 27.

Griffith, P., and J. D. Wallis. 1960. The role of surface conditions in nucleate boiling. *Chem. Eng. Progr. Symp. Ser.* 56(49):49–63.

Haley, K. W., and J. W. Westwater. 1966. Boiling heat transfer from single fins. *Proc. 3rd Int. Heat Transfer Conf.* 3:245–53.

Harvey, E. N. 1944. *J. Cell. Comp. Physiol.* 24:1,23.

Harvey, E. N. 1945. *J. Am. Chem. Soc.* 67:156.

Hasegawa, S., R. Echigo, and S. Irie. 1975. Boiling characteristics and burnout phenomena on a heating surface covered with woven screens. *J. Nucl. Sci. Tech.* 12(11): 722–24.

Hitachi Cable. 1984. Hitachi high-performance heat-transfer tubes. Hitachi Cat. No. EA-500, Japan.

Hsu, Y. Y. 1962. On the size range of active nucleation cavities on a heating surface. *J. Heat Transfer* 84:207–16.

———. and R. W. Graham. 1961. *An analytical and experimental study of the thermal boundary layer and ebullition cycle in nucleate boiling* (NASA paper TND-594).

Hummel, R. L. 1965. Means for increasing the heat transfer coefficient between a wall and a boiling liquid. U. S. Patent 3,207,209, September 21.

Inoue, K. 1974. Heat exchanger. U. S. Patent 3,825,064, July 23.

Jacob, M., and W. Fritz. 1931. *Forsch. Geb. Ing.* 2:435.

Janowski, K. R., M. S. Shum, and S. A. Bradley. 1978. Heat transfer surface. U. S. Patent 4,129,181, December 12.

Jones, W. 1941. Cooler and condenser heat transfer with low pressure Freon refrigerant. *Refrig. Eng.* 41(6):413–18.

Katz, D. L., J. E. Myers, and E. H. Young. 1955. Boiling outside finned tubes. *Petrol. Refiner* 34(2):113–16.

Kemnal, J. 1930. Air heater, U. S. Patent 1,770,208, July 8.

Kotchaphakdee, P., and M. C. Williams. 1969. Enhancement of nucleate boiling with polymeric additives. *Int. J. Heat Mass Transfer* 13:835–48.

Kubanek, G. R., and D. L. Miletti. 1979. Evaporative heat transfer and pressure drop performance of internally finned tubes with refrigerant-22. *J. Heat Transfer* 101:447–52.

Kun, L. C., and A. M. Czikk. 1969. Surface of boiling liquids. U. S. Patent 3,454,081, July 8.

Kurihari, H. M., and J. E. Myers. 1960. Effects of superheat and roughness on boiling coefficients. *AIChE J.* 6(1):83–91.

Lea, E. S. 1941. Finned condenser tube. U. S. Patent 2,241,209, March 11.

Lorentz, J. J., B. B. Mikic, and W. M. Rohsenow. 1974. The effect of surface conditions on boiling characteristics. *Proc. 5th Int. Heat Transfer Conf.* 4:35–39.

Ma, T., X. Liu, J. Wu, and H. Li. 1986. Effects of geometrical shapes and parameters of reentrant grooves on nucleate pool boiling heat transfer from porous surfaces. *Proc. 8th Int. Heat Transfer Conf.* 4:2013–18.

Marto, P. J., and B. Hernandez. 1983. Nucleate pool boiling characteristics of a Gewa-T surface in Freon-113. *AIChE Symp. Ser.* 79(225):1–10.

Marto, P. J., and J. Lepere. 1981. Pool boiling heat transfer from enhanced surfaces to dielectric fluids. In *Advances in Enhanced Heat Transfer*, HTD Vol. 18, 93–102. New York: American Society of Mechanical Engineers.

Marto, P. J., J. A. Moulson, and M. D. Maynard. 1968. Nucleate pool boiling of nitrogen with different surface conditions. *J. Heat Transfer* 90:437–44.

Marto, P. J., and W. M. Rohsenow. 1965. *The effect of surface conditions on nucleate pool boiling heat transfer to sodium* (M. I. T. Mechanical Engineering Department report MIT-3357-1).

———. 1966. Effects of surface conditions on nucleate pool boiling of sodium. *J. Heat Transfer* 88:196–204.

Marto, P. J., A. S. Wanniarachchi, and R. J. Pulido. 1985. Augmenting the nucleate pool boiling characteristics of Gewa-T finned tubes in R-113. In *Augmentation of heat transfer in energy systems*, ed. P. J. Bishop, HTD Vol. 52, 67–73. New York: American Society of Mechanical Engineers.

Mascone, C. F. 1986. CPI strive to improve heat transfer in tubes. *Chem. Eng.* February.

Milton, R. M. 1968. Heat exchange system. U. S. Patent 3,384,154, May 21.

———. 1970. Heat exchange system. U. S. Patent 3,523,577, August 11.

———. 1971. Heat exchange system with porous boiling layer. U. S. Patent 3,587,730, June 28.

———. and C. F. Gottzmann. 1972. High efficiency reboilers and condensers. *Chem. Eng. Progr.* 68(9):56–61.

Myers, J. E., and D. L. Katz. 1952. Boiling coefficients outside horizontal plain and finned tubes. *Refrig. Eng.* 60(1):56–69.

———. 1953. Boiling coefficients outside horizontal tubes. *Chem. Eng. Progr. Symp. Ser.* 49(5): 107–14.

Nakayama, W., T. Daikoku, H. Kuwahara, and T. Nakajima. 1979. Dynamic model of enhanced boiling heat transfer on porous surfaces. In *Advances in Enhanced Heat Transfer*, 31–43. New York: American Society of Mechanical Engineers.

Nishikawa, K., T. Ito, and K. Tanaka. 1983. Augmented heat transfer by nucleate boiling at prepared surfaces. *Proc. ASME-JSME Therm. Eng. Joint Conf.* 1:387–93.

Oktay, S. 1982. Departure from natural convection (DNC) in low-temperature boiling heat transfer encountered in cooling micro-electronic LSI devices. *Proc. 7th Int. Heat Transfer Conf.* 4:113–18.

———. and A. Schmeckenbecher. 1972. Method for forming heat sinks on semiconductor device chips. U. S. Patent 3,706,127, December 19.

Phillip, L. A. 1933. Refrigerating system. U. S. Patent 1,931,268, October 17.

Ragi, E. G. 1972. Composite structure for boiling liquids and its formation. U. S. Patent 3,684,007, August 15.

Robinson, D. B., and D. L. Katz. 1951. Effect of vapor agitation on boiling coefficients. *Chem. Eng. Progr.* 6:317–24.

Rogers, J. S. 1949. Method of making externally and internally finned tubes. U. S. Patent 2,463,997, March 8.

Saier, M., H. W. Kastner, and R. Klockler. 1979. Y- and T-finned tubes and methods and apparatus for their making. U. S. Patent 4,179,911, December 25.

Sauer, E. T. 1935. Master's thesis, Massachusetts Institute of Technology.

Schade, S. S., and E. L. Park. 1979. Effect of a plasma deposited polymer coating on the nucleate boiling behavior of Freon-113. In *Multiphase transport fundamentals, reactor safety, applications*, 771–26. Washington, D.C.: Hemisphere.

Schmittle, K. V., and K. E. Starner. 1978. Heat transfer in pool boiling. U. S. Patent 4,074,753, February 21.

Shih, C. C., and J. W. Westwater. 1974. Spheres, hemispheres and discs as high-performance fins for boiling heat transfer. *Int. J. Heat Mass Transfer* 17:125–34.

Shum, M. S. 1980. Finned heat transfer tube with porous boiling surface and method for producing same. U. S. Patent 4,182,412, January 8.

Starner, K. E., and R. A. Cromis. 1977. Energy savings using High Flux evaporator surface in centrifugal chillers. *ASHRAE J.* 19(12):24–27.

Still, W. J. 1929. Heat transmitting tube. U. S. Patent 1,716,743, June 11.

Stone, R. H., and E. F. Tilley. 1938. Finned tube. U. S. Patent 2,118,060, May 24.

Szumigala, E. T. 1971. Manufacturing method for boiling surfaces. U. S. Patent 3,566,514, March 2.

Taborek, J. 1980. Cited in S. Yilmaz and J. W. Westwater, Effect of commercial enhanced surfaces on the boiling heat transfer curve. In *Advances in Enhanced Heat Transfer*, 73–91. New York: American Society of Mechanical Engineers.

Thome, J. R. 1978. Bubble growth and nucleate pool boiling in liquid nitrogen, liquid argon, and their mixtures. Ph.D. diss., Oxford Unversity.

Trepp, C., and T. V. Hoffman. 1980. Boiling heat transfer from structured surfaces to liquid nitrogen. *Warme Stoffubertrag.* 14:15–22.

Union Carbide. 1979. Advanced heat transfer technology with High Flux tubing. Product bulletin F-4063 87-0117, October.

Vachon, R. I., G. H. Nix, G. E. Tanger, and R. E. Cobb. 1969. Pool boiling heat transfer from Teflon-coated stainless steel. *J. Heat Transfer* 91:364–70.

Venart, J. E. S., A. C. M. Sousa, and D. S. Jung. 1986. Nucleate and film boiling heat transfer in R-11: The effects of enhanced surfaces and inclination. *Proc. 8th Int. Heat Transfer Conf.* 4:2019–24.

Ware, C. D. 1967. Heat transfer surface. U. S. Patent 3,326,283, June 20.

Webb, R. L. 1970. Heat transfer surface which promotes nucleate ebullition. U. S. Patent 3,521,708, July 28.

———. 1972. Heat transfer surface having a high boiling heat transfer coefficient. U. S. Patent 3,696,861, October 10.

———. 1981. The evolution of enhanced surface geometries for nucleate boiling. *Heat Transfer Eng.* 2(3–4):46–69.

———. A. E. Bergles, and G. H. Junkhan. 1983. *Bibliography of U. S. patents on augmentation of convective heat and mass transfer II* (Department of Mechanical Engineering report HTL-32, Iowa State University, Ames).

Webber, W. O. 1960. Under fouling conditions—Finned tubes can save money. *Chem. Eng.* March:149–52.

Wolverine Tube. 1985. *Turbo-B an improved evaporator tube, product bulletin*, June.

Young, R. K., and R. L. Hummel. 1965. Improved nucleate boiling heat transfer. *Chem. Eng. Progr.* 60(7):53–58.

Yukiteru, T., and O. Kiyoshi. 1986. Development of high performance heat transfer tube Thermoexcel. Paper read at Tomorrow's Tube, 1986 International Conference and Exhibition, 10–12 June, Birmingham, England (International Tube Association, P.O. Box 84, Leamington Spa, Warwickshire, England).

Zeiman, W. E., and D. L. Katz. 1947. Boiling coefficients for finned tubes. *Petrol. Refiner* 26(8): 78–82.

CHAPTER

FOUR

COMMERCIAL ENHANCED BOILING TUBES

4-1 INTRODUCTION

Many types of enhanced boiling tubes have become commercially available for shell-side boiling and intube boiling, several with enhancement for the heating fluid side also. In addition, a number of intube inserts have been commercially developed specifically to augment boiling. This chapter documents the various geometries of these tubes and inserts together with their physical characteristics for the benefit of practicing engineers and heat transfer researchers. Boiling data for many of these tubes and inserts are presented elsewhere, especially in Chapters Seven and Ten; the photographs and diagrams presented here are intended to serve as a quick reference for the reader. Many commercial enhancements are unavoidably left out of the review; there are many still in the development stage, others for which a photograph or diagram could not be obtained, and yet others that are manufactured only for in-house use.

The boiling enhancements are classified into six categories: externally finned tubes, modified externally finned tubes, porous layer–coated tubes, internally finned tubes, doubly enhanced tubes, and tube inserts.

4-2 EXTERNALLY FINNED TUBES

Integral externally finned tubes are manufactured with various fin densities, profiles, and heights. It is only the low-finned versions that are used for boiling applications because the large boiling heat transfer coefficients for medium- and high-finned tubes would produce an unacceptably low fin efficiency. External fins that are welded, brazed, or wrapped on the outside of a tube are not used for boiling applications. External longitudinally finned tubes are also not typically utilized in boiling services except in some double-pipe heat exchangers. The integral low-finned tubes are considered enhanced boiling tubes because they provide substantially higher thermal duties per unit length than plain, drawn tubing.

Figure 4-1 is a schematic diagram of a typical low-finned tube. The basic characteristics of low-finned tubing are (1) the flat end that facilitates rolling of the tube into a tube sheet; (2) the transition zone from the unfinned flat to the finned section, where the fin height increases and the inside tube diameter decreases; (3) the finned portion made up of helical fins (tip diameter slightly less than or equal to that of the flat) and a thin wall underneath the fins; and (4) the intermediate land or flat, which is sometimes used at the baffle locations. On the S/T Trufin tube of Wolverine Tube, shown in Fig. 4-2, these characteristics can be seen (except for the baffle land). The inside of low-finned tubes may be rippled by the external finning process, as shown in Fig. 4-3 for the Hitachi low-fin and middle-fin tubes.

Standard fin densities for low-finned tubes are 11, 19, 26, 30, and 40 fins per inch (fpi), that is 434, 750, 1,026, 1,184, and 1,579 fins per meter, respectively. As can be discerned from Fig. 4-3, the higher the fin density, the lower the fin height and thickness. Fin densities in copper as high as 60 fpi (2,369 fins per meter) can be made but are rarely used commercially because the fins are thin and thus structurally weak. The shape of the fin is dependent on each manufacturer's tooling.

Integral low-finned tubes are made in many different metals. The fins are plastically formed by pinching the tube between a plug inside the tube and three sets of planetary rings of increasing diameter that compress the outer tube wall and raise the fins. The more ductile materials such as copper, copper alloys, and aluminum are easily finned, as are Admiralty, cupro-nickels, and low-carbon steels. Corrosion-resistant materials

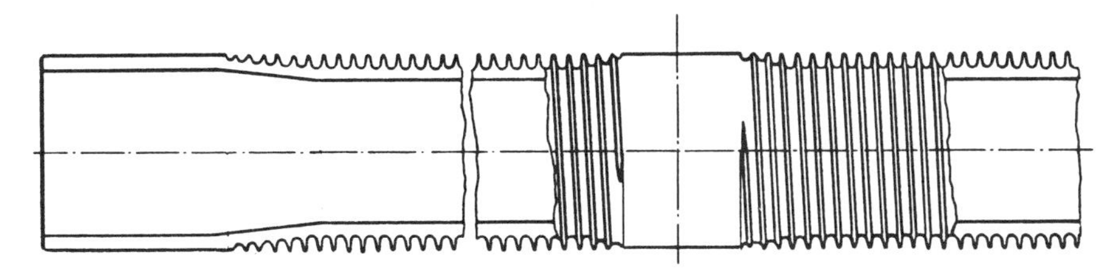

Figure 4-1 Schematic diagram of a low-finned tube.

Figure 4-2 S/T Trufin tube (courtesy Wolverine Tube).

Figure 4-3 Low- and middle-fin tubes (courtesy Hitachi).

such as stainless steel, titanium, and Inconels can also be finned. Mascone (1986) reports that the materials zirconium and Hastelloy C22, which previously could not be finned by conventional techniques because of their low ductility, can now be made into fins by a patented extrusion method under pressure developed by High Performance Tube Inc. Therefore, the design engineer can now choose from a wide selection of low-finned tube materials that can handle just about any service application.

4-3 MODIFIED EXTERNALLY FINNED TUBES

The externally finned tubes described above can be further modified by cutting, knurling, notching, or rolling a low-finned tube to form new complex fin geometries with higher boiling perfomances than those of the original tube.

The process can begin by knurling a plain tube to produce fine fins before cutting and rolling the projections to obtain a new geometry. This process is used to manufacture the Thermoexcel-E tube of Hitachi, shown in Fig. 4-4 and described by Tanno and Ooizumi (1986). Shallow grooves with a pitch of 0.7 mm (1,430 grooves per meter or 36 grooves per inch) at an angle of 45° are produced by the rotating knurling tools. The projections are then cut by the triangular cutting tool, which plows up the surface without detaching it, to produce the saw-toothed fins of the Thermoexcel-C enhanced condensation surface. Finally, these fins are rolled to bend them over to form reentrant channels to produce the Thermoexcel-E geometry shown schematically in Fig. 4-5; a photograph of the Thermoexcel-E tube is shown in Fig. 4-6. In this process only the

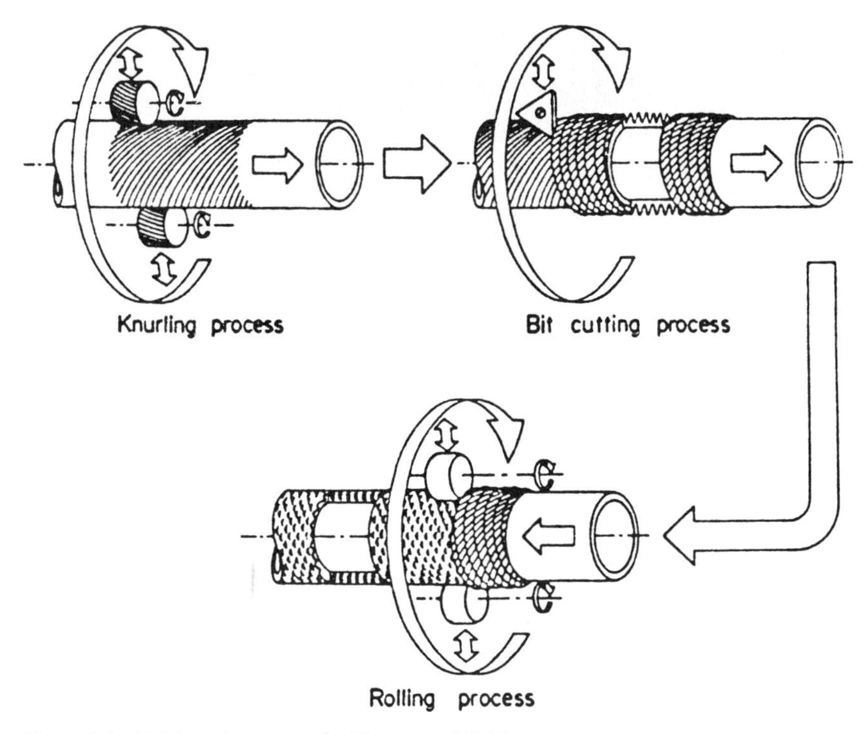

Figure 4-4 Metalworking process for Thermoexcel-E tube.

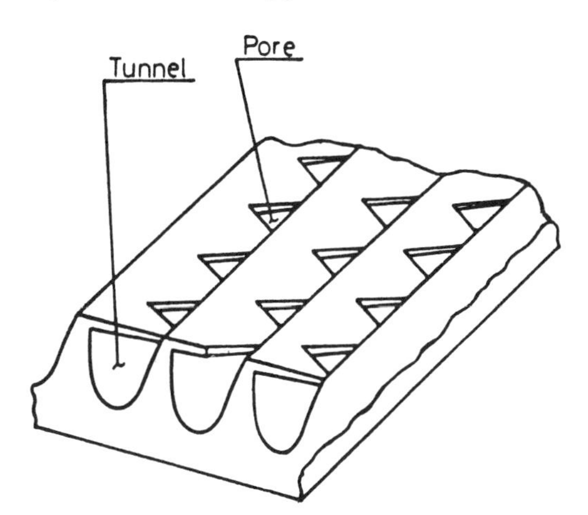

Figure 4-5 Schematic diagram of Thermoexcel-E tube.

surface layer of the tube is cut by the triangular tool; the body of the tool is not deformed. Therefore, Tanno and Ooizumi (1986) reported that hard materials such as stainless steel, aluminum brass, and titanium can be used to make Thermoexcel tubes.

Another version of the above tube is the Thermoexcel-HE tube, which undergoes another operation to raise fine projections just beneath the triangular openings to lessen the influence of subcooling on boiling performance.

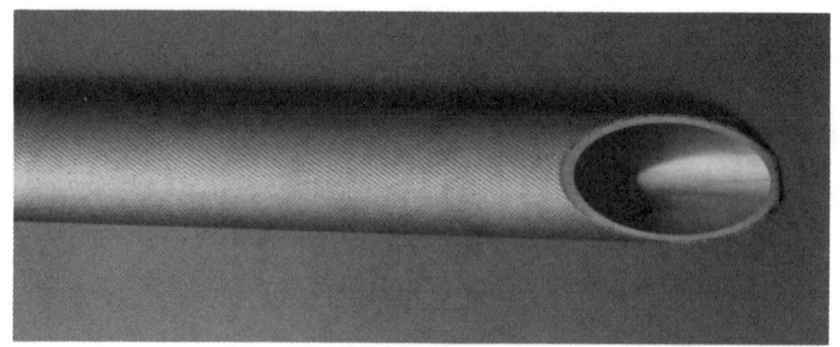

Figure 4-6 Photograph of Thermoexcel-E tube (courtesy Hitachi).

A modified low-finned tube named the Gewa-T (Fig. 4-7) is produced by Wieland-Werke. This tube is made from an integral low-finned tube by splitting the fins into a Y shape and then rolling these fins to obtain T-shaped fins and a uniform outside surface with a prescribed gap size between adjacent fin tips. Thus a continuous reentrant cavity is formed around the circumference of the tube between the T-shaped fins. An advanced version of this tube is the Gewa-TX, which has periodic notches inside the grooves as shown in Fig. 4-8. These tubes can be made in a wide range of materials, such as copper, cupro-nickel, aluminum, and low-carbon steel. T-fin densities can be either 741 or 1,000 fins per meter (18.8 and 25.3 fpi, respectively).

The Turbo-B tube of Wolverine Tube has a different enhancement geometry from the above two. The exterior boiling enhancement is made by raising integral low fins,

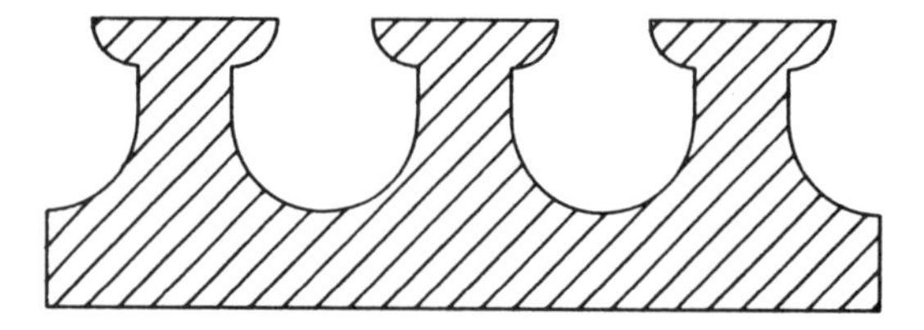

Figure 4-7 Gewa-T tube.

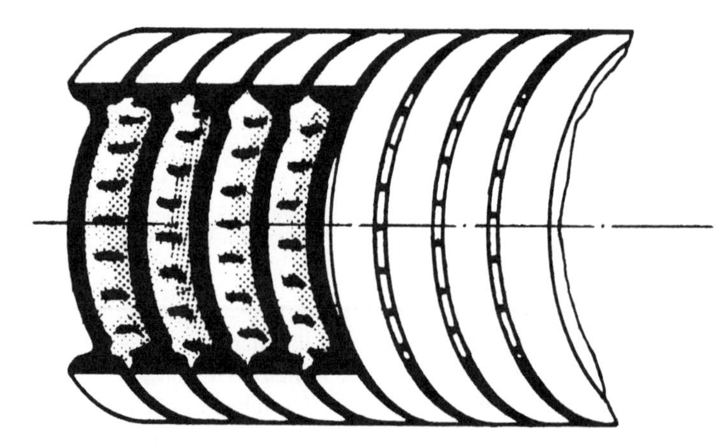

Figure 4-8 Gewa-TX tube (courtesy Wieland-Werke).

cutting diagonally across these fins, and then rolling the fins to compress them to form mushroomlike pedestals. Reentrant passageways are thus formed in a rectangular crosshatch pattern. Figure 4-9 depicts a cross-section of the tube wall, and Fig. 4-10 is a photograph of the surface (this tube can be made without the internal fins and is therefore included here). The shape of the fins when viewed from above is close to rectangular. This tube is currently available in copper, cupro-nickel, and low-carbon steel.

Another enhanced boiling tube similar in geometry to the Turbo-B tube is the Ever-Fin ECR-40 of Furukawa Electric. This tube is made from a 40 fpi (1,579 fins per meter low-finned tube. A schematic diagram is shown in Fig. 4-11. The fins are

Figure 4-9 Cross-sectional view of Turbo-B tube (courtesy Wolverine Tube).

Figure 4-10 Photograph of Turbo-B tube (courtesy Wolverine Tube).

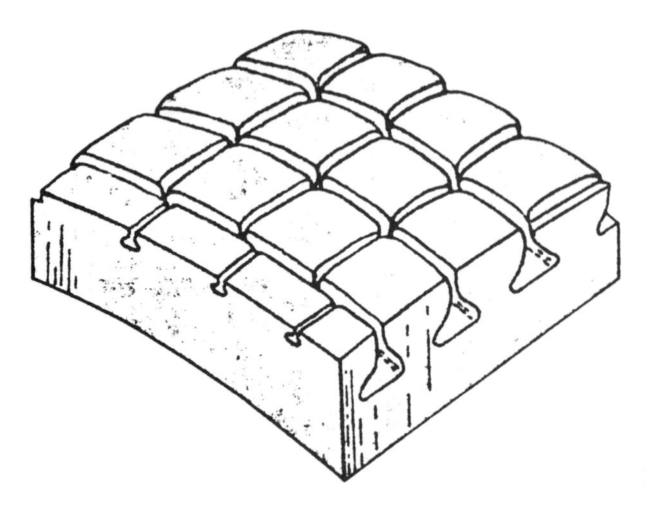

Figure 4-11 ECR-40 tube.

described by the manufacturer to be a concave saw-toothed shape. This tube can be made in copper and 90/10 cupro-nickel.

4-4 POROUS LAYER–COATED TUBES

Various methods for producing porous layers on the outside of tubes have been developed and patented, as described in Chapter Three. Some of these techniques are flame spraying of wire or particles onto the surface, plating of powders fixed to the tube wall or to the cavities of foam wrapped around a tube, and sintering or brazing of screens to a surface. Apparently only one of these methods has produced a commercially viable enhanced boiling tube, the High Flux tube of Union Carbide (now of UOP). The Korotex II tube of Wolverine Tube was available for a short period of time but proved to be too expensive to manufacture.

The High Flux tube was the first enhanced boiling tube after low-finned tubes to become widely used in industry. Figure 4-12 is a photograph of various types of High Flux tubing. This tube is produced by spraying a specially developed coating made up of a binder, a metallic powder, and a brazing powder on its exterior surface to form a thin film. The coated tube is then heated in an oven to melt the brazing powder and to burn off the binding material, leaving behind a thin, porous metallic matrix that is several layers of particles thick and has a multitude of random, interconnected passageways. The outside diameter of the tube's boiling section is therefore slightly greater than that of its plain, uncoated ends by about 0.07 to 0.3 mm. The process can also be applied to the inner surface of tubes to enhance intube boiling. The porous layers can be made of copper, copper-iron alloy, cupro-nickel, or steel particles (an aluminum version is made for in-house use in Union Carbide's air separation plants). The particle size range, thickness, and material are varied to match different service applications.

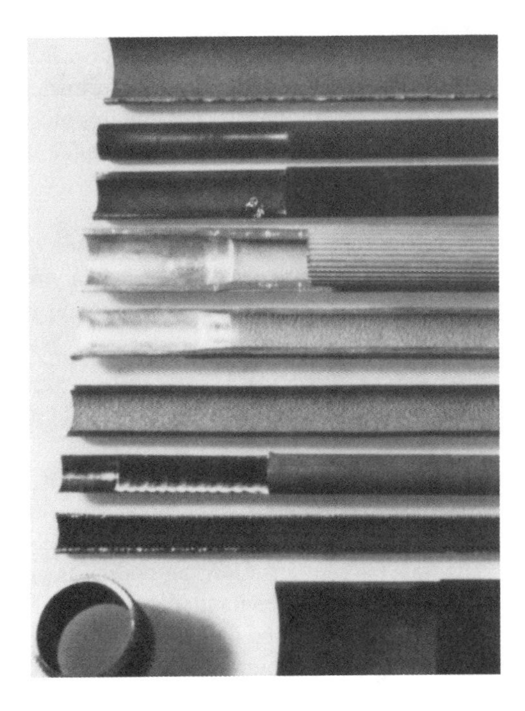

Figure 4-12 High Flux tube configurations (courtesy Union Carbide).

4-5 INTERNALLY FINNED TUBES

A wide variety of integral internally finned tubes is available commercially. Nearly all these are for single-phase flows, however, such as the Turbo-B's internal enhancement shown in Fig. 4-10 for chilled water. In recent years, tubes with special internal geometries for augmenting boiling in refrigeration systems have been developed as an alternative to aluminum star-shaped inserts (described in Sec. 4-7). Each of these tubes has its own optimized fin contour or cross-section, number of fins around the inside circumference of the tube, and helix angle. The fins are produced by drawing plain tubing over a die somewhat similar to the knurling tool shown in Fig. 4-4. The tubes are available in copper and have diameters as small as 7.94 mm (0.31 inch).

Figure 4-13 depicts the different internal fin shapes of the Thermofin tubes of Hitachi, the first three of which are described by Shinohara and Tobe (1985) and the last by W. Nakayama (personal communication 1987). The Thermofin shape A is nearly trapezoidal in cross-section, and the B shape is triangular. A newer version, the Thermofin-EX tube, also has triangular fins that are higher and have a smaller apex

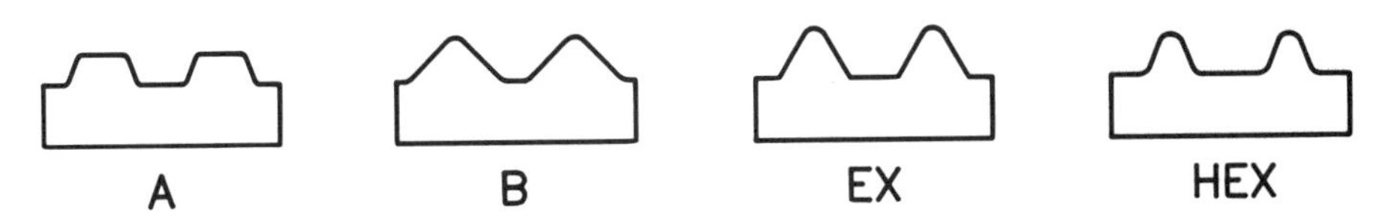

Figure 4-13 Thermofin tubes A, B, EX, and HEX.

angle than the older B version. This newer shape increases the fin efficiency while reducing the tube's weight per unit length by 3%. The newest and highest-performance version is the HEX, which has still higher and narrower fins profiles but requires more manufacturing skill to produce. Figure 4-14 shows the Thermofin-EX tube installed in an aluminum louvered-fin heat exchanger.

Internally ribbed steel tubes are manufactured for use in steam generators to increase the critical heat flux rather than the boiling heat transfer coefficient. Figure 4-15 (from Kitto and Albrecht [1988]) depicts one such tube produced by a subsidiary of Babcock and Wilcox for their fossil-fuel steam generators. The fin cross-section is trapezoidal and is produced with a helical spiral.

4-6 DOUBLY ENHANCED TUBES

The heating fluid's thermal resistance almost always becomes controlling when an enhanced boiling tube is used. Therefore, a doubly enhanced tube may be economically justifiable to increase further the overall heat transfer coefficient. When boiling is on the outside, the heating-side enhancement is usually integral longitudinal or helical fins on the inside of the tube. When boiling is on the tube side, the outside enhancement may be longitudinal or fluted fins to augment vertical condensation, for example, or aluminum plates force-fit or brazed to the outside of the tube in air-conditioning systems, as shown in Fig. 4-14. The Gewa-TX tube is also available with internal helical fins

Figure 4-14 Thermofin-EX tubes in aluminum louvered-fin heat exchanger (courtesy Hitachi).

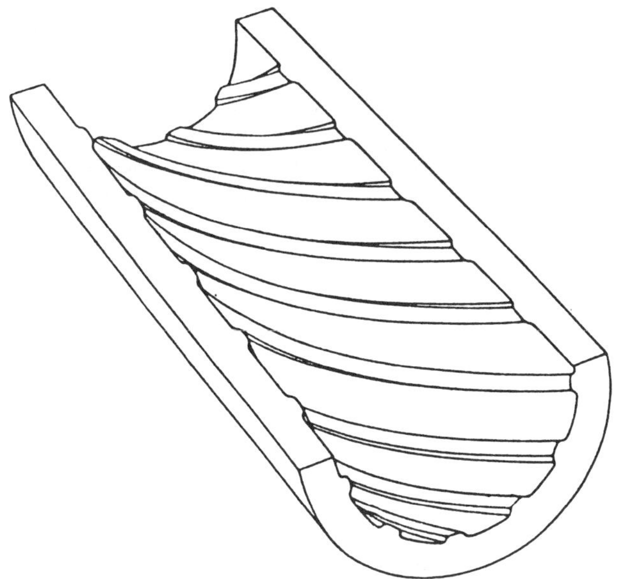

Figure 4-15 Ribbed tube of Babcock and Wilcox.

for chilled water; such a tube is designated the Gewa-TWX tube. The Thermoexcel tube has a doubly enhanced version called the Thermoexcel-HEC, which has shallow corrugations as shown in the photograph in Fig. 4-16. The standard tube material for doubly enhanced tubes is copper.

The S/T Turbo-Chil tube of Wolverine Tube is an integral low-finned tube with helical fins on the inside. This tube is depicted in Fig. 4-17. It has the same exterior as the S/T Trufin and fin densities from 750 to 1,579 fins per meter (19 to 40 fpi). It is available in copper, 90/10 and 70/30 cupro-nickel, stainless steel, and titanium.

The High Flux tube has several doubly enhanced versions. One has external fluted fins with the porous layer applied to the inside wall of the tube. Internally finned tubes can be coated on the outside to produce a doubly enhanced version for shell-side boiling.

Doubly enhanced tubes can also be made by deforming the tube wall to augment

Figure 4-16 Thermoexcel-HEC tube (courtesy Hitachi).

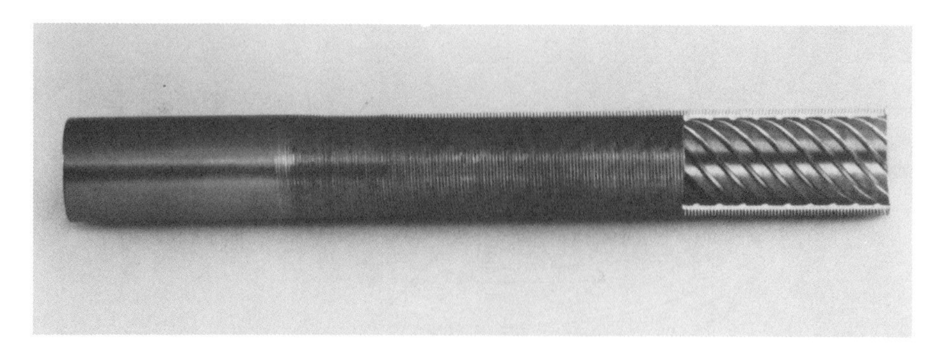

Figure 4-17 S/T Turbo-Chil tube (courtesy Wolverine Tube).

Figure 4-18 Spirally fluted tube (courtesy GA Technologies).

the heat transfer processes both inside and outside. The aluminum, spirally fluted tube of GA Technologies (Fig. 4-18) is an example. The fluting is produced in sheets in a rolling mill. Fluted strips are then driven into a closed cylindrical die to form a cylinder whose helical contact is continuously welded.

Another type of doubly enhanced tube is the corrugated tube. This geometry is made by various manufacturers. A corrugated tube of Wolverine Tube, shown in Fig. 4-19, is primarily used for water chillers in which water is cooled on the outside of the

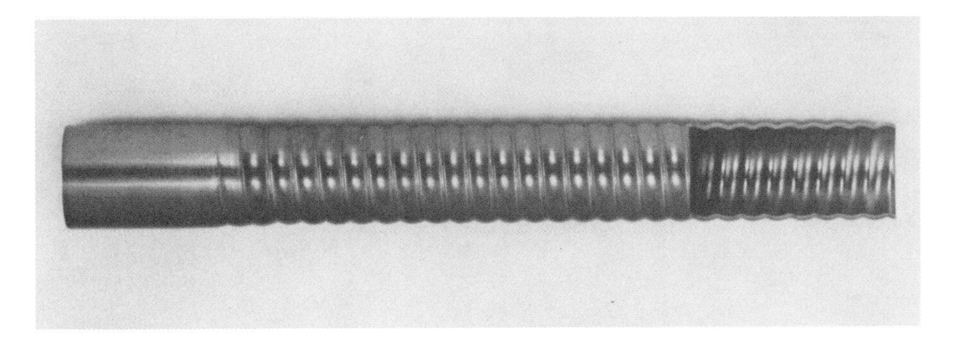

Figure 4-19 Corrugated tube (courtesy Wolverine Tube).

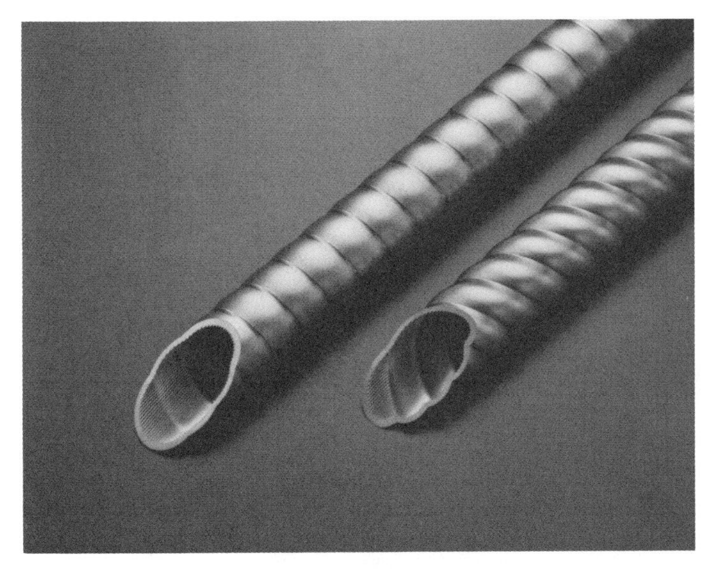

Figure 4-20 Thermofin-CR tube (courtesy Hitachi).

tube by a refrigerant expanding and evaporating on the inside. The principal geometric characteristics of corrugated tubes are the depth, the pitch, and the helix angle of the corrugation. The corrugations can be produced without thinning the tube wall and with the same outside diameter over the corrugations as that of the plain flats at the ends of the tube. Lands can also be made periodically along the tube. Corrugated tubes are typically available in copper and copper alloys.

An enhanced version of the corrugated tube is the Thermofin-CR tube of Hitachi, shown in Fig. 4-20. This tube is the internally finned Thermofin tube described above (Fig. 4-13) corrugated for further improvement of its performance.

4-7 TUBE INSERTS

Inserts can be placed inside smooth tubes to augment the boiling heat transfer coefficient. The most widely used one is the extruded aluminum star insert. These are

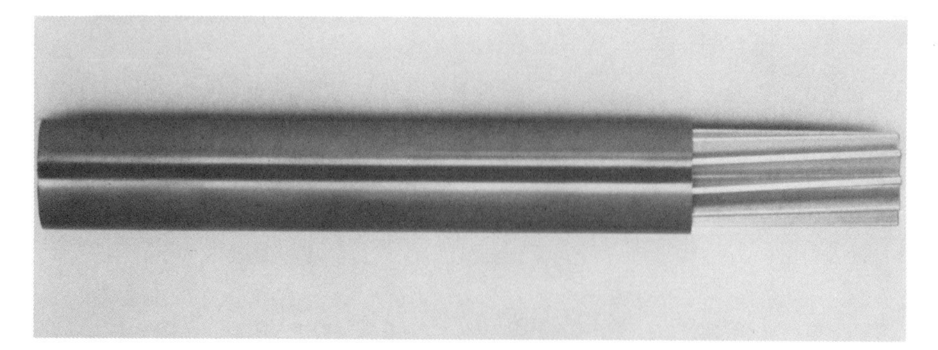

Figure 4-21 Water Chiller tube with aluminum insert (courtesy Wolverine Tube).

Figure 4-22 Aluminum Inner Fin tube (courtesy Hitachi).

typically made with five to twelve splines. They are available from numerous manufacturers. Figure 4-21, for example, depicts the Wolverine Water Chiller tube, which has an insert with a helical twist and plain splines. The splines can also be made with a wavy contour, as shown in Fig. 4-22 for the Hitachi Inner Fin tube. These inserts are installed inside copper tubes by mechanically stretching the tubing to create a solid line of contact between the spline tips and the inner tube wall. The integrity of this contact is important so that heat can be effectively conducted into the insert. Star inserts are available in aluminum and aluminum alloys, which are readily extruded.

Twisted steel tapes are also used as inserts; the number of twists per unit length can be varied to optimize performance. These tapes are typically installed to improve the performance of existing evaporators that are operating below expectations.

The last common type of insert is the coiled wire or brush insert. These inserts are typically made of copper or steel wires wound around a central core. They can be installed in existing evaporators, even inside U-tube bundles, to increase their heat duty without retubing the bundle. The wire coil is spring-loaded against the inner tube wall to obtain good thermal contact and can be removed for cleaning of the tube.

REFERENCES

Kitto, J. B. Jr., and M. J. Albrecht. 1988. Elements of two-phase flow in fossil boilers. In *Two-phase flow heat exchangers*, eds. S. Kakac, A. E. Bergles, and E. D. Fernandes, NATO ASI Series E, Vol. 143, 495–551. Dordrecht: Kluwer Academic Publishers.

Mascone, C. F. 1986. CPI strive to improve heat transfer in tubes. *Chem. Eng.* February.

Shinohara, Y., and M. Tobe. 1985. Development of an improved, "Thermofin Tube." *Hitachi Cable Rev.* 4:47–50.

Tanno, Y., and K. Ooizumi. 1986. Development of high performance heat transfer tube "Thermoexcel." *Tomorrow's tube*. Leamington Spa, Warwickshire: International Tube Association.

CHAPTER

FIVE

BOILING NUCLEATION ON ENHANCED BOILING SURFACES

5-1 INTRODUCTION

The inception of boiling is the dividing point between single-phase convection and two-phase boiling. Under pool boiling conditions, the commencement of boiling is usually referred to as boiling nucleation or boiling incipience. The wall superheat at which boiling initiates is called the nucleation superheat or the boiling activation superheat. For boiling inside tubes, the most commonly used term to describe the commencement of boiling is the onset of nucleate boiling. The wall superheat at which this occurs is also called the nucleation superheat. Curiously, research on the phenomenon of onset of nucleate boiling on the outside of tube bundles has been nearly completely neglected.

Boiling nucleation is of fundamental importance to enhanced boiling, as described in Chapter Three. It is also a key factor involved in the proper operation of evaporators that use enhanced boiling surfaces. Startup of an evaporator, for instance, requires knowledge of the minimum superheat that must be available under the process conditions. On the other hand, it is sometimes necessary for the nucleation superheat to be smaller than a certain value so as not to damage the equipment to be cooled, such as electronic chips. The deactivation superheat, the point at which boiling ceases to occur on a heated surface, is another significant parameter involved in the proper functioning of plain and enhanced boiling surfaces because it is the minimum possible operating wall superheat necessary to sustain nucleate boiling after startup.

Most thermal design methods do not include these parameters as limitations, which can lead to poor thermal operation. With enhanced boiling tubes, it is imperative to include these parameters in the design process because their operation in low-temperature approaches in heat integration schemes may not provide sufficient superheat unless particularly specified. Thus boiling nucleation is not only a fundamental problem of research interest but is also a significant factor to weigh during design and operation.

In this chapter, results of two visualization studies on the enhanced boiling nucleation process are described. Experimental data for boiling activation and deactivation are then surveyed. Finally, the factors affecting the onset of nucleate boiling in flow boiling are discussed. The reader is referred to Chapter Two for a brief review of boiling nucleation on ordinary plain surfaces and to Chapter Three for a survey of the early fundamental research on boiling nucleation in pits, grooves, and reentrant cavities. Boiling nucleation in liquid mixtures is addressed in Chapter Nine.

5-2 VISUALIZATION STUDIES

Understanding of the physical aspects of a process is always easier when the process can be observed. Several studies have therefore attempted to obtain a view of the nucleation process inside enhanced boiling geometries to see what is actually occurring.

Nakayama and Colleagues

Nakayama et al. (1979) were apparently the first to perform a visualization study on the boiling nucleation process in an enhanced boiling geometry. They studied a reentrant channel that was connected to an outside pool of liquid by a series of small-diameter openings, a geometry that is representative of many of the integral extended surfaces described earlier (see Chapter Three's discussion of integral extended surfaces and Chapter Four, Sec. 4-3). Figure 5-1 depicts their experimental setup. The boiling apparatus consisted of a base block, two glass side plates, end plates, and a thin metal lid with a series of holes drilled along its center. These parts formed a rectangular cross-section channel connected to a liquid pool of R-11 by the holes in the cover plate. In one series of experiments heat was provided from below by a cartridge heater in the base block; in other tests the lid was electrically heated by a dc current and was used as a resistance thermometer. Boiling activation was recorded by a high-speed cine film camera or a still camera oriented for a side view of the channel. Channel heights varied between 0.5 and 1.0 mm, and the width was apparently held fixed at 1.0 mm. The diameters of the holes were uniform, ranging from 0.05 to 0.5 mm in different lids tested.

A composite sequence of the events that they observed is as follows:

1. In an unheated state vapor existed in the reentrant channel, and its volume was dependent on the temperature of the room. At low temperatures the vapor nucleus was smaller in diameter than the channel itself.
2. On addition of heat the vapor region expanded along the channel, forcing liquid out through the holes in the cover plate. Liquid remained in the corners of the channel, however, and thus must have been replenished by liquid entering at inactive holes in the cover plate to make up for the liquid evaporated.

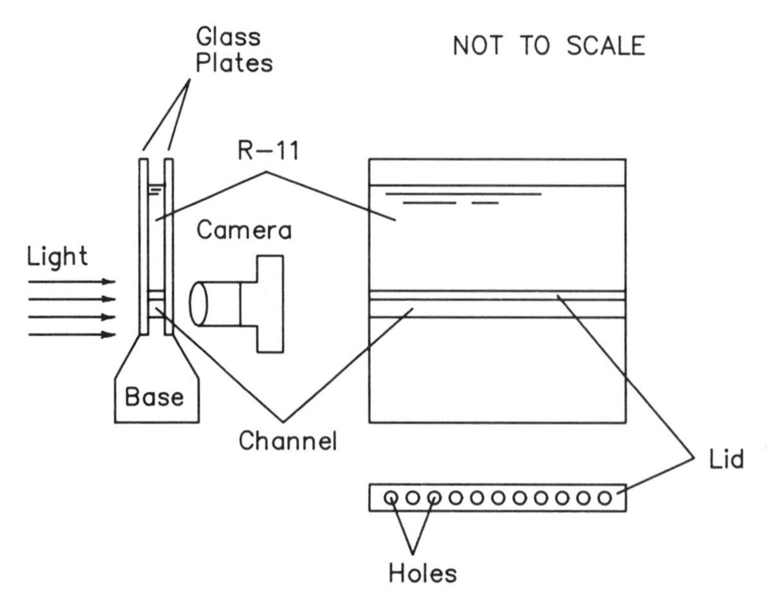

Figure 5-1 Visualization apparatus used in the experiments of Nakayama et al. (1979).

3. At wall superheats less than 0.6 K vapor occupied the channel as described in event 2 above, but no bubbles emerged from the holes in the lid.
4. At wall superheats greater than 0.6 K, bubbles began to emerge from several of the holes at a frequency of about one every 8 seconds. Pulsation of the liquid menisci in the corners of the channel was observed to be in phase with the departure of the bubbles.
5. Incrementally increasing the wall superheat increased the bubble departure frequency and also the number of active holes. The essential features of the process remained unchanged, however.

These characteristics were observed for all test geometries except for the lid with the largest diameter holes (0.5 mm). In this case, expulsion of the liquid from the channel was never complete, and liquid could be seen sloshing around in the channel while occupying roughly 10% to 50% of the channel's volume. Therefore, it can be concluded that the process of activating an enhanced boiling surface is different from that involving a nucleus in a pit in a plain surface.

Arshad and Thome

Arshad (1982) and Arshad and Thome (1983) carried out a visualization study with water using an experimental apparatus similar to the one described above. Figure 5-2 depicts their experimental apparatus, which was designed to provide observation along the axis of a channel to view the nucleation process in the corners and to see the cross-section of the liquid film on the channel walls. Their setup consisted of a brass block

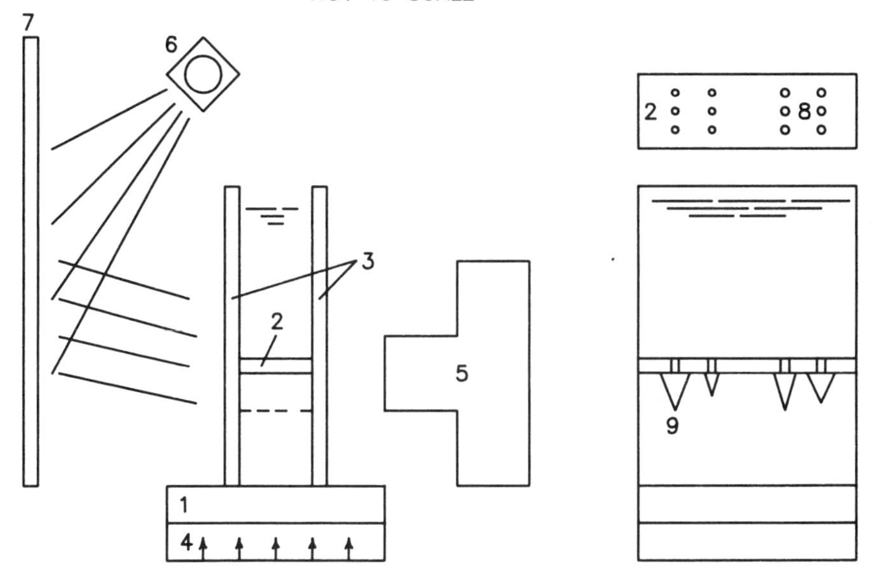

Figure 5-2 Arshad and Thome (1983) visualization setup. Legend: 1, brass base block; 2, copper lid; 3, glass plates; 4, heater; 5, camera; 6, florescent light source; 7, diffusively reflecting screen; 8, holes; 9, groove.

with grooves in its top surface. Triangular, rectangular, and circular cross-sectional grooves were tested. Very thin copper shimming stock was used to make the cover plates. Holes were drilled in the cover plate for each groove; hole diameters were 0.15 and 0.25 mm, sometimes used in combination. Heat was provided by an electrically heated Nichrome ribbon attached to the bottom of the base block. The ends of the channels were covered with thin glass plates. The process inside the channels was recorded through a magnifying lens with either a video tape camera or a motor-driven still camera.

Figure 5-3 depicts the activation of a triangular groove with hole diameters of 0.15 mm and a groove height of about 1 mm. Nucleation occurred at the top right corner in this sequence, but in other tests it also occurred at the bottom or the top left corner. Once the vapor reached a hole in the cover plate, a bubble immediately grew in the liquid pool at this hole, and a thin liquid film could be observed covering the channel walls, taking their shape. At large heat fluxes, the liquid film progressively dried out. Figure 5-4 shows this process schematically.

Figure 5-5 depicts the nucleation process in a rectangular channel. In this instance, the vapor front reached the opposite wall before reaching the hole in the cover plate. (Because of the short depth perception obtained when using magnification, the bubble emerging from the cover plate was not in focus and is not shown.

Figure 5-6 shows the curious instance of multiple nucleation sites inside a circular channel about 1.0 mm in diameter. Activation of circular channels was more difficult compared to the other two geometries. Liquid films formed on the channel walls in all three geometries and for all sizes of channels and holes.

(*a*) Nucleation

(*b*) expansion of vapor nucleus

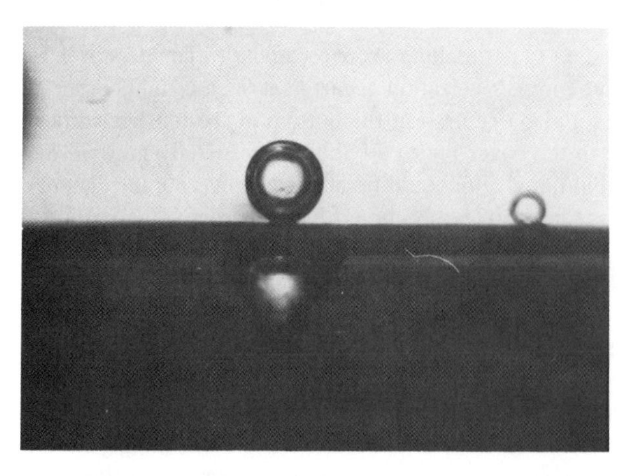

(*c*) departure of first bubble and formation of liquid film.

Figure 5-3 Boiling activation in a triangular reentrant channel.

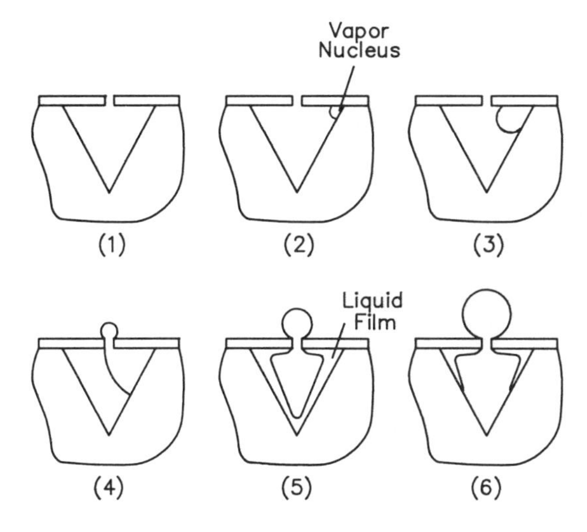

Figure 5-4 Schematic representation of the boiling nucleation and film formation process shown in Fig. 5-3.

Flooding (boiling deactivation) of a triangular channel is shown in Fig. 5-7. Liquid rushed in through the holes in the cover plate and trapped vapor in the bottom of the channel. Afterwards vapor was sometimes no longer visible; this was the case even shortly after cessation of boiling but especially for the circular channels. The circular channels tended to be flooded with liquid more easily than the others.

Summary

In summary, the above two studies reached the following conclusions about boiling nucleation and deactivation processes in reentrant channels:

1. The radius of the hole in the cover plate did not control the boiling nucleation process, which occurred at a vapor nucleus trapped inside the channel. When visible nuclei were present they were situated in corners, not at the holes in the cover plate.
2. Corners were the preferred place for nucleation to occur.
3. The liquid film in the reentrant channel was formed as the liquid-vapor interface advanced in the channel and reached a hole to the outside liquid pool.
4. The liquid film tended to take the shape of the channel but was thicker in the corners as a result of surface tension forces. It partially or completely dried out at high heat fluxes.
5. The diameter of the holes in the cover plate and the channel geometry were crucial parameters affecting the flooding process. Vapor nuclei were sometimes visible in the reentrant channel, depending on the amount of subcooling and the cavity geometry.

5-3 NUCLEATION SUPERHEAT EXPERIMENTS

Experimental studies to measure boiling activation and deactivation superheats have been performed for many different fluids. As in previous experiences with plain sur-

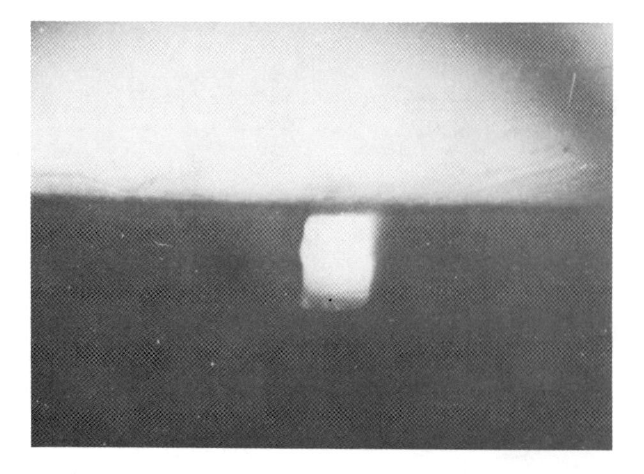

(*a*) Flooded channel

(*b*) nucleation in corner

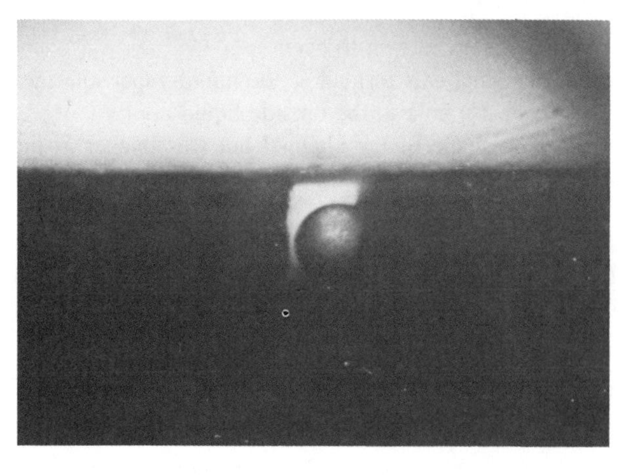

(*c*) advancing vapor front.

Figure 5-5 Nucleation in a rectangular reentrant channel.

Figure 5-6 Multiple nucleation within a circular reentrant channel.

(*a*) Advancing liquid front

(*b*) last view of vapor nucleus.

Figure 5-7 Flooding of a triangular reentrant channel.

faces, the manner in which heat is applied to an enhanced surface and the operating history of the surface have an impact on the size of the activation superheat. For example, large steps in the wall heat flux applied by a cartridge heater typically cause nucleation to occur at lower wall superheats than a slow gradual increase.

Bergles and Chyu (1981) performed a comparative study on the effect of operating history on boiling nucleation with several different High Flux tubes and a plain tube. They tested both water and refrigerant-113. They observed that the High Flux tubes activated at slightly smaller superheats than the smooth surface for water but at much smaller superheats for R-113. Some of their results were contrary to expectations. For instance, the nucleation superheat for smooth surface 2A with R-113 was found to decrease with increasing subcooling. The pool boiling curves for the High Flux tube, however, were not affected by this.

Marto and Lepere (1981) conducted boiling nucleation studies using different operating histories with three different enhanced boiling surfaces: High Flux (with a 0.08-mm coating and the same particle size distribution as that used by Bergles and Chyu), Thermoexcel-E, and Gewa-T tubes. Refrigerant-113 and Fluorinert FC-72 (a dielectric produced by 3M Company) were tested at their normal boiling points. The results showed that the nucleation superheat increased with increasing subcooling, as would be expected from plain surface studies. The wall superheat required to initiate boiling on the High Flux tube was the lowest, and the Gewa-T tube required the highest. These investigators did not, however, test a plain tube to determine the relative difference in the nucleation superheats. Nevertheless, the ratio of the nucleation superheats for the two different fluids on each tube was well predicted by boiling nucleation theory (e.g., Eq. [3-3]).

Shakir and Thome (1986) measured boiling activation and deactivation superheats for water, ethanol, *n*-propanol, methanol, and benzene for a commercial R-11 High Flux tube (with a 0.075-mm thick coating made of a 1% copper-iron alloy) and several different smooth surfaces and tubes. They used only one standard type of operating history: boiling immediately before the nucleation test with no subcooling. The boiling nucleation wall superheats were attained by means of fixed step intervals of heat flux up until the surface activated and were shown to be reproducible. The High Flux tube consistently activated at lower superheats than the plain surfaces. The margin of improvement depended primarily on the variation in the plain surface values from one test surface to another.

In these tests water had the lowest measured boiling activation superheat of all the fluids, both on the High Flux tube and on the various smooth plates and plain tubes. Yet water has the highest surface tension of the tested fluids and would therefore be expected to have the largest nucleation superheat. To determine the reason for this, the investigators measured contact angles at 25°C on copper and brass surfaces. Table 5-1 shows that water has a much larger contact angle than the other fluids. Thus water's larger contact angle apparently increased the size of its trapped vapor nuclei [e.g., the radius R in Eq. (3-3)] and reduced the nucleation superheat. The deactivation wall superheats for the High Flux tube were lower than those for rough and polished tubes and also those for their own respective activation superheats; the deactivation superheats varied from fluid to fluid.

Table 5-1 Contact angles of pure fluids

Fluid	Contact angles; degrees*	
	Copper	Brass
Water	78	94
Ethanol	14, 19†	14, 18†
n-Propanol	13	8
Methanol	25	22
Benzene	25	24

*An average of 36 droplets was used to obtain each mean value.
†Same surface with two different roughnesses.

Figure 5-8 Dendritic surface (courtesy IBM).

Oktay (1982) experimentally investigated the boiling nucleation characteristics of a dendritic surface with FC-86 (a dielectric produced by 3M Company). A photograph of a dendritic surface is shown in Fig. 5-8. Superheats of about 18 K were required to initiate boiling.

To compare these and other results, Table 5-2 gives a compilation of boiling nucleation wall superheats measured for nucleate pool boiling on several commercial enhanced boiling tubes and on smooth surfaces and tubes. The enhanced boiling tubes are shown to activate consistently at the lowest wall superheats. The Gewa-T tube,

Table 5-2 Boiling nucleation superheats on smooth and enhanced boiling surfaces

	Boiling nucleation superheat, K			
Liquid	Smooth surfaces	High Flux	Gewa-T tube	Thermoexcel-E tube
Water	2–4[*]	1–3[*]		
	5–8[†]	2–3[†]		
Ethanol	6–23[†]	6[†]		
n-Propanol	8[†]	4[†]		
Methanol	9–10[†]	4[†]		
Benzene	10–15[†]	7[‡]		
R-11			14[§]	
R-12		1[‖]		
R-113	12–15[*]	2–9[*]	2–19[#]	1–12[#]
	19–21[**]	0.8–9[#]	7–18[**]	12[††]
	19–22[‡‡]			
	12[§§]		6–9[§§]	
FC-72		1–5[#]	1.5–8[#]	1–7[#]

[*]Bergles and Chyu (1981).
[†]Shakir and Thome (1986).
[‡]Ali and Thome (1984).
[§]Stephan and Mitrovic (1981).
[‖]Czikk, O'Neill, and Gottzmann (1981).
[#]Marto and Lepere (1981).
[**]Marto and Hernandez (1983).
[††]Hitachi Cable (1978).
[‡‡]Bahhuth and Genetti (1983).
[§§]Ayub and Bergles (1988).

with its 0.25-mm opening between fin tips, probably allows natural convection to occur within its reentrant channels, which affects its nucleation characteristics. The spread of 6 to 23 K for ethanol on smooth surfaces represents mean values of 6, 10, and 23 K for a crocus paper roughened tube, a highly polished tube, and a roughened disk (silicon carbide 320 emery paper), respectively, all tested with the same operating history before nucleation.

One aspect of boiling activation not illustrated by these data is the degree to which the surface becomes activated on nucleation of the first one or two boiling sites. For instance, Ayub and Bergles (1988) observed that the Gewa-T tube tended to initiate boiling in isolated patches, apparently because its reentrant channels do not communicate with one another.

5-4 ONSET OF NUCLEATE BOILING

The area of enhanced boiling nucleation under forced convection conditions has attracted little research interest, even though the onset of nucleate boiling is an important aspect of thermal design and operation. For instance, a porous boiling enhancement is not effective when only sensibly heating a subcooled liquid. For internally finned tubes, single-phase heat transfer may or may not be effective depending on the type of fin. The interested reader is referred to Carnavos (1979) for a comparative study of internally finned tubes with single-phase forced convection.

The operating history of the tube or tube bundle may affect the wall superheat at the onset of nucleate boiling and its thermal performance for a period of time after startup. Bergles and Chyu (1981) ascribed the somewhat erratic heat transfer behavior of a 279-tube High Flux bundle for boiling ammonia reported by Lewis and Sather (1978) to nucleation effects. Altering the startup procedure was seen to change the thermal performance of the bundle over the first 40 hours of operation. The overall heat transfer coefficients were about 15% lower than the steady-state values when the unit was started up flooded with subcooled ammonia. The steady-state overall heat transfer coefficient could be obtained immediately on startup by charging the ammonia after the heating medium began flowing in the tubes, however. For intube boiling, Murphy and Bergles (1972) reported that porous boiling surfaces eliminated the boiling hysteresis for refrigerants that they had observed for plain tubes.

Antonelli and O'Neill (1981) commented that boiling wall superheat temperature differences in excess of 0.5 K in High Flux heat exchangers were usually sufficient to initiate boiling. In some cases 0.2 K was enough, according to O'Neill and Gottzmann (1980). Czikk, O'Neill, and Gottzmann (1981) reported on their involvement with more than 400 High Flux heat exchangers in commercial operation (refrigerated water chillers, olefin plant reboiler condensers, and various refinery reboilers). They noted that special startup procedures were not used and not required, even for climate control and air conditioning systems for which operation is intermittent. They suggested that the following factors were responsible for the absence of transient boiling behavior in these various units:

1. Operating temperatures of refrigerant evaporators are often less than ambient air temperature, making subcooling or startup with subcooled liquid feed unlikely.
2. A large wall superheat is usually available temporarily to activate the surface in reboilers attached to distillation columns.
3. A fluid heating medium is used as the heat source rather than electrical heating.
4. Heat fluxes are typically greater than 3,100 W/m^2 and thus beyond the normal single-phase natural convection regime.

In addition, in refrigeration systems a two-phase fluid enters the evaporator with a typical vapor quality of about 20%. This may promote boiling nucleation by "seeding" cavities with vapor. This inlet condition can also take the process directly into the annular flow regime inside a tube, bypassing the nucleation process altogether, or into the liquid film flow regime on the outside of a tube bundle. Uniform vapor distribution then becomes an important design parameter.

Accurate prediction of boiling nucleation superheats or deactivation superheats is still not possible. For boiling on the outside of tubes, one must guess the probable radius R of trapped vapor nuclei to evaluate the nucleation superheat with Eq. (3-3). For flow boiling, Eq. (2-23) for smooth tubes will give an approximate estimate.

REFERENCES

Ali, S. M., and J. R. Thome. 1984. Boiling of ethanol-water and ethanol-benzene mixtures on an enhanced boiling surface. *Heat Transfer Eng.* 5(3–4):70–81.

Antonelli, R., and P. S. O'Neill. 1981. Design and application considerations for heat exchangers with enhanced boiling surfaces. Paper presented at International Conference on Advances in Heat Exchangers, September, Dubrovnik, Yugoslavia.

Arshad, J. 1982. Enhanced nucleate boiling. Master's thesis, Michigan State University, East Lansing.

———. and J. R. Thome. 1983. Enhanced boiling surfaces: Heat transfer mechanism and mixture boiling. *Proc. ASME-JSME Therm. Eng. Joint Conf.* 1:191–97.

Ayub, Z. H., and A. E. Bergles. 1988. Nucleate pool boiling curve hysteresis for Gewa-T surfaces in saturated R-113. In *Advances in Enhanced Heat Transfer*, HTD Vol. 97. New York: American Society of Mechanical Engineers.

Bahhuth, A., and W. E. Genetti. 1983. Nucleate boiling heat transfer enhancement with heat transfer surface cavities. *AIChE Symp. Ser.* 79(225):28–33.

Bergles, A. E., and M. C. Chyu. 1981. Characteristics of nucleate pool boiling from porous metallic coatings. In *Advances in Enhanced Heat Transfer*, HTD Vol. 18, 61–71. New York: American Society of Mechanical Engineers.

Carnavos, T. C. 1979. An experimental study: Pool boiling R-11 with augmented tubes. In *Advances in Enhanced Heat Transfer*, 103–108. New York: American Society of Mechanical Engineers.

Czikk, A. M., P. S. O'Neill, and C. F. Gottzmann. 1981. Nucleate boiling from porous metal films: Effect of primary variables. In *Advances in Enhanced Heat Transfer*, HTD Vol. 18, 109–22. New York: American Society of Mechanical Engineers.

Hitachi Cable. 1978. High Flux boiling and condensation tube. Production Bulletin. Thermoexcel, Tokyo.

Lewis, L. G., and N. F. Sather. 1978. *OTEC performance tests of the Union Carbide flooded-bundle evaporator* (Argonne National Laboratory report ANL-OTEC-PS-1).

Marto, P. J., and B. Hernandez. 1983. Nucleate pool boiling characteristics of a Gewa-T surface in Freon-113. *AIChE Symp. Ser.* 79(225):1–10.

Marto, P. J., and J. Lepere. 1981. Pool boiling heat transfer from enhanced surfaces to dielectric fluids. In *Advances in Enhanced Heat Transfer*, HTD Vol. 18, 93–102. New York: American Society of Mechanical Engineers.

Murphy, R. W., and A. E. Bergles. 1972. Subcooled flow boiling of fluorocarbons—Hysteresis and dissolved gas effects on heat transfer. *Proc. Heat Transfer Fluid Mech. Inst.* 400–16.

Nakayama, W., T. Daikoku, H. Kuwahara, and T. Nakajima. 1979. Dynamic model of enhanced boiling heat transfer on porous surfaces. In *Advances in Enhanced Heat Transfer*, 31–43. New York: American Society of Mechanical Engineers.

Oktay, S. 1982. Departure from natural convection (DNC) in low-temperature boiling heat transfer encountered in cooling micro-electronic LSI devices. *Proc. 7th Int. Heat Transfer Conf.* 4:113–18.

O'Neill, P. S., and C. F. Gottzmann. 1980. Improved air plant main condenser. Paper read at ASME Century 2 Meeting, Emerging Technology Conferences (cyrogenic processes and equipment session), August, San Francisco.

Shakir, S., and J. R. Thome. 1986. Boiling nucleation of mixtures on smooth and enhanced boiling surfaces. *Proc. 8th Int. Heat Transfer Conf.* 4:2081–86.

Stephan, K., and J. Mitrovic. 1981. Heat transfer in natural convective boiling of refrigerants and refrigerant-oil mixtures in bundles of T-shaped finned tubes. In *Advances in Enhanced Heat Transfer*, HTD Vol. 18, 131–46. New York: American Society of Mechanical Engineers.

CHAPTER

SIX

MECHANISMS OF ENHANCED NUCLEATE POOL BOILING

6-1 INTRODUCTION

Nucleate pool boiling heat transfer coefficients for enhanced boiling surfaces can be up to 20 times greater than those for boiling on a polished, bare tube when compared at the same heat flux. Conversely, at a fixed wall superheat the enhanced boiling performance can be as much as 100 times greater. Therefore, the boiling heat transfer process and its thermal mechanisms have to be substantially modified by the enhancement geometry to make these levels of augmentation possible. This chapter reviews current understanding of enhanced boiling phenomena for nucleate pool boiling. First, the heat transfer mechanisms responsible for the transport of heat are discussed. The fundamental studies on the enhanced boiling process and phenomena are then described. Finally, some of the miscellaneous effects on enhanced boiling are reviewed.

6-2 HEAT TRANSFER MECHANISMS AND PROCESSES

Boiling on Plain Surfaces

Hsu and Graham (1976) summarized the heat transfer mechanisms responsible for heat transport in the nucleate pool boiling process on plain, smooth surfaces. Because heat transfer coefficients for nucleate pool boiling are much larger than those for single-phase natural convection from the same surface, these mechanisms represent the changes in the physical process that produce the augmentation. These investigators identified the principal mechanisms as follows (refer to Fig. 6-1 for schematic illustrations):

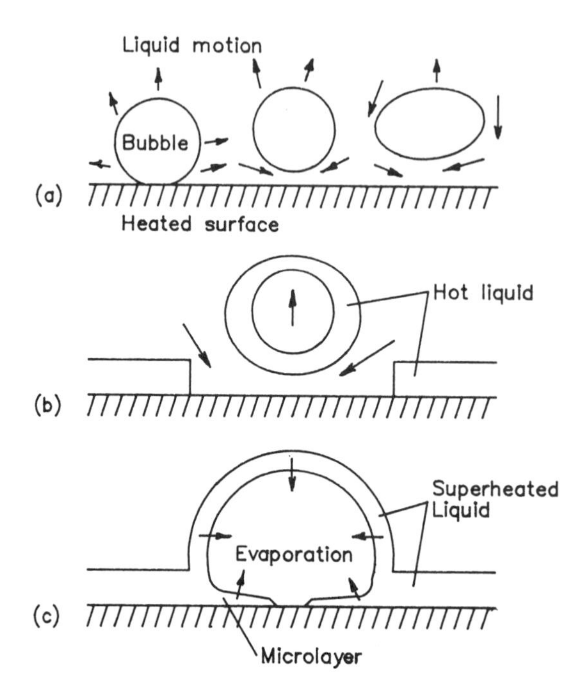

Figure 6-1 Nucleate boiling mechanisms. (*a*) Bubble agitation, (*b*) vapor-liquid exchange, (*c*) evaporation.

Bubble agitation (Fig. 6-1a). Improved liquid-phase convection results from the motion imparted to the liquid from the growth and departure of vapor bubbles. Essentially, this means that the natural convection process has been transformed into a "forced" convection process, in which the velocity of the liquid near the heated wall is comparable to that of the growing bubbles. Thus convection to the liquid is primarily controlled by the intensity of the boiling process, and heat is transported from the surface in the form of superheated liquid.

Vapor-liquid exchange (Fig. 6-1b). Convection to the liquid is augmented by the quenching of the heated wall by fresh liquid rushing in after the departure of a vapor bubble. This process is characterized by the cyclic removal of the thermal boundary layer covering the bubble and the adjacent heated wall. The sensible heat transport rate is controlled by the quantity of superheated liquid removed from the thermal boundary layer, which is proportional to the thermal boundary layer thickness and its average temperature, the bubble departure diameter and frequency, and the number of boiling sites.

Evaporation (Fig. 6-1c). Vapor bubbles grow by virtue of the heat conducted into the liquid from the heated wall and then to the bubble interface, where phase change occurs. This process can be thought of as a thermal "short circuit" because large amounts of latent heat can be efficiently carried away in the bubbles, which rise much more rapidly than buoyant liquid convection currents. Vaporization occurs from (1) a liquid microlayer trapped between the heated wall and the bottom of a rapidly growing bubble and (2) the original thermal boundary layer covering the top of the bubble. The quantity of latent heat transported is determined by the bubble departure volumes and departure frequencies and by the number of boiling sites on the surface.

The actual boiling process is a combination of these heat transfer mechanisms. The heat transported is not the summation of these three individual contributions, however, because these mechanisms interact and there is an "overlap" in the heat carried away by each process; that is, some of the heat entering the thermal boundary layer leaves as latent heat in the bubbles rather than as sensible heat in the superheated liquid.

Enhanced Boiling

The heat transfer mechanisms contributing to enhanced boiling process are affected by the particular geometry of the enhancement. Thus each different enhanced boiling surface, be it a porous layer or a mechanically deformed low-finned tube, augments heat transfer in a unique way. Consequently, the better performance of one enhanced boiling surface compared to another should in theory be able to be traced to its geometry's relative effectiveness in promoting these mechanisms.

Heat can leave a surface in several forms. Because of the relatively large radii of departing bubbles, the vapor leaves in its saturated state and little or no heat is transported as superheated vapor. In addition, the temperature difference between the heated wall and the saturated bulk liquid is too small for thermal radiation to be significant. Therefore, energy can leave a boiling surface in only two modes, either as latent heat in the departing vapor bubbles or as sensible heat in the superheated liquid rising from the surface.

Evaporation and convection to the liquid can occur both on the exterior of an enhanced boiling surface (as for conventional boiling) and inside its passageways. Hence there are four possible paths by which heat can leave an enhanced boiling surface:

1. as latent heat in vapor formed within the enhancement matrix;
2. as latent heat in bubbles growing on the exterior surface, both in bubbles originating on the exterior and by additional evaporation in bubbles emerging from the enhancement matrix;
3. superheating of the liquid drawn into and driven out of the enhancement by the pumping action of the bubbles; or
4. superheating of the liquid flowing over the exterior surface.

The mechanisms that contribute to these heat flows are described below. Any particular enhanced boiling surface may encourage the occurrence of several of these heat transfer mechanisms.

Nucleation superheat. Although not actually a heat transfer mechanism, the nucleation superheat is of paramount importance because it controls the activation of boiling on the surface, without which the heat transfer mechanisms cannot exist (with the exception of single-phase natural convection). The smaller nucleation superheats enjoyed by many enhanced boiling surfaces thus provide augmentation relative to a smooth surface because the former is able to operate in the more effective boiling regime while the latter is still in the natural convection regime.

Evaporation. The evaporation process can be subdivided into three types: thin film evaporation, capillary evaporation, and external evaporation. These are depicted schematically in Fig. 6-2.

Thin film evaporation refers to evaporation resulting from the conduction or convection (or both) of heat across liquid films formed in the enhancement passageways. The heat reaching the free vapor-liquid interface causes the liquid to flash and form vapor, which is then transported away by virtue of its buoyancy relative to the liquid above it. The existence of these films has been verified visually by both Nakayama et al. (1979, 1980) and Arshad and Thome (1983) (the formation of a liquid film is depicted in Figs. 5-3 and 5-4).

Thin film evaporation is similar to film condensation and is an effective heat transfer mechanism because the only resistance to heat flow is the thin liquid film, which is typically on the order of 0.1 mm thick. For steady-state conduction across the film, the local thin film heat transfer coefficient can be calculated as

$$\alpha_{\mathrm{tf}} = \frac{\lambda_{\mathrm{L}}}{t_{\mathrm{f}}} \tag{6-1}$$

where t_f is the film thickness. The larger surface area that can be covered by a thin film in the enhancement passageways relative to the nominal external surface area further multiplies the augmentation provided by this heat transfer process.

Capillary evaporation is characterized by the vaporization of liquid at menisci in the enhancement passageways. It is distinct from thin film evaporation in that heat is conducted from the wall to a meniscus rather than across a continuous film, phenomena that are quite dissimilar. For instance, for capillary evaporation the liquid meniscus is superheated relative to the local saturation temperature; in thin film evaporation the film's interface is nearly planar or has such a large radius that no significant superheating

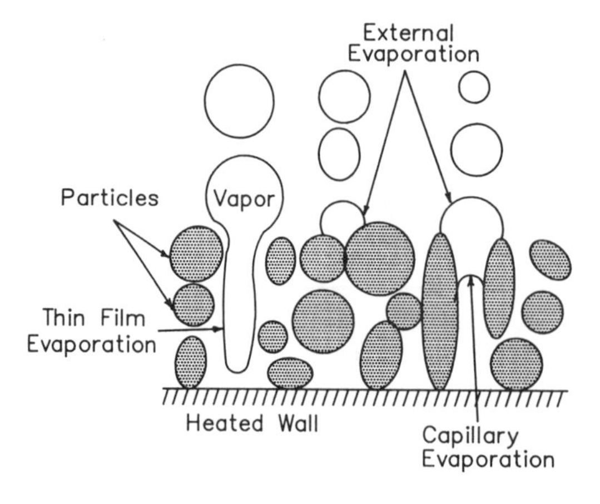

Figure 6-2 Evaporation processes on an enhanced boiling surface.

of the liquid is required for evaporation to occur, even though capillary forces may be acting to supply liquid to the film from a corner or an inactive pore.

According to Soloviyov (1986), the meniscus can be subdivided into three regions as shown in Fig. 6-3. In region I on the upper part of the meniscus the adhesive forces impede the evaporation of liquid, and virtually no heat is transferred. In region II the influence of the adhesive forces is less, and some evaporation occurs. Liquid is drawn into this region by the variation of the pressure along the meniscus. The thickness of the liquid film in this region was estimated to be on the order of 1,000 Å for water on glass. Region III at the lower portion of the meniscus is not influenced by the adhesive forces in any significant manner, and evaporation also occurs. Liquid influx occurs as a result of the phase interface curvature gradient. Heat transfer in this region has been visualized by Mirzamoghadam and Catton (1985) by means of holographic interferometry, and layer thicknesses have been measured optically by Tung, Muralidhar, and Wayner (1982).

For a liquid meniscus in a wedge as depicted in Fig. 6-4, Ma et al. (1986) showed theoretically that the existence of a capillary force to supply liquid to the film is dependent on the contact angle of the fluid and the wedge angle. The condition for the existence of a capillary force in a wedge was found to be

$$\beta < \frac{\pi}{2} - \frac{\theta}{2} \tag{6-2}$$

Therefore, fluids with small contact angles tend to produce larger capillary forces and better wet corners and wedges in enhancement geometries.

External evaporation refers to the same process that occurs on plain smooth surfaces. The most important difference here is that at low superheats the plain surface will be inactive, whereas on the enhanced surface the bubbles formed by the vapor emerging from the enhancement can continue to grow in the external thermal boundary

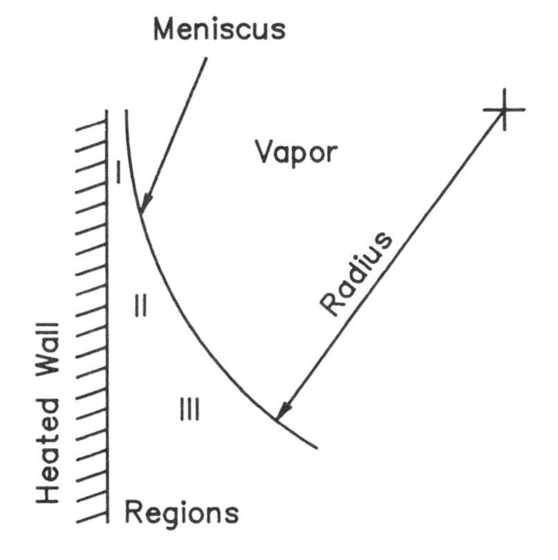

Figure 6-3 Heat transfer regions of an evaporating liquid meniscus (not to scale).

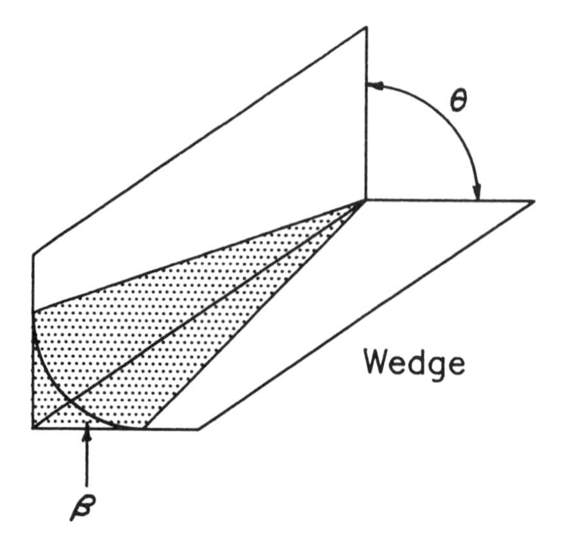

Figure 6-4 Liquid meniscus in a wedge.

layer. In addition, the boiling site density is much larger on the enhanced surface, enhancing the microlayer evaporation process. The wall superheat driving the external evaporation process is less than that for a plain surface operating at the same heat flux, however.

Convection to the liquid. The convection process is influenced by the following heat transfer mechanisms: two-phase flow and laminar flow with entrance effects in the enhancement itself, and bubble agitation and vapor-liquid exchange on the exterior surface. The superheated liquid formed in the enhancement can either provide latent heat for the thin film evaporation process or be carried out of the enhancement by the pumping action of the vapor escaping to the outside.

The two-phase flow mechanism was apparently first proposed by Stephan and Mitrovic (1981) to describe the boiling process inside the reentrant channels formed by the T-shaped fins of the Gewa-T tube (refer to Fig. 4-7 for a diagram of this enhancement) and in the space between adjacent fins of a low-finned tube. Essentially, bubbles grow inside the channel and then rise up and around the channel before escaping near the top. The effect of the T-shaped fins was to retain the bubbles within the channel to the top of the horizontal tube.

The internal two-phase flow process was investigated photographically by Fath and Gorenflo (1986). Figure 6-5 is a photograph of R-22 boiling on a Gewa-T tube with a gap opening of about 0.45 mm. The tube had been chemically etched before the test, and the surface is clearly smooth without many nucleation centers. Small diameter bubbles can be seen inside the reentrant channel formed by the T fins. Because the tube was mounted horizontally, the buoyancy force pushes the bubbles against the root of the fins in the lower half of the tube and against the fin tips in the top half of the tube. These bubbles thus augment the liquid-only convection contribution to heat transfer by increasing the liquid's velocity inside the reentrant channel; this is similar to the F factor's effect in the Chen correlation for plain tubes [see Eq. (2-25)].

Figure 6-5 Boiling of R-22 on a Gewa-T tube at 5.0 bar and 52 kW/m^2 (courtesy Laboratorium für Wärme- und Kältetechnik, Universität-GH-Paderborn).

The passageways in enhanced boiling surfaces have small hydraulic diameters and are often short. Thus the liquid flows are laminar and have entrance effects, and the local heat transfer coefficients are large. A conservative estimate of their values can be obtained from the fully developed heat transfer coefficient for laminar flow. For water flowing through a passageway with a hydraulic diameter of 0.1 mm (characteristic of a porous costing), the heat transfer coefficient is about 30 $kW/(m^2 \cdot K)$. For a Gewa-T channel the hydraulic diameter is about 10 times larger, approximately 1 mm, and its local liquid-only heat transfer coefficient is about 3 $kW/(m^2 \cdot K)$. The average value in a never fully developed entrance region can increase by an order of magnitude.

Laminar entrance region flow is also important to the thin film evaporation process. The liquid entering a film from an inactive opening represents a type of laminar entrance region flow that is much more effective than steady-state heat conduction. Hence the convective heat transfer coefficients of the film are large, and the liquid in the film is quickly superheated.

The external convection process is controlled by the bubble agitation and vapor-liquid exchange mechanisms. Both these mechanisms are augmented on an enhanced surface by the larger boiling site density and the increased intensity of the boiling process and also perhaps by the larger surface roughness. The average superheat of the thermal boundary layer is lower on an enhanced boiling surface than on a plain surface, however, because of the former's higher heat transfer coefficient. Thus a higher flow rate of superheated liquid away from the surface is required to carry away the same amount of sensible heat as from a plain surface.

Wetted surface area. Part of the increase in the heat transfer performance of enhanced boiling surfaces can be attributed to the larger wetted surface areas available for heat transfer. For example, an integral low-finned tube has two to three times the heat transfer surface area of a plain tube with a diameter equal to that of the tip of the fins. A porous coated surface with one layer of uniform spherical particles packed in a square layout produces a surface area ratio of $\pi + 1$, two layers produces a ratio of $2\pi + 1$, three layers $3\pi + 1$, and so forth. Mechanically deformed surfaces, such as the Thermoexcel-E shown schematically in Fig. 4-5, produce ratios ranging from about 3 to 4. Consequently, if effectively used, this additional surface area can produce large enhanced boiling heat transfer coefficients because the coefficients for these surfaces are normally defined in terms of the nominal exterior surface area.

6-3 FUNDAMENTAL EXPERIMENTAL STUDIES

This section surveys studies on the fundamental physical phenomena and processes affecting the performance of enhanced boiling surfaces. Emphasis is placed on those studies that resulted in an enlarged understanding of how the particular type of enhancement geometry augments heat transfer and also on those in which a wide range of characteristic dimensions was tested to determine an optimum configuration.

Low-Finned Surfaces

There are two factors of foremost importance to boiling on integral low-finned tubes. These are the increase in wetted surface area and the effect of the fins on the boiling process relative to that on a plain tube. Both are related to one another.

Photographic studies of boiling on finned surfaces have been performed to obtain a better understanding of the physical process. Westwater (1973) summarized the many investigations that he and co-workers undertook on various types of fins. For low-finned tubes, he suggested that the spacing between adjacent fins should be about equal to the bubble departure diameter of the particular fluid. Apparently, bubbles much larger than the interfin spacing create a thermally resistive vapor film at the root of the fins, and bubbles that are too small cause the tube to function as a plain surface.

An extensive photographic study on boiling heat transfer of R-22 on a low-finned tube was described by Fath and Gorenflo (1986), Fath (1986), and Gorenflo and Fath (1987). The principal factor studied was the effect of pressure on boiling performance. They tested a Gewa-K tube (manufactured by Wieland-Werke) that was 15.81 mm in outer diameter with a root diameter of 13.01 mm. The fins were 1.40 mm high and 0.30 mm thick with a tapering of 3°. The fin density was 741 fins per meter (18.8 fins per inch [fpi]). This trapezoidal geometry produced a 1.05-mm gap between the fins and a surface area ratio of 2.51 relative to a 15.81-mm diameter plain tube.

Figure 6-6 compares boiling at pressures of 5, 25, and 40 bar, representing reduced pressures of 0.1, 0.5, and 0.8, respectively, at about the same heat flux (based on the nominal area of a plain tube with a diameter of 15.81 mm). At 5 bar the bubbles are observed to be about the same diameter as the interfin spacing and were reported to have a sweeping effect on other bubbles growing on the opposite fin. At 25 and 40 bar, the bubbles leaving the top of the tube are much smaller than the fin spacing. They are more numerous and produce a bubble wake at the top of each fin. In addition, the bubbles at high pressures were noted to rise much more slowly than at 5 bar because of their smaller buoyancy force. Thus at 5 bar it appears that the boiling process is affected by the presence of adjacent fins and that at reduced pressures greater than 0.5 the process is similar to that on a plain surface. This was partially confirmed by the heat transfer results, which showed that the heat transfer augmentation of the finned tube tended to decline with increasing heat flux and pressure to that of a plain tube.

Interpreting the photographs in Fig. 6-6 in terms of the heat transfer mechanisms described in the section above on enhanced boiling, it appears that for a low-finned tube the following mechanisms are most important to their performance at low heat fluxes: (1) "external" type evaporation on the entire wetted surface area (primarily a reflection of the increased wetted surface area available for nucleate boiling), and (2) two-phase flow convection between the fins. Thin film evaporation may exist at large heat fluxes at low pressures, but complete suppression of individual boiling sites is unlikely for a single tube because the vapor formed at the bottom side of the tube tends to disengage once reaching the sides of the tube. The thin film process could be created intermittently at fin spacings smaller than the bubble departure diameters, but this process would probably be more similar to slug flow than to true film flow.

Theory and practice of condensation on low-finned tubes and tube bundles are fairly well advanced, and it is generally accepted that for fluids with large surface tensions, such as water and ammonia at low pressures, low fin densities should be used and that higher densities can be used for fluids with "normal" surface tension values. The advantage of using tubes with larger fin densities is their larger ratio of external wetted surface area to inside surface area. For nucleate pool boiling, however, no such rule of thumb has been presented in the literature.

For low-finned tube bundles, the bundle boiling heat transfer coefficient is augmented relative to that for a single tube at low to medium heat fluxes (see Chapter Ten's discussion of bundle boiling studies of low-finned tubes). Thus it is not sufficient

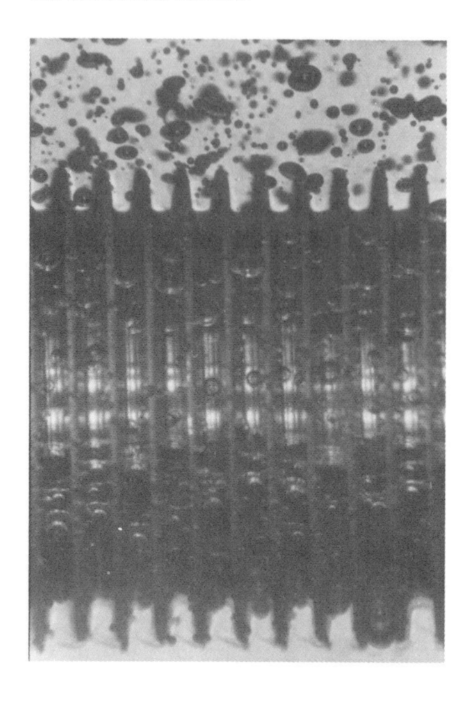

(a) 5 bar, 4.1 kW/m^2

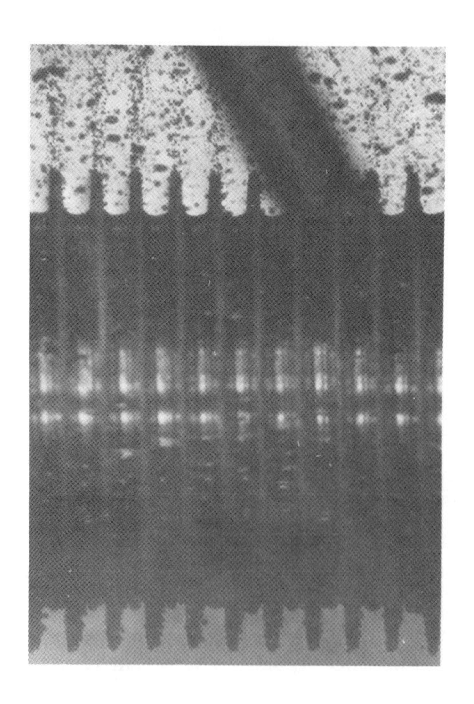

(b) 20 bar, 1.2 kW/m^2

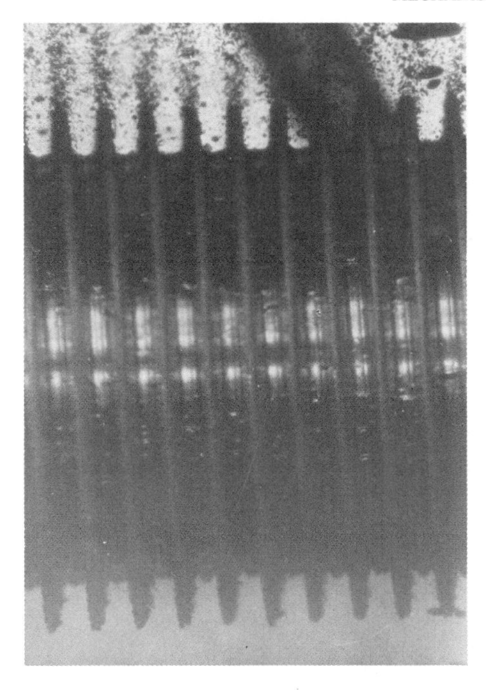

(*c*) 40 bar, 1.2, kW/m^2

Figure 6-6 Boiling of R-22 on a low-finned tube (courtesy Laboratorium für Wärme- und Kältetechnik, Universität-GH-Paderborn).

to determine the optimum fin density in single tube tests because a comparison of bundle boiling performance may yield different results. For example, Danilova and Dyundin (1972) attempted to determine an optimum low-finned tube for boiling R-12 and R-22 using single tube tests. They found that a 27.7-fpi tube (1,095 fins per meter) outperformed a 12.5-fpi tube (493 fins per meter) by about 20%. The subsequent bundle tests that they ran, however, showed only marginal improvement, about 5%. In addition, the optimum fin density for one fluid at one pressure is probably not the optimum at another pressure or for a different fluid. Therefore, experimental studies to determine an optimum fin density of a low-finned tube have less importance than similar studies of the higher-performance boiling tubes, which tend to have high single tube performances that are similar to their bundle boiling values.

Modified Finned Surfaces

These types of surfaces are shown in Chapter Four. All their enhancement geometries have been optimized in industrial research laboratories operated by the particular tube's manufacturer, where sometimes hundreds of variations of the particular geometry were tested. The objective of a tube manufacturer is not necessarily to obtain the best boiling geometry but rather a geometry that produces substantial augmentation and can be produced cheaply and within manufacturing tolerances that do not significantly affect the tube's boiling performance. In addition, for water chiller applications for example, a tube manufacturer may guarantee that a particular tube will perform up to or exceed specifications set by the boiling curve established as the product standard. For crucial applications, such as the refrigeration system in nuclear submarines, the refrigeration company may require the tube manufacturer to provide test data for a tube sample taken every so many tubes to document the quality of the tubing.

In the open literature, only a few enhancement geometries in modified finned surfaces have been extensively investigated: surfaces with T-shaped fins, bent-over fins, and channels covered by perforated plates.

T-shaped fins. The boiling process on a Gewa-T tube is shown in Fig. 6-5 for R-22 at 5 bar. For comparison purposes, boiling on the same tube (in an unetched condition with a gap opening of 0.35 mm) at 32.5 bar is shown in Fig. 6-7 (from Fath [1986]). The bubbles outside the enhancement are much smaller at the larger reduced pressure but much more numerous because of the larger surface roughness. The bubbles inside

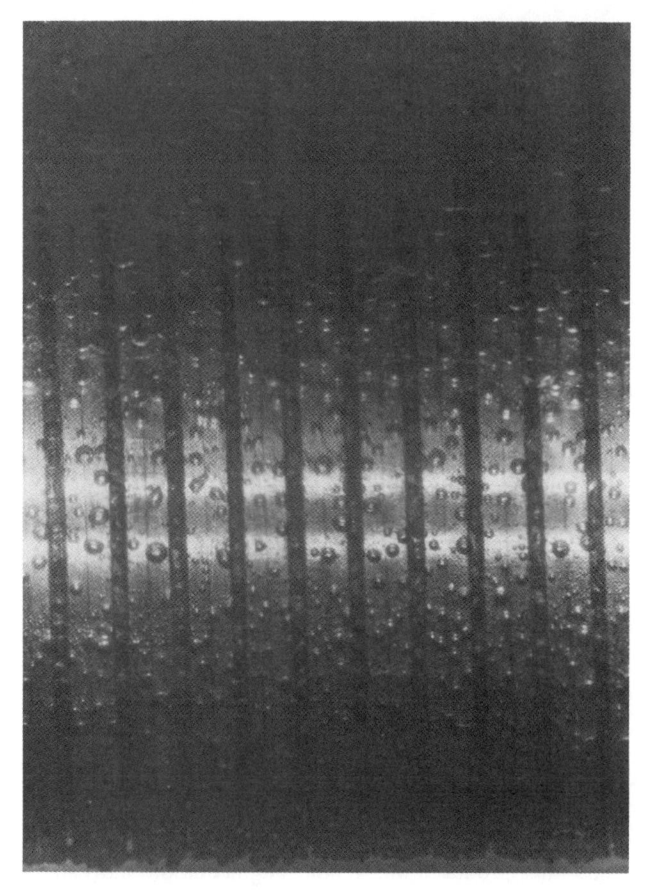

Figure 6-7 Boiling of R-22 on a Gewa-T tube at 32.5 bar and 52 kW/m^2 (courtesy Laboratorium für Wärme- und Kältetechnik, Universität-GH-Paderborn).

the reentrant channels appear to be about the same size, irrespective of the larger reduced pressure, but are again more numerous at the higher pressure. The flow inside the reentrant channel appears to be a slug flow regime. Unexpectedly, thin film evaporation does not appear to occur even at the heat flux of 52 kW/m^2.

Ayub and Bergles (1985) investigated the effect of gap size between fin tips of tubes with T-shaped fins. They ran tests with water and R-113 for copper tubes with gap sizes of 0.15, 0.25, 0.35, and 0.55 mm while holding the other dimensions fixed (e.g., a fin density of 740 fins per meter [18.8 fpi], a fin height of 1.1 mm, and tube diameters from 25.4 to 26.0 mm). Figure 6-8 shows their results plotted for several heat fluxes. For R-113 a maximum in performance occurred at a gap width of about 0.25 mm. For water the maximum performance was obtained with a larger opening of about 0.4 mm, perhaps because water forms larger diameter bubbles. By injecting a blue dye into the liquid around the circumference of the tube with a hypodermic needle and syringe, they investigated the liquid flows around the tube and observed that there

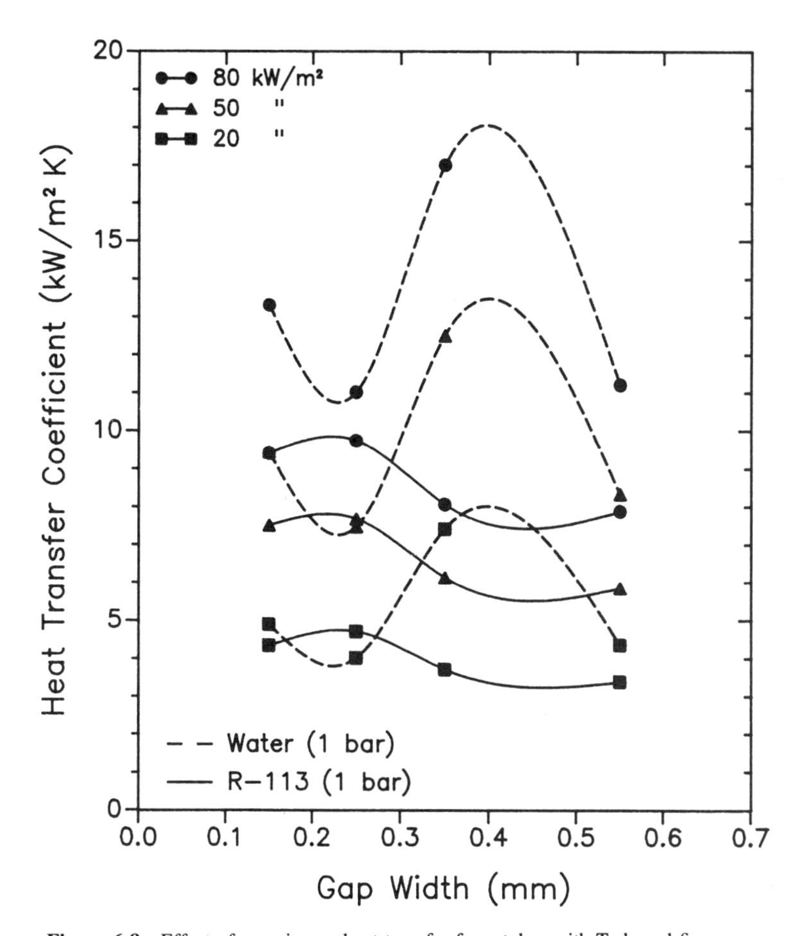

Figure 6-8 Effect of gap size on heat transfer from tubes with T-shaped fins.

were distinct locations where liquid tended to enter the reentrant channel, namely at the bottom, at both sides, and at the top dead center.

In a similar study with R-113, Marto, Wanniarachchi, and Pulido (1985) investigated the effect of gap width, fin density, wrapping wires inside the reentrant channel, and placing a metal shroud on tubes with T-shaped fins. For tubes with fin densities of 980 and 1,333 fins per meter (24.8 and 33.8 fpi) and fin heights of about 1 mm, they found that a gap width of 0.25 mm produced the best performance, as in the above study. For tubes with the same gap width, the higher fin density version gave slightly better heat transfer performance, probably because of the reentrant channel's smaller hydraulic diameter. The use of three wires wrapped inside the reentrant channel with a 0.25-mm gap increased the boiling heat transfer coefficient substantially at heat fluxes less than 50 kW/m^2 (see Fig. 3-19).

Still further improvement was attained by placing a tightly fitting shroud over the tube with openings at the top and bottom, by which liquid entered at the bottom opening and vapor and liquid exited from the top. With the shroud in place boiling only occurred inside the reentrant channel at low to medium heat fluxes, and hence external enhancement was not much of a factor. The augmentation with the shroud can be explained by two factors: (1) the shroud retains the bubbles inside the reentrant channel until they reach the top of the tube, and (2) the shroud reduces the intake of liquid into the reentrant channel, which increases the vapor quality and the two-phase velocity and hence the factor F multiplying the liquid-only convective heat transfer coefficient.

These results with the wires and shroud thus tend to confirm that the two-phase flow heat transfer mechanism inside the reentrant channel is responsible for heat transfer augmentation on tubes with T-shaped fins.

Xin and Chao (1985) produced T-shaped fins in flat plates and tested them with ethanol, R-113, and water. The surfaces had the following geometric specifications: gap widths ranging from 0.09 to 0.24 mm, fin heights of either 0.5 or 1.0 mm, and fin densities of either 833 or 1,250 fins per meter (21.1 or 31.7 fpi). For R-113, the gap widths of 0.09 and 0.11 mm produced better performance than a width of 0.20 mm, which is contrary to the results of the above two studies. The fin heights of the present surfaces were much smaller (0.5 mm), however.

For water, a gap width of 0.17 mm yielded the optimum performance, which is also contrary to the Ayub and Bergles (1985) results given above even though the fin heights and densities were similar. Thus the boiling process is speculated to be different for a horizontal plate and a horizontal tube with T-shaped fins. The buoyancy force of the vapor formed in the reentrant channels of a horizontal surface cause the vapor to escape upward such that less two-phase flow enhancement is produced.

For ethanol, a gap width of 0.11 mm gave much better heat transfer than widths of 0.09, 0.14, and 0.20 mm while all other dimensions were held fixed. Most of the heat transfer augmentation was attributed to thin film evaporation, but it was also suggested that the external convection process was important (a correlation that Xin and Chao developed for these surfaces is described in Chapter Eight's section on modified externally finned tubes).

Bent-over fins. A simple but effective method for augmenting the boiling performance of low-finned tubes was developed and patented by Webb (1972). By rotating a low-finned tube in a chuck and running a bending or rolling tool over the fins, the fins were deformed to make a continuous reentrant channel as depicted in Fig. 3-21c. The tip of the tool was designed to produce the desired degree of bending and a controlled gap size from the tip of one fin to the back of the next one. From experimental studies with R-11, Webb observed that the heat transfer performance dropped off rapidly with a gap size beyond the range from 0.0254 to 0.127 mm (0.001 to 0.005 inch) and that the preferable gap size was between 0.038 and 0.089 mm (0.0015 and 0.0035 inch). He also noted that a tube with 1,303 fins per meter (33 fpi) whose fins were 0.76 mm high and 0.254 mm thick (0.03 and 0.01 inch, respectively) produced essentially the same performance as another tube with twice the fin density and fins half as high and half as thick. A boiling comparison between his tube (used by the Trane Company) with bent-over fins to an ordinary low-finned tube is shown in Fig. 6-9.

On the basis of physical observations of the boiling process and the resulting tube

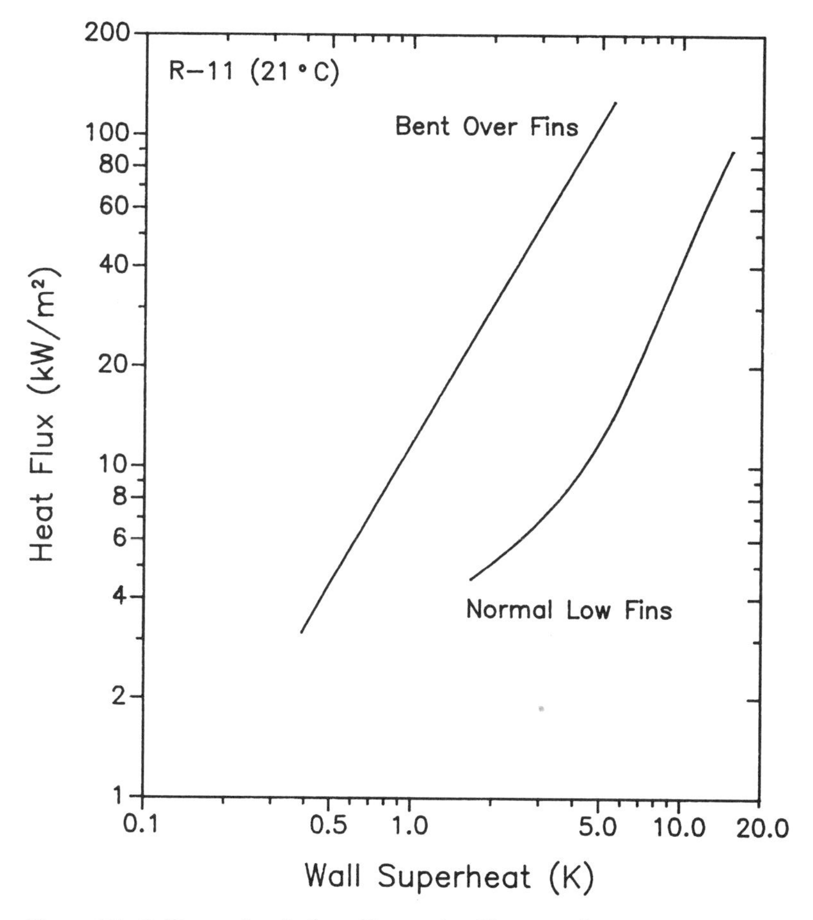

Figure 6-9 Boiling on finned tubes with normal and bent-over fins.

performances with various gap sizes, Webb proposed a possible physical explanation for the substantial boiling performance that he obtained. First of all, the augmentation could partly be attributed to a modification of the boiling nucleation process as follows. With ordinary plain surfaces or low-finned tubes, "cold" saturated liquid flowing in from the bulk tends to prevent activation of potential boiling sites, and thus a larger heat flux is required to bring this fluid to its incipience point. In contrast, for bent-over fins the inrushing liquid first passes through the narrow gap before flowing along the interior to the nucleation site. Thus the nucleation sites are shielded from the cold bulk liquid and instead are furnished with superheated liquid. Consequently nucleation can occur at lower wall superheats and heat fluxes, and the nucleate boiling process begins at conditions under which plain and ordinary finned surfaces are still in the single-phase natural convection regime. In addition, by specifying that the spacing between the base of the fins be larger than the gap at the fin tip, the channel is reentrant and permits the radius of the vapor-liquid interface to be relatively large, so that the superheat required to sustain the boiling process is reduced. Thus the gap size is the most crucial parameter of this enhancement because the nucleation superheat is increased if it is too small and cold liquid can enter if it is too large.

Webb also hypothesized that the evaporation process was accentuated by the presence of a large quantity of superheated liquid within the channel. For instance, the departure of a bubble carries away the locally superheated liquid when it is boiling on an ordinary surface, which temporarily deactivates the site until the inrushing cold liquid is again superheated. In contrast, the continuous supply of superheated liquid within the reentrant channel allows the process to proceed without interruption. On departure of a bubble, Webb concluded; the inrushing liquid pushed the liquid interface along the channel because the sites of vapor emanating from the gap opening tended to move. At higher heat fluxes, closely spaced bubble columns were formed that produced bubbles of similar size but out of phase with one another, which also tended to confirm Webb's oscillating vapor-liquid interface theory. Therefore, a second important feature of the channel geometry was to provide thermal and hydraulic communication between adjacent boiling columns, which improved their combined performance.

In summary, the bent-over fins augment heat transfer primarily by (1) reducing the nucleation superheat by preventing cold bulk liquid from reaching the nucleation sites, (2) facilitating the existence and continuous functioning of the thin film evaporation mechanism within the reentrant channels, and (3) superheating the liquid passing through the narrow gap by means of the laminar entrance region convection mechanism.

Channels with perforated outer surfaces. Nakayama et al. (1979, 1980) and Nakayama, Daikoku, and Nakajima (1981) investigated the geometric effects on boiling of grooved surfaces covered with perforated plates, similar in geometry to the Thermoexcel-E surfaces shown in Figs. 4-5 and 4-6. They ran tests with liquid nitrogen, R-11, and water to obtain boiling curves, boiling site densities, and bubble departure diameters and frequencies for various pore and channel sizes (their nucleation and thin film visualization work is described in Chapter Five, Sec. 5-2).

In tests with R-11 on two identical enhancements (differing only in pore diameter) and on four plain surfaces, these investigators observed that the plain surfaces required wall superheats an order of magnitude larger than those of the enhanced surfaces to produce the same boiling site density and that the different pore diameters (0.10 and 0.04 mm) had little effect on the boiling site density. The pore diameter was defined as that of a circle tangent to the three sides of the triangular pores. Figure 6-10 shows that pore size had a minimal effect on R-11 boiling performance, which was much higher than that of the plain surface. It is also interesting to note that the boiling performance of these three enhancements were nearly identical to that of Webb's bent-over fin design, which has a gap size similar to the pore sizes used here.

Similar results were obtained in tests with the other liquids, although boiling performances varied much more with pore size. For water, a pore diameter of 0.20 mm performed much more poorly than smaller ones of 0.08 and 0.14 mm. For liquid nitrogen, pore diameters of 0.03 and 0.06 mm produced boiling curves with small slopes that tended quickly toward the plain surface boiling curve, apparently because the openings were too small for liquid to pass and adequately wet the channels; the

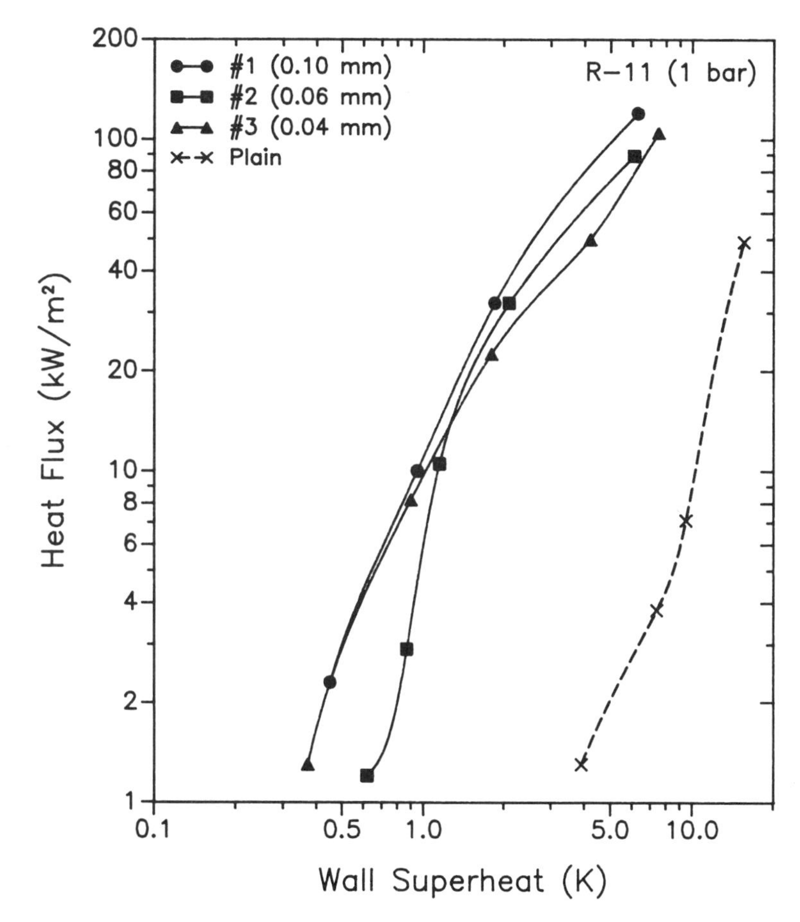

Figure 6-10 Effect of pore size on Thermoexcel-E performance.

optimal pore size was found to be 0.09 mm. Thus for three fluids with widely differing physical properties, a pore size of about 0.10 mm was found to be the best.

Using an optical probe specially developed to measure bubble departure diameters and frequencies and to count visually boiling site densities, Nakayama and co-workers determined the latent heat flux from a surface from the expression

$$q_{\text{lat}} = \left(\frac{N}{A}\right)\left(\frac{\pi D^3}{6}\right) f \Delta h_v \rho_v \tag{6-3}$$

where N/A is the boiling site density, D is the average bubble departure diameter, and f is the average departure frequency.

Figure 6-11 depicts their ratios of the latent heat flux to the total heat flux for three different surfaces (the sensible heat flux ratio can be obtained by subtracting the latent heat flux ratio from 1.0). At a total heat flux of 0.7 kW/m^2, about 90% of all heat left the two enhanced surfaces in the departing bubbles; the remaining 10% was thus carried away as superheated liquid. As the total heat flux increased, however, the percentage leaving as latent heat dropped substantially, reaching about 25% at 20 kW/m^2. In contrast, the plain surface did not begin to boil until reaching a heat flux of 3 kW/m^2,

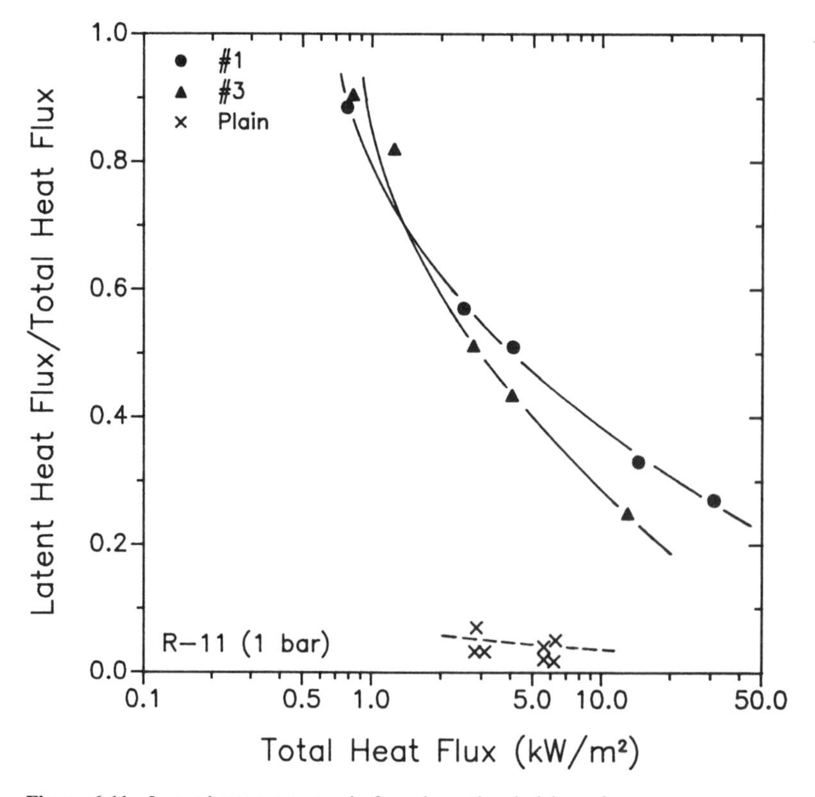

Figure 6-11 Latent heat transport ratio for enhanced and plain surfaces.

and then only a small amount of heat left in the vapor. The high performance of the Thermoexcel-E type of surface was therefore attributed to the low nucleation superheat required to activate the surface, the thin film evaporation process observed inside the channels, and augmentation to the external convection process by the increased boiling site density.

Ali and Thome (1984) also evaluated the data in Figs. 6-10 and 6-11. They pointed out that, at heat fluxes of practical interest in refrigeration water chillers, sensible heat remains the dominant heat transfer mode for the enhanced surfaces, as in conventional boiling. For instance, at a total heat flux of 10 kW/m^2, sensible heating is seen to be responsible 70% and 98% of the total for enhanced and smooth surfaces, respectively, in Fig. 6-11. Referring to Fig. 6-10, the corresponding wall superheats are 1.0 and 10 K, respectively. Thus it can be deduced that the evaporation heat transfer coefficients, (i.e., those characterizing the evaporation heat transfer mechanisms and thus determined from q_{lat} divided by the wall superheat) are 3 and 0.02 kW/m^2 K and that the convective heat transfer coefficients, determined from $(q_{total} - q_{lat})$ divided by the wall superheat, are 7 and 0.98 kW/m^2 for the enhanced and plain surfaces, respectively. Consequently, their enhanced surface geometry augmented the evaporation process by a factor of 150 and the convective heat transfer process by a factor of 7!

In the above tests, the pores on the surface were all of uniform size. Optimization of performance was also attempted by using pores of two different diameters on the same surface, the small ones for growing bubbles with the vapor produced inside the channels and the large, inactive ones for supplying liquid to the channels. In Nakayama, Daikoku, and Nakajima (1981), two otherwise identical enhanced surfaces were constructed with rectangular channels 0.25 by 0.4 mm in area, a channel pitch of 0.55 mm, and a pore pitch along the channel of 0.7 mm. Their surface identified as C15-10 had pores 0.15 mm in diameter (8.3%) and 0.10 mm in diameter (91.7%). Surface C10-5-1 had smaller pores, 0.10 mm in diameter (8.3%) and 0.05 mm in diameter (91.7%).

Using the boiling data that they obtained, one can estimate that the surface with small pores produced about 50% more vapor than the other at a heat flux of 2.6 kW/m^2. Thus this surface transferred 50% more energy in the form of latent heat and correspondingly less energy as sensible heat. The wall superheat for the small-pore surface was 0.45 K, and that for the large-pore surface was only 0.06 K. Therefore, even though the smaller pores were better at producing vapor, the heat transfer coefficient was only ⅛ that of the large-pore surface. Hence the high performance of channels with perforated outer surfaces is only partially controlled by their enlarged vapor-producing capacity relative to that of a plain surface.

Evaluating these results further, it is noted that most of the heat still leaves both these surfaces as sensible heat. Consequently, the most important aspect in their design appears to be their ability to augment the convective heat transfer process for superheating the liquid. The surface C15-10 had a smaller boiling site density and a much smaller wall superheat driving the convection process compared to C10-5-1. Thus, C10-5-1 would be expected to provide more augmentation to the external convection process. By a process of elimination of the heat transfer modes described above (under enhanced boiling), the only heat transfer mechanism that could be responsible for the better performance of surface C15-10 is apparently the internal convection process. It does not appear to be possible for a two-phase flow process similar to that shown in Figs. 6-5 and 6-7 to occur in the present surfaces because the outlets were quite small and because the test sections were flat and mounted horizontally. Superheating of the liquid entering the small triangular pores and flowing in the liquid films by entrance region laminar flow would be effective, however. The location of the pores appears to favor this because they were not located at the center of the channel but at one of the channel's walls, so that liquid could flow directly onto that wall or from that wall out through the pore. Thus it is hypothesized here that the pumping action of the bubbles draws part of the liquid film out of the active pores after departure of bubbles, or draws liquid in and then forces it out of the inactive pores, or both. The larger pore diameters would facilitate this process and thus explain surface C15-10's performance improvement relative to C10-5-1.

Ma et al. (1986) tested fourteen different flat, horizontal surfaces for boiling water and methanol at atmospheric pressure. These surfaces had various channel cross-sectional geometries and were covered by sintering a screen or a perforated plate (or both) to the outer surface. For rectangular channels, screens with pore sizes of 0.068 and 0.091 mm gave much better performance than one with 0.145 mm for both fluids. The investigators concluded from this that the large pore size was too big and allowed the channels to be flooded with liquid, reducing effectiveness. Indeed, the surface with the largest pore size performed like a low-finned tube. In another series of tests in which the size of the rectangular channel was varied and the 0.145-mm pore screen was used, the smaller channel size (0.3 by 0.4 mm) gave slightly better results than channels up to twice as large.

Ma and co-workers made another set of test sections with rectangular, triangular, and U-shaped channels. These channels were covered by a fine mesh screen topped with a perforated plate. Different test sections had different diameters of laser-drilled pores in the cover plate, ranging from 0.080 to 0.220 mm. The triangular and rectangular grooves tended to give the best performance in both fluids, although optimal configurations were different for water and methanol. For instance, the best surface for water at heat fluxes less than 100 kW/m^2 was the worst surface for methanol. Thus heat transfer augmentation was shown to be sensitive to the physical properties of the fluid for this type of geometry.

Ma and co-workers attributed some of the above findings to the pore diameter's effect on the boiling process. They hypothesized that each pore served two main functions: (1) to provide for the outlet flow of vapor, and (2) to allow liquid to enter the grooves without flooding them. These processes, however, produce a conflicting requirement on the size of the pore. In the first process, the vapor has to overcome the surface tension holding it to the mouth of the pore for the bubble to depart. Thus the larger the radius, the greater the surface tension impeding bubble departure and the vapor transport process. The flow of vapor through the pore into the bubble encounters less hydraulic resistance with increasing pore diameter, however. On the other hand, after a bubble departs the pressure inside the channel is lower than the external pressure, facilitating the inflow of liquid. The liquid influx, however, is resisted by the capillary pressure of the vapor nucleus, which increases with decreasing pore size. Thus the pore radius must be small enough to resist flooding and to assist bubble departure but large enough to allow sufficient inflow of fresh liquid to the evaporating films. Consequently, for a particular fluid there should exist an optimum pore size that properly balances these conflicting requirements.

In summary, pore diameters of about 0.10 mm and grooves forming wedges (triangular and rectangular cross-sections) provide the best performance for surfaces with grooves covered by a perforated cover. The additional geometric effects of pore spacing and channel density are not well understood. Comparing channels with perforated outer surfaces and bent-over fins, such as the Thermoexcel-E, the Turbo-B, the ECR-40, and a tube patented by Webb (1972), to surfaces with T-shaped and straight fins, one finds that the former tend consistently to give better heat transfer performance. Therefore, it can be concluded that the thin film evaporation mechanism, the laminar developing flow mechanism, and the nucleation characteristics of these channels are more effective than the two-phase flow mechanism in larger channels between fins.

Porous Layers

Porous layers have been studied thoroughly because of this type of enhancement's importance to heat exchangers, heat pipes, and electronic cooling. The various geometric parameters that have been studied include particle diameter or size range of particles, layer thickness, particle material, and porosity. The actual geometry of the enhancement is dependent on the type of process used to produce it, be it sintering, plating, or flame spraying of particles.

Various forms of boiling curves can be produced for boiling on surfaces coated with porous layers and, to a lesser extent, on surfaces with grooves and perforated cover plates. Many investigators have noted different boiling curves for the former, including Afgan et al (1985), Kovalev, Solov'yev, and Ovodkov (1987), Styrikovich et al. (1987), and a host of other eastern European investigators. The boiling curves for nitrogen in Nakayama et al. (1979, 1980) also display this type of trend for Thermoexcel-E geometries, for instance. In this type of geometry, Arshad and Thome (1983) observed that the channel walls progressively dry out from the bottom up with increasing heat flux, which causes the boiling curve to flatten out.

Figure 6-12 shows a composite of these different shapes (from Kovalev, Solov'yev, and Ovodkov [1987]). The plain surface boiling curve is that normally observed when the wall temperature is the independently controlled experimental parameter. When the porous material has a high thermal conductivity and is highly permeable, the heat transfer performance is much better than that of a plain surface, especially at low heat fluxes. As the heat flux rises from low values along curve 1 after boiling nucleation has occurred, the slope is large. At higher heat fluxes the curve tends to flatten out before rising again at the highest heat fluxes. On decreasing the heat flux, either curve 1 or curve 2 may be followed. In contrast, another porous coating can have a boiling curve similar to 3, which represents a good performance at low heat fluxes that deteriorates quickly at higher ones, producing a lower "peak" nucleate heat flux than the plain surface. For porous coated surfaces the transition from the nucleate boiling to film boiling regime can be attained without passing through the characteristic minimum

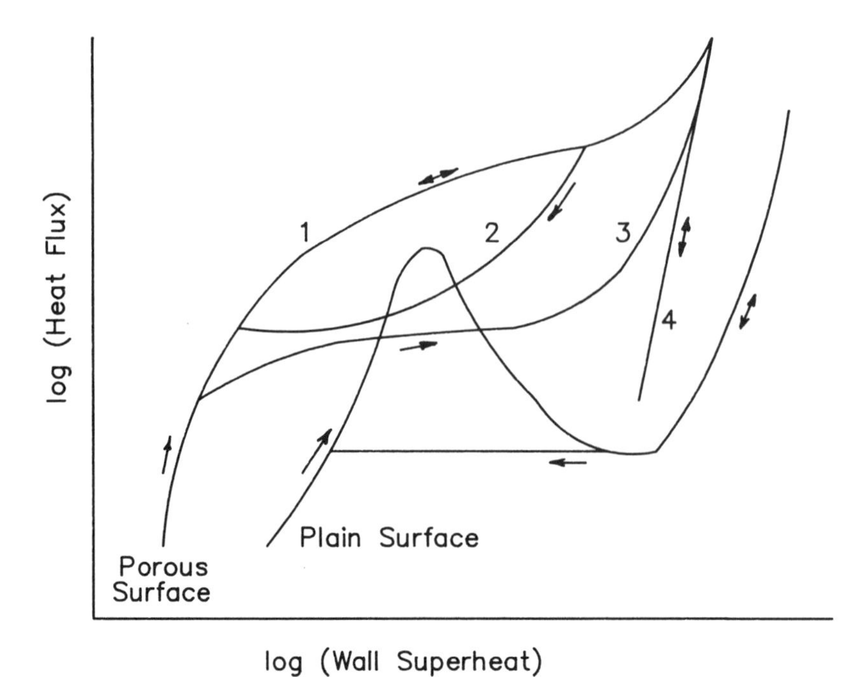

Figure 6-12 Boiling curves for surfaces with and without porous coatings.

observed for plain surfaces. The film boiling curve 4 can also be at lower superheats for a surface with a porous layer than for a plain surface if the porous layer is thick. This probably occurs because the temperature gradient through the layer maintains nucleate boiling at the outer surface, in many respects like the Westwater (1973) studies on long fins with film boiling at the base and nucleate boiling at the tip.

Figure 6-13 (from Afgan et al. [1985]) compares the boiling processes occurring on a plain surface to those occurring on a porous coated surface. For a plain surface, evaporation only occurs externally at cavities in the surface (Fig. 6-13*a*). A porous surface promotes evaporation within the layer. This can occur as shown in Fig. 6-13*b*, where the base heated wall is wetted by liquid, or as shown in Fig. 6-13*c*,

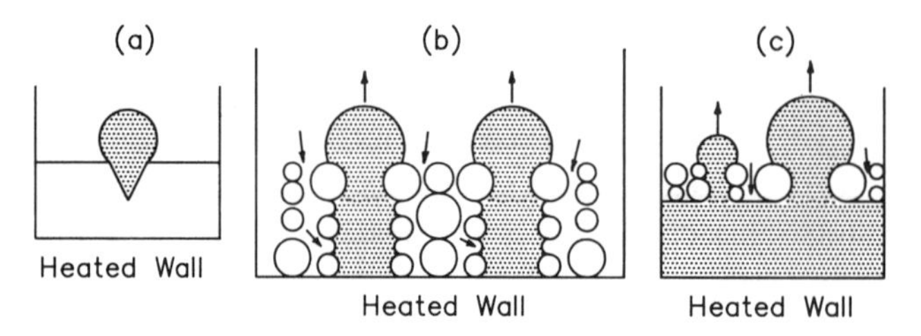

Figure 6-13 Vapor generation processes. (*a*) Bubble growing on a plain surface, (*b*) evaporating menisci in a porous layer, (*c*) evaporation at a continuous vapor film in the porous layer.

where the bottom of the porous layer is filled with a continuous vapor region and the heated wall is not wetted by liquid. Afgan and colleagues hypothesized that the processes in Fig. 6-13, *b* and *c* are similar to those in a heat pipe, where vapor is generated in the porous matrix and condensation occurs at the top of the bubble in the bulk liquid.

The shapes of the porous layer boiling curves in Fig. 6-12 can be explained in terms of these two processes. Curve 1 is produced by layers that are wetted all the way to the base wall at low to medium heat fluxes but form a continuous vapor region near the heated wall at high heat fluxes. Curve 2 may occur if the heat flux was originally raised past the critical value required to form the continuous vapor region. Hysteresis may then occur if the continuous vapor region remains intact on lowering of the heat flux. Boiling curves of type 3 occur for layers with low permeability or that are thick and thus form a continuous vapor region as in Fig. 6-13*c* at low heat fluxes.

Kovalev, Solov'yev, and Ovodkov (1987) performed experiments to measure flow rates of vapor, liquid, and opposing vapor and liquid flows through porous layers that could result from small pressure gradients. They found that the hydraulic resistance to gas flow and, to a greater extent, liquid flow occurs when there is a countercurrent flow, similar to that which occurs for vapor trying to escape and liquid trying to enter a porous layer during boiling. Solving some hydrodynamic and heat conduction differential expressions describing the above process numerically, they obtained boiling curves of shapes 1 and 3 depending on the thermal conductivity, permeability, and thickness of the coating.

Czikk and O'Neill (1979) surveyed Union Carbide's experience during the development of the High Flux surface. This surface's particles are not of uniform size or shape. It was determined that the heat transfer coefficient for a given fluid and operating condition (pressure and heat flux) varied with the porosity and pore size produced by the particles. For example, if several test surfaces were prepared with metallic powders sieved though screens of different mesh sizes, typically a maximum in performance was found for a powder producing an intermediate pore size. In general, high–surface tension fluids such as water and ammonia benefited from a coarse matrix, whereas low–surface tension fluids such as light hydrocarbons and cryogens performed better when fine-grain powders were used to obtain smaller pores.

Bukin, Danilova, and Dyundin (1982) tested porous coated tubes in R-12 and R-22 produced by three different methods: electric-arc spraying of copper, sintering of spherical stainless steel particles, and jacketing with glass or stainless steel clothes. They observed a structural effect on the heat transfer performance. When spraying molten metal, the droplets struck the surface and flattened out, producing a surface with a lower porosity than that achievable by sintering particles to the surface. Hence the sprayed surfaces, even though made of copper, performed worse than the sintered stainless steel ones. They also observed an optimum layer thickness for the sintered coatings. This was explained as follows: when the layer thickness was increased, the liquid supply to the innermost active layers of the coating was decreased, reducing performance; when the thickness was decreased to the bubble departure diameter, the enhancing effect on the growth of the vapor nuclei decreased, also reducing performance. A thickness of 0.5 mm gave the best results, but the optimum layer thickness was rather broad, and layers from 0.3 to 1.0 mm gave nearly equal heat transfer coefficients.

When covering a tube with a tightly wrapped porous glass cloth, the vapor was primarily generated underneath the cloth and had difficulty in escaping, hampering the heat transfer process. By loosening the cloth this problem was alleviated, and performance similar to that of the best metallic sintered surfaces was attained. Consequently, the low thermal conductivity of glass compared to stainless steel did not adversely affect the process. Surprisingly, using a stainless steel gauze with a 0.04-mm mesh size and 0.04-mm wire size produced inferior performance compared to the glass cloth.

Nishikawa, Ito, and Tanaka (1979, 1983) ran parametric studies on the effect of particle size, layer thickness, and particle material on boiling in R-11 and R-113. Using spherical copper particles 0.25 mm in diameter, they observed a large maximum in heat transfer performance with layer thickness, as shown in Fig. 6-14. The optimal thickness was found to be 1 mm, which corresponded to a ratio of layer thickness to particle diameter of 4. At the optimum thickness the heat transfer coefficient was a

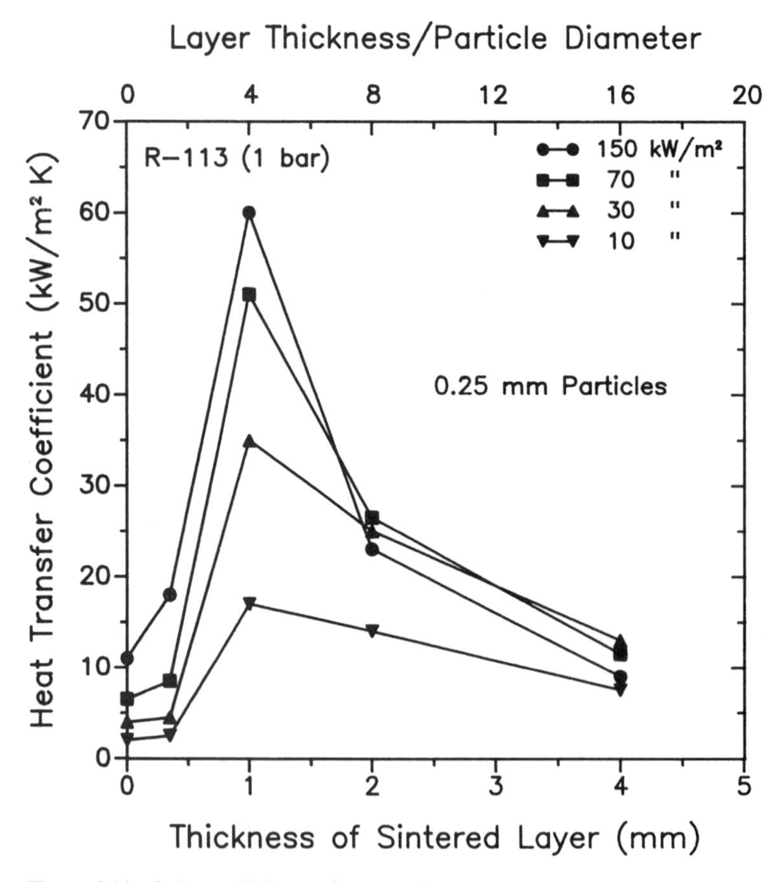

Figure 6-14 Optimum thickness of a porous layer.

strong function of heat flux, whereas at nonoptimal values much less dependency was evident. Some of the boiling curves obtained by these investigators were similar to the ones shown in Fig. 6-12.

In another set of tests, Nishikawa and co-workers varied the particle diameter and material (copper and bronze) for boiling R-11 and R-113. The optimum ratios of layer thickness to particle diameter for copper and bronze particles 0.25 mm in diameter were nearly the same, about 4. For bronze with R-113, this ratio varied from 3 for 0.3-mm particles to about 11 for 0.1-mm particles. The optimal porous layers gave performance similar to that of the High Flux surface but were much thicker and had a much larger particle size (the High Flux coatings tend to be about 0.07 to 0.3 mm thick).

For porous layers made with sintered or plated particles, Webb (1983) developed a method for categorizing their geometric characteristics by means of four parameters: particle size or size distribution, particle shape, porous coating thickness, and particle packing arrangement. The last parameter establishes the porosity of the layer, as can be deduced from Fig. 6-15. For instance, the porosity of a triangular layout of tightly packed round particles of uniform size is only 8.9%, whereas for an in-line "square" layout the porosity is 47.6%. In contrast, an in-line triangular layout (not considered in Webb's study) has a porosity of 39.5%, or $[1 - (\pi/3)\sqrt{3}] \times 100\%$, which is much larger than the tightly packed triangular layout.

Webb's measurements of High Flux samples showed porosities ranging from 50% to 65%, and those obtained by Nishikawa and co-workers were reported to vary from 38% to 71%. Bukin, Danilova, and Dyundin (1982) reported porosities from 46% to 52% for their sintered stainless steel coatings and from 23.7% to 37% for the copper arc-sprayed ones. The plasma sputtering of Nichrome by Styrikovich et al. (1987) produced porosities ranging from 40% to 50%, but the boiling performances were much lower than those for the sintered or plated layers. Webb explained the differences in performance of porous layers of differing or similar porosities on the basis of the types of pores produced:

1. active pores with stable nucleation sites;
2. intermittent pores, which can be filled with either liquid or a vapor meniscus depending on whether annular or plug flow exists in the matrix, respectively;
3. liquid-filled pores that act to supply superheated liquid to the active and intermittent pores; and

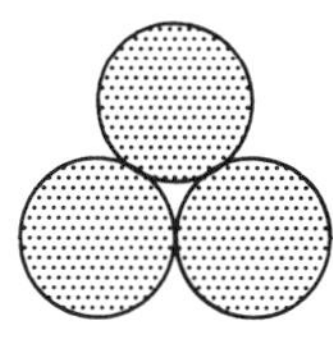

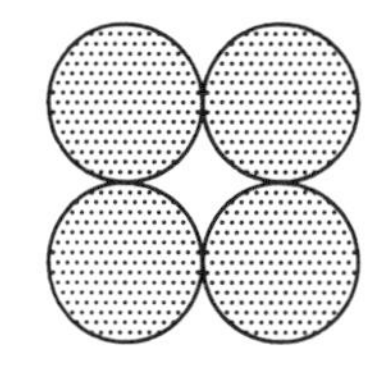

Figure 6-15 Particle packing arrangements.

4. nonfunctional pores, that is closed voids containing neither boiling liquid nor vapor.

Fujii, Nishiyama, and Yamanaka (1979) and Fujii (1984) tested a wide range of particle sizes and layer thicknesses for spherical particles electroplated to a heated wall. From cutaway photographs of their surfaces, they noted that some of the particles became larger in diameter than others during the plating process. Layers with two particle thicknesses tended to form an in-line triangular arrangement, and three or more layers tended to form an in-line square layout. They compared performances by defining a boiling enhancement ratio as the enhanced heat transfer coefficient divided by the smooth surface value at the same heat flux. Figure 6-16 depicts their results. For R-113 the layer 0.115 mm thick with particles 0.46 mm in diameter gave the best performance; for water one layer of 0.53-mm diameter particles was superior. This is different from the R-113 results shown in Fig. 6-14, where the optimal thickness was 1.0 mm for smaller diameter particles. Figure 6-16 is also a good indication of the difficulty that would be met in trying to correlate porous layer boiling curves by using plain tube boiling curve parameters or by only including additional surface factors.

In summary, within a specific set of test geometries it is possible to determine an

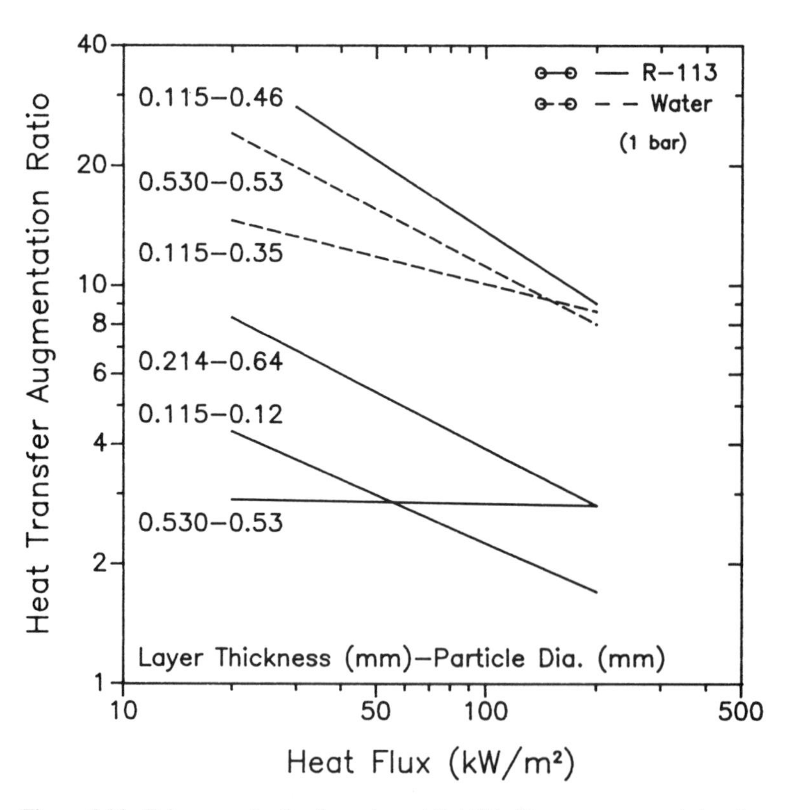

Figure 6-16 Enhancement ratios for water and R-113 boiling on porous coated surfaces.

optimum layer thickness, particle diameter, or method for producing a porous layer. In general, however, the optimal values found in one study only vaguely correspond to those found in others. Thus it is not possible to recommend a priori any particular method or layer geometry to obtain optimal performance. Ideally, one should attempt to attain large porosities with many interconnecting and reentrant passageways to promote the influx of liquid and the outflow of vapor. Layers more than 1.5 to 2.0 mm thick probably operate as shown in Fig. 6-13*c* over most of their nucleate pool boiling range. Hence layers thicker than these should be avoided because they add another thermal resistance to the process (i.e., conduction through the particles and vapor) that is unnecessary.

No visualization studies of the actual boiling process within porous matrices of these small dimensions have apparently been made. Therefore, one can only hypothesize that the important heat transfer mechanisms appear to be thin film and capillary evaporation and entrance region laminar flow effects within the matrix together with improved external evaporation (a larger vapor-liquid surface area) and external convection (a larger boiling site density). These surfaces also have good nucleation characteristics that "lift" them out of the poor single-phase heat transfer regime at low wall superheats and heat fluxes.

6-4 MISCELLANEOUS EFFECTS ON BOILING PERFORMANCE

Large Reduced Pressure

The effect of pressure on enhanced boiling is of fundamental and practical interest. For instance, the boiling curves for plain surfaces move to the left with increasing pressure. Thus conventional boiling performance increases with rising pressure, most rapidly above reduced pressures of about 0.1 to 0.2. Therefore, at high reduced pressures the relative advantage of using enhanced boiling surfaces may be less. As was seen earlier in Figs. 6-5 through 6-7, the size of departing vapor bubbles decreases with increasing pressure. In addition, the nucleation superheat predicted by Eq. (2-1) decreases with increasing pressure and facilitates the activation of cavities and scratches on ordinary surfaces. Thus the thermal mechanisms of an enhancement with its gap or pore sizes optimized for reduced pressures less than 0.1 may not be effective at high reduced pressures.

To investigate these effects, Fath (1986), Fath and Gorenflo (1986), and Gorenflo and Fath (1987) obtained boiling curves in R-22 for a plain tube at reduced pressures from 0.1 to 0.93, for a low-finned tube from 0.064 to 0.93, and for a Gewa-T tube from 0.064 to 0.97 (the low-finned and Gewa-T tubes were those described in Sec. 6-3).

Figure 6-17 shows a comparison of their boiling data over the reduced pressure range from 0.1 to 0.93. The boiling curves (as is the convention throughout this book) are based on the nominal outside areas of tubes with the same diameter as these tubes' external diameters. Performances of all three surfaces are noted to increase substantially with increasing pressure. At a reduced pressure of 0.10 and low heat fluxes, the low-finned tube performs like the Gewa-T tube, yielding a performance about 2.5 times

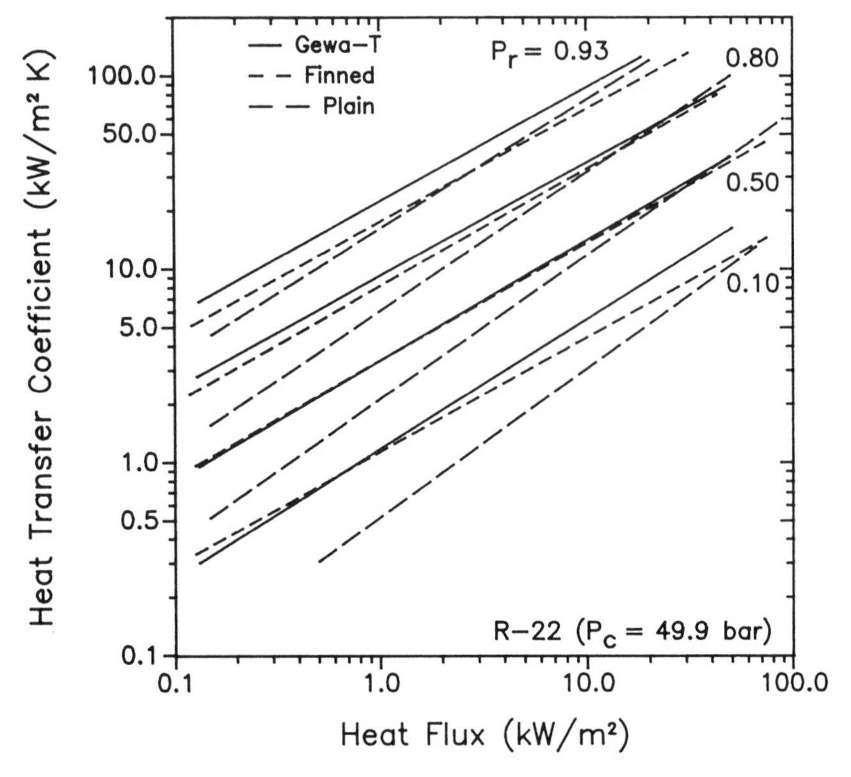

Figure 6-17 Boiling comparison over a wide range of reduced pressures.

that of the plain tube. With increasing pressure, however, the level of augmentation drops off until at a reduced pressure of 0.93 the low-finned tube performs within experimental error of the plain tube, notwithstanding that its surface area is 2.51 times that of an equal-sized plain tube. The Gewa-T provides some augmentation, but much less than its surface area ratio of 2.84. Consequently, for these two enhancement geometries the boiling performance per unit wetted surface area decreases relative to a plain surface with rising pressure. For low-finned tubes, the use of higher fin densities with increasing pressure may be effective when the spacing between the fins is maintained at about the same size as the departing bubbles.

Another extensive study of the effect of pressure on enhanced boiling was described by Czikk, O'Neill, and Gottzmann (1981), who compared the boiling heat transfer coefficients of various High Flux porous coatings to those of smooth surfaces with a machine or mill finish for R-11, R-22, R-113, benzene, and water. They observed that the performance of the porous layers increased at the same rate or more rapidly with increasing pressure than plain surfaces at pressures ranging from 0.19 to 13 bar. The marked improvement in the porous coating's coefficients with increasing pressure was attributed to a reduction in the nucleation superheat required for activation. These investigators found that coatings with pores smaller than 0.030 mm performed best at high pressures and that larger pore sizes were more suitable at subatmospheric pressures. They also noted that the porous surface's heat transfer coefficient decreased at high pressures approaching the critical value. No reason was offered, but one can speculate that as the buoyancy force exerted by the bubbles decreases near the critical point the driving force for removing vapor from the porous matrix is decreased, and the circulation through the matrix suffers.

Subcooling

The effect of subcooling on enhanced boiling has not been investigated thoroughly. For boiling on plain surfaces (see Fig. 2-7), the boiling heat transfer coefficient drops off rapidly with increasing subcooling. The performance of enhanced boiling surfaces would also be expected to decrease with subcooling. Yet, because enhancement geometries (other than low-finned tubes) are thought to improve the resistance of trapped vapor nuclei to subcooling by virtue of their reentrant cavities and tend to demonstrate lower nucleation superheats once saturation conditions have been restored, perhaps their performance will diminish less. In addition, the evaporation process inside the enhancement will be isolated from the subcooled bulk liquid because incoming liquid will be preheated as it passes though the narrow passageways or inactive pores. On the other hand, wall superheats of high performance enhanced boiling surfaces are much smaller than those of smooth surfaces, typically on the order of 1 to 3 K in many applications. Thus small subcoolings of only a few kelvins can quickly eliminate the temperature driving force for evaporation, that is $T_w - T_{sat}$, when the wall temperature is fixed by the heating-side process.

Bajorek (1988) investigated the effect of subcooling on boiling on a copper low-finned tube. He obtained data for ethanol, water, and several ethanol-water mixtures at 1.01 bar for subcoolings ranging from 0 to 40 K. His test section was a Gewa-K tube (manufactured by Wieland-Werke) that was 19.05 mm in diameter over the fins. The fin density was 741 fins per meter (18.8 fpi). The fins were 1.55 mm high and 0.46 and 0.30 mm thick at their base and tip, respectively. Figure 6-18 depicts the results for water at various heat fluxes. The largest drop in the heat transfer coefficient is seen to occur with the first 10 K of subcooling,where performance is reduced by 40% to 50%. Bajorek's results were similar to those obtained by Hui and Thome (1985) with the same fluids for a flat vertical disk.

Surface Orientation

Fujii, Nishiyama, and Yamanaka (1979) studied the effect of surface orientation on boiling from a porous coated disk in R-113, primarily with electronic cooling applications in mind. Their surface was made by electroplating copper particles 0.115 mm in average diameter to a disk to form a layer two to three particles thick. The results for this surface and a smooth surface are shown in Fig. 6-19, in which the angles of 0°, 45°, and 90° represent a horizontal disk facing upwards, an inclined disk, and a vertical disk, respectively. For a plain surface these investigators observed a small increase in performance with angle; for the porous layer the opposite effect was found. For the

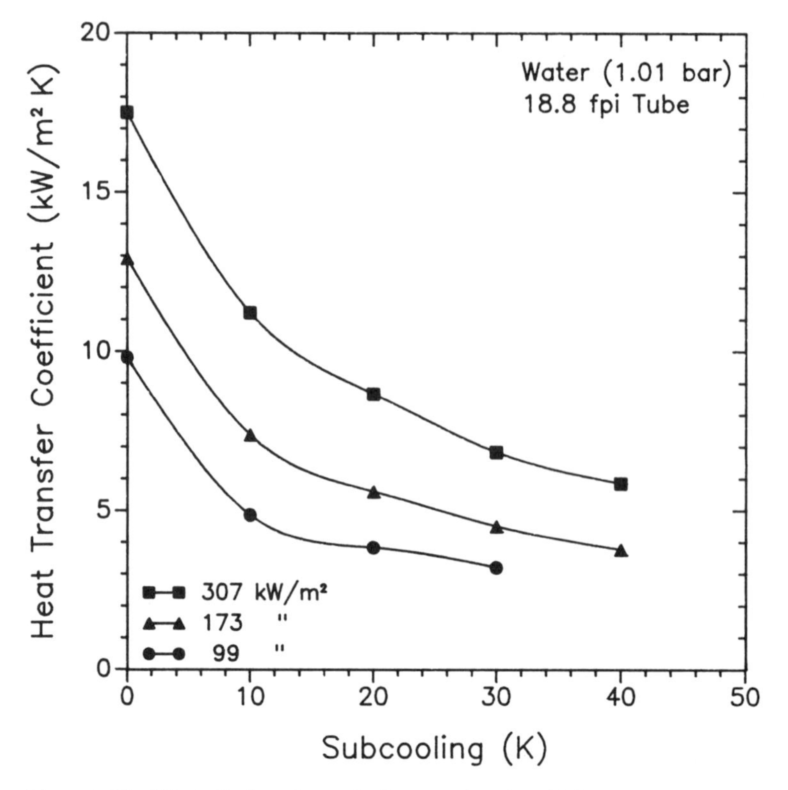

Figure 6-18 Effect of subcooling on boiling on a low-finned tube.

plain surface at low heat fluxes, they explained the enhancement as resulting from the more intense agitation of the thermal boundary layer by departing bubbles rising along the surface as the angle of inclination increased. At large heat fluxes the boiling process is already intense, and little change with angle occurred. The decrease in performance of the porous surface with increasing inclination was thought to result from trapping of vapor inside the matrix (similar to that shown in Fig. 6-13*c*), which would increase the thermal resistance.

Czikk, O'Neill, and Gottzmann (1981) noted a similar trend for the High Flux porous surface. Venart, Sousa, and Jung (1986) also found that the nucleation superheat decreased with increasing angle of inclination from a horizontal facing-upwards position to a horizontal facing-downwards position for R-11 on a porous metallic sprayed surface.

Marto, Wanniarachchi, and Pulido (1985) tested a Gewa-T tube with a 0.25-mm gap between fin tips in both a horizontal and a vertical position with R-113. Somewhat surprisingly, the vertical orientation gave better performance, even though this position would inhibit the two-phase convection heat transfer mechanism inside the reentrant channels that is speculated to control the performance of this tube. For instance, at a heat flux of 36 kW/m^2 the heat transfer coefficient increased by 40%. The difference diminished at a heat flux of 108 kW/m^2 to 5%, however. The vertical orientation would

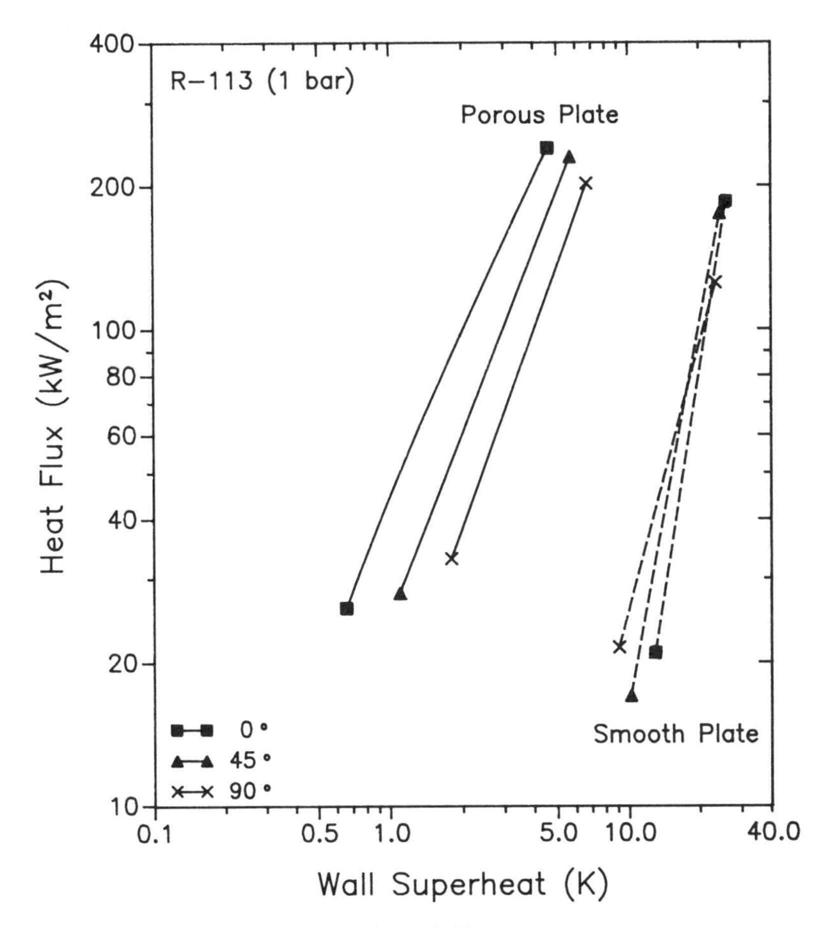

Figure 6-19 Effect of surface orientation on boiling.

be expected to reduce the circumferential flow around the reentrant channel that occurs in the horizontal position, but this may have induced thin film evaporation to occur, which would augment heat transfer. Curiously, this means that in a U-tube heat exchanger the boiling process will be more effective on the vertical bends than on the straight horizontal tubes.

Peak Nucleate Heat Flux

Much of the practical interest in enhanced boiling surfaces rests in their operation at low to medium heat fluxes. Startup of a heat exchanger, heat pipe, or electronic chip, however, may occur under circumstances that produce initially large heat fluxes or temperature differences. Thus knowledge of the effect of the surface enhancement on the peak nucleate heat flux and the wall superheat at which it occurs is of some importance (the maximum heat flux for tube bundles is discussed in Chapter Eleven's section on special thermal constraints and operational limits).

Table 6-1 Peak heat flux ratios relative to a smooth tube*

Tube	$(\alpha_{max})_{enh}/(\alpha_{max})_{sm}$	$(\Delta T_{max})_{enh}/(\Delta T_{max})_{sm}$
Gewa-K	0.643	0.873
Gewa-T	1.468	0.915
Thermoexcel-E	1.470	1.541
CSBS	0.898	0.432
High Flux	1.968	0.687
ECR-40	1.404	0.730

*Yilmaz and Westwater (1981).

Yilmaz, Hwalek, and Westwater (1980) and Yilmaz and Westwater (1981) completed an broad comparative study of commercial enhanced boiling tubes that included measurement of their peak nucleate heat fluxes. The test sections were steam heated, and the peaks were approached gradually. Isopropanol and *p*-xylene were the test fluids. Table 6-1 summarizes their results for isopropanol at 1.0 bar; the peak nucleate heat flux and the wall superheat at the peak nucleate heat flux are given as ratios relative to the smooth tube values. The tubes in Table 6-1 are described in Chapter Four except for the CSBS, which was produced by flame spraying copper wire to the tube.

The Gewa-T, Thermoexcel-E, and ECR-40 tubes exhibited about a 40% to 50% increase in peak heat flux, and the High Flux tube demonstrated a value about twice that of the smooth tube. The Gewa-K low-finned tube with 1,200 fins per meter (30 fpi) and the flame sprayed CSBS surface perform worse than the smooth tube by about 36% and 10%, respectively. The peak nucleate heat flux occurs at smaller wall superheats for all the surfaces except Thermoexcel-E. Most of the surfaces displayed fairly broad peaks relative to the plain tube, somewhat similar to those in Fig. 6-12, but with definite maxima before the drop off in heat flux associated with transition boiling.

A higher peak heat flux for a High Flux surface has also been observed by O'Neill, Gottzmann, and Terbot (1972) relative to a smooth surface for trichloroethylene at 0.7 bar. The peak heat flux of the High Flux surface was about 650 kW/m^2 compared to 375 kW/m^2 for the plain surface. The wall superheat at the peak heat flux was also lower for the High Flux tube, as in the Yilmaz and Westwater (1981) study: 21 K compared to 33 K.

Czikk, O'Neill, and Gottzmann (1981) presented critical heat flux data in tabular form for various pore sizes of High Flux coatings. The increases in peak nucleate heat flux tended to be about 15% to 20% relative to a smooth surface for R-113, R-114, and FC-88. No consistent trend in the peak nucleate heat flux with pore size (fine, medium, and coarse) was evident. For instance, with FC-88 five different porous layers, ranging from fine to coarse, showed only a 10% deviation from their combined average value of 235 kW/m^2. The most dramatic change was the wall superheat at the peak nucleate heat flux, which occurred at a wall superheat as low as 2.6 K for a fine pore size in R-113.

6-5 NOMENCLATURE

D	bubble departure diameter (m)
f	bubble departure frequency (s^{-1})

Δh_v	latent heat of vaporization (J/kg)
N/A	boiling site density (m^{-2})
P_c	critical pressure (bar)
P_r	reduced pressure
q_{lat}	latent heat flux (W/m^2)
q_{total}	total heat flux (W/m^2)
$(q_{max})_{enh}$	enhanced tube peak nucleate heat flux (W/m^2)
$(q_{max})_{sm}$	smooth tube peak nucleate heat flux (W/m^2)
t_f	thickness of liquid film (m)
T_{sat}	saturation temperature (K)
T_w	wall temperature (K)
$(\Delta T_{max})_{enh}$	wall superheat at $(q_{max})_{enh}$ (K)
$(\Delta T_{max})_{sm}$	wall superheat at $(q_{max})_{sm}$ (K)
α_{tf}	thin film evaporation heat transfer coefficient [$W/(m^2 \cdot K)$]
β	contact angle
θ	wedge angle
λ_L	liquid thermal conductivity [$W/(m \cdot K)$]
ρ_v	vapor density (kg/m^3)

REFERENCES

Afgan, N. H., L. A. Jovic, S. A. Kovalev, and V. A. Lenykov. 1985. Boiling heat transfer from surfaces with porous layers. *Int. J. Heat Mass Transfer* 28:415–22.

Ali, S. M., and J. R. Thome. 1984. Boiling of ethanol-water and ethanol-benzene mixtures on an enhanced boiling surface. *Heat Transfer Eng.* 5(3–4):70–81.

Arshad, J., and J. R. Thome. 1983. Enhanced boiling surfaces: Heat transfer mechanism and mixture boiling. *Proc. ASME-JSME Therm. Eng. Joint Conf.* 1:191–97.

Ayub, Z. H., and A. E. Bergles. 1985. Pool boiling from Gewa surfaces in water and R-113. In *Augmentation of heat transfer in energy systems*, ed. P. J. Bishop, HTD Vol. 52, 57–66. New York: American Society of Mechanical Engineers.

Bajorek, S. M. 1988. An experimental and theoretical investigation of multicomponent boiling on smooth and finned surfaces. Ph.D. diss., Michigan State University, East Lansing.

Bukin, V. G., G. N. Danilova, and V. A. Dyundin. 1982. Heat transfer from Freons in a film flowing over bundles of horizontal tubes that carry a porous coating. *Heat Transfer Sov. Res.* 14(2):98–103.

Czikk, A. M., and P. S. O'Neill. 1979. Correlation of nucleate boiling from porous metal films. In *Advances in Enhanced Heat Transfer*, 53–60. New York: American Society of Mechanical Engineers.

———. and C. F. Gottzmann. 1981. Nucleate boiling from porous metal films: Effect of primary variables. In *Advances in Enhanced Heat Transfer*, 109–22. New York: American Society of Mechanical Engineers.

Danilova, G. N., and V. A. Dyundin. 1972. Heat transfer with Freons 12 and 22 boiling at bundles of finned tubes. *Heat Transfer Sov. Res.* 4(4):48–54.

Fath, W. 1986. Wärmeübergangsmessungen an flatt- und rippernrohren in einer standardapparatur für siedeversuche. Ph.D. diss., University TH Paderborn, West Germany.

———. and D. Gorenflo. 1986. Zum einsatz von rippenrohr in uberfluteten verdampfern bei hohen siededrucken. *Dtsch. Kälte. Klimatechn. Verien.* 315–32.

Fujii, M. 1984. Nucleate pool boiling heat transfer from a porous heating surface (optimum particle diameter). *Heat Transfer Jap. Res.* 13(1):76–91.

———. E. Nishiyama, and G. Yamanaka. 1979. Nucleate pool boiling heat transfer from micro-porous heating surface. In *Advances in Enhanced Heat Transfer*, 45–51. New York: American Society of Mechanical Engineers.

Gorenflo, D., and W. Fath. 1987. Pool boiling heat transfer on the outside of finned tubes at high saturation pressures. *Proc. 17th Int. Congr. Refrig.* B:955–60.

Hsu, Y. Y., and R. W. Graham. 1976. *Transport processes in boiling and two-phase systems*. Washington, D.C.: Hemisphere.

Hui, T. O., and J. R. Thome. 1985. A study of binary mixture boiling: Boiling site density and subcooled heat transfer. *Int. J. Heat Mass Transfer* 28:919–28.

Kovalev, S. A., S. L. Solov'yev, and O. A. Ovodkov. 1987. Liquid boiling on porous surfaces. *Heat Transfer Sov. Res.* 19(3):109–20.

Ma, T., X. Liu, J. Wu, and H. Li. 1986. Effects of geometrical shapes and parameters of reentrant grooves on nucleate pool boiling heat transfer from porous surfaces. *Proc 8th Int. Heat Transfer Conf.* 4:2013–18.

Marto, P. J., A. S. Wanniarachchi, and R. J. Pulido. 1985. Augmenting the nucleate pool boiling characteristics of Gewa-T finned tubes in R-113. In *Augmentation of heat transfer in energy systems,* ed. P. J. Bishop, Vol. HTD-52, 67–73. New York: American Society of Mechanical Engineers.

Mirzamoghadam, A. V., and I. Catton. 1985. Holographic interferometry investigation of meniscus behavior. In *Multiphase flow and heat transfer,* ed. V. K. Dhir, J. C. Chen, and O. C. James. HTD Vol. 47, 49–56. New York: American Society of Mechanical Engineers.

Nakayama, W., T. Daikoku, H. Kuwahara, and T. Nakajima. 1979. Dynamic model of enhanced boiling heat transfer on porous surfaces. In *Advances in Enhanced Heat Transfer*, 31–43.

———. 1980. Dynamic model of enhanced boiling heat transfer on porous surfaces—Part I: Experimental investigation. *J. Heat Transfer* 102:445–50.

Nakayama, W., T. Daikoku, and T. Nakajima. 1981. Effects of pore diameters and system pressure on nucleate boiling heat transfer from porous surfaces. In *Advances in Enhanced Heat Transfer*, HTD Vol. 18, 147–53. New York: American Society of Mechanical Engineers.

Nishikawa, K., T. Ito, and K. Tanaka. 1979. Enhanced heat transfer by nucleate boiling at sintered metal layer. *Heat Transfer Jap. Res.* 8(2):65–81.

———. 1983. Augmented heat transfer by nucleate boiling at prepared surfaces. *Proc. ASME-JSME Therm. Eng. Joint Conf.* 1:387–93.

O'Neill, P. S., C. F. Gottzmann, and J. W. Terbot. 1972. Novel heat exchanger increases cascade cycle efficiency for natural gas liquefaction. *Adv. Cryogenic Eng.* 17:420–37.

Soloviyov, S. L. 1986. Liquid evaporation heat transfer on a porous surface. *Heat Transfer Sov. Res.* 18(3):58–64.

Stephan, K., and J. Mitrovic. 1981. Heat transfer in natural convective boiling of refrigerants and refrigerant-oil mixtures in bundles of T-shaped finned tubes. In *Advances in Enhanced Heat Transfer*, HTD Vol. 18, 131–46. New York: American Society of Mechanical Engineers.

Styrikovich, M. A., S. P. Malyshenko, A. B. Andrianov, and I. V. Talaev. 1987. Investigation of boiling on porous surfaces. *Heat Transfer Sov. Res.* 19(1):23–29.

Tung, C. V., T. Muralidhar, and P. C. Wayner, Jr. 1982. Experimental study of evaporation in the contact line region of a mixture of decane and 2% tetradecane. *Proc. 7th Int. Heat Transfer Conf.* 4:101–106.

Venart, J. E. S., A. C. M. Sousa, and D. S. Jung. 1986. Nucleate and film boiling heat transfer in R-11: The effects of enhanced surfaces and inclination. *Proc. 8th Int. Heat Transfer Conf.* 4:2019–24.

Webb, R. L. 1972. Heat transfer surface having a high boiling heat transfer coefficient. U. S. Patent 3,696,861, October 10.

———. 1983. Nucleate boiling on porous coated surfaces. *Heat Transfer Eng.* 4:(3–4)71–82.

Westwater, J. W. 1973. Development of extended surfaces for use in boiling liquids. *AIChE Symp. Ser.* 69(131):1–9.

Xin, M., and Y. Chao. 1985. Analysis and experiment of boiling heat transfer on T-shaped finned surfaces. Paper read at 23rd National Heat Transfer Conference (enhanced heat transfer equipment session); August 4–7, Denver.

Yilmaz, S., J. J. Hwalek, and J. W. Westwater. 1980. *Pool boiling heat transfer performance for commerical enhanced tube surfaces* (ASME paper 80-HT-41).

Yilmaz, S., and J. W. Westwater. 1981. Effect of commercial enhanced surfaces on the boiling heat transfer curve. In *Advances in Enhanced Heat Transfer*, HTD Vol. 18, 73–91. New York: American Society of Mechanical Engineers.

CHAPTER

SEVEN

PERFORMANCE COMPARISONS OF ENHANCED BOILING TUBES

This chapter presents pure fluid boiling performances of commercial enhanced boiling tubes to provide comparisons relative to one another and also in reference to boiling on plain, smooth tubes. The comparisons are made for nucleate pool boiling on the outside of a single tube. Ideally, selection of the optimum enhanced boiling tubes should be made on a basis of their overall heat transfer performances as tube bundles. A simple method for performing this type of comparison for particular applications is presented later in Chapter Eleven, Sec. 11-9.

The following classes of fluids are surveyed: water, refrigerants, cryogens, hydrocarbons, alcohols, solvents, and dielectrics. Representative fluids in each class were selected, although the choices of fluids were limited by the availability of boiling data. The enhanced boiling surfaces represented were restricted to commercially available surfaces so as to compare those that are viable in an industrial setting and not only in the laboratory. When data from more than one source were available for a particular combination of enhanced boiling tube and fluid, these independent data sets have been plotted to serve as a double-check. It is important to keep in mind that nearly all the experimental data for enhanced boiling tubes are for copper tubes and that in the petroleum and chemical processing industries boiling data for steel and alloy tubes are required. Thus additional data should be sought from the manufacturers for boiling performances with these materials, which may differ greatly from those obtained with copper.

The experimental wall superheat in nucleate pool boiling can be measured with an accuracy of 0.2 to 0.5 K. Errors can be caused by fluctuations in the temperature within the pool of test liquid, thermocouple calibration errors, thermocouple purity tolerances, inaccuracies in thermocouple electromotive force measurement, and distortion of the heat flow path by thermocouples embedded in the wall of the test section. In some cases, the wall thermocouples have been installed in tubes tightly fitted inside thin-walled enhanced boiling tubes, which can create thermal contact resistance and yield a larger wall superheat than actually exists (but giving a conservative performance curve). The error in the cited heat fluxes is normally 5% or less. These errors derive from inaccuracies in the heat transfer surface area of the test section, errors in the electrical power supplied to the cartridge heater for electrically heated test sections, and errors in the inlet and outlet temperatures (or vapor quality) and flow rate for hot liquid–heated test sections (or a condensating medium). Thus these error bands should be kept in mind when comparing the relative performances of the different enhanced boiling tubes and when selecting an over-design safety factor, if any, to use in their thermal design.

Experimentally obtained nucleate pool boiling curves for smooth tubes (and plates) are also depicted on many of the graphs. The surface finish can influence their boiling curves by a factor of 2, and hence these curves are only intended as approximations of the performance of commercial drawn tubing. As a further reference, the Mostinski (1963) pool boiling correlation, familiar to many thermal designers, has also been plotted in many cases. The standard version given by Eqs. (2-8) and (2-9) has been used for nonrefrigerants. For refrigerants R-11, R-12, R-22, and R-113, the values for coefficient A^* listed in Collier (1983) and Bell and Mueller (1984) for the following Borishanski expressions have been used:

$$\alpha_{nb} = A^* q^{0.7} F_P \tag{7-1}$$

$$F_P = 0.7 + 2P_r \left[4 + \frac{1}{(1 - P_r)} \right] \tag{7-2}$$

Straight lines and, where appropriate, splines have been used to represent the published data points. The interested reader is referred to the specific reference given for the individual data sets. The boiling curves have not been extrapolated past the test conditions given in the original reports. Thus tubes tested over only a small range of heat flux are shown only for the range actually tested. All heat fluxes refer to the outside nominal area of the enhanced boiling section of the tubes. Heat fluxes for finned tubes were *not* calculated from their total outside wetted surface areas because these are not used to evaluate the complex geometry enhanced boiling surfaces. The nominal outside surface area is defined as

$$A = \pi DL \tag{7-3}$$

where D refers to the diameter over the outside of the enhancement or fin tips and L to the heated test length. The external diameter was chosen over the root diameter of the enhancement for defining heat flux because this is essentially the same diameter of tubing that would enter the tube sheet hole when an existing plain tube heat exchanger is being retubed.

The wall superheats refer to the difference between the wall temperature at the base of the enhancement or fins and the bulk saturated liquid temperature. These wall superheats are the values obtained experimentally, and recalculating them to the outside surface of the enhancement for a thermal definition consistent with the heat flux is not feasible.

7-1 WATER

Figure 7-1 depicts the boiling data for water at atmospheric pressure that were obtained for three different enhanced boiling surfaces: High Flux, Thermoexcel-E, and Gewa-K low-finned tubing. The specifications of the High Flux tube tested by Gottzmann, O'Neill, and Minton (1973) were not given. The High Flux tube supplied to Bergles and Chyu (1981) was the copper R-11 version according to O'Neill (person communication 1986). It had a porous layer thickness of 0.38 mm and was made from copper particles

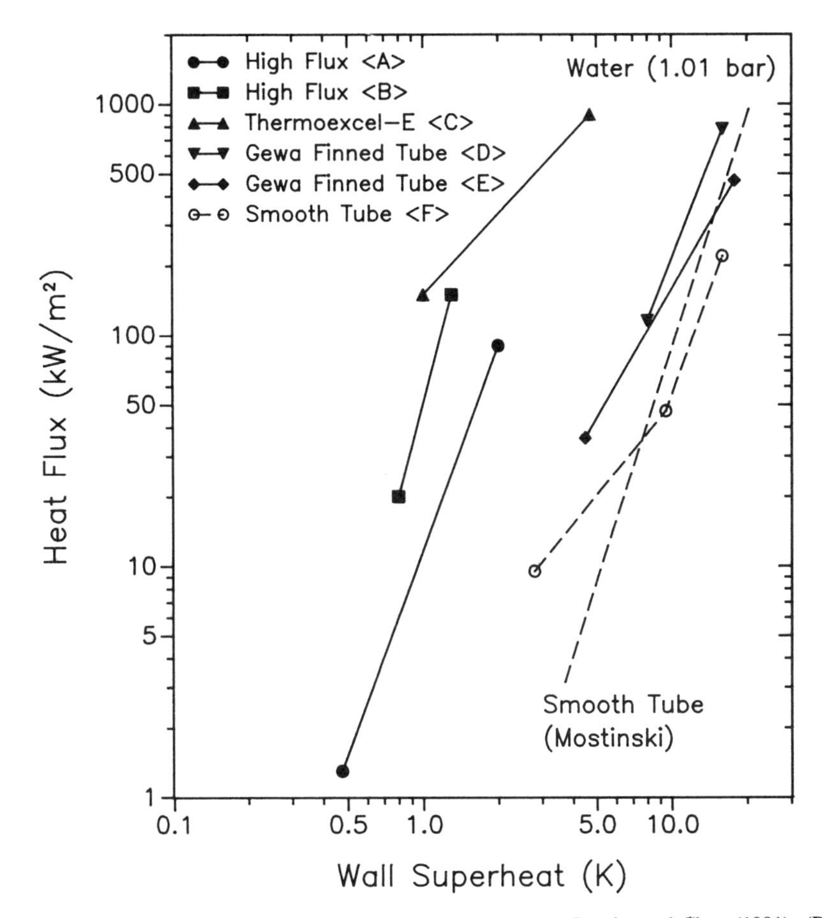

Figure 7-1 Boiling performances in water. References: (A) Bergles and Chyu (1981), (B) Gottzmann, O'Neill, and Minton (1973), (C) Nakayama et al. (1979, 1980), (D) Yilmaz and Westwater (1981), (E) Bajorek (1988), and (F) Shakir, Thome, and Lloyd (1985).

composed of about 45% 200- to 325-mesh size, with the remainder being even finer. The porosity of this version typically ranges from 50% to 65% according to the manufacturer. The copper Thermoexcel-E boiling curve shown corresponds to surface W-3 in Nakayama et al. (1979, 1980). This surface had pore diameters of 0.08 mm and a ratio of wetted surface area to nominal outside surface area of 3.7. The copper finned tube tested by Yilmaz and Westwater (1981) had 30.7 fins per inch (fpi) (1,213 fins per meter), an outside diameter of 12.57 mm, a root diameter of 10.94 mm, a fin height of 0.81 mm, and a wetted surface area 2.36 times that of a plain tube of the same outside diameter. The Gewa-K copper low-finned tube of Bajorek (1988) was 19.05 mm in diameter with 18.8 fpi (741 fins per meter). The fins were 1.55 mm high and 0.46 and 0.30 mm thick at their base and top, respectively. The plain copper tube tested by Shakir, Thome, and Lloyd (1985) was 22.2 mm in diameter and had a surface finish prepared with crocus paper to resemble drawn commercial copper tubing.

At a fixed wall superheat of 1.0 K, the Thermoexcel-E tube out-performs the High Flux tube by several times, and at a heat flux of 160 kW/m^2 their performances are essentially identical. The two finned tubes perform similarly, irrespective of their different fin densities, and are only marginally better than a plain tube (a lower fin density would probably yield better performance for water). The boiling data sets for the two High Flux test sections are within normal experimental error of one another. The plain tube data correlated well with the Mostinski correlation.

7-2 REFRIGERANTS

Refrigerant-11

Refrigerant-11 has been the most widely studied fluid with enhanced boiling surfaces because of its widespread use in refrigeration systems in the United States. Therefore, the enhanced boiling data for this fluid are represented in three different graphs: Fig. 7-2 for integral low-finned tubes, Fig. 7-3 for complex-geometry enhanced boiling tubes at subatmospheric pressures, and Fig. 7-4 for complex-geometry tubes at or above atmospheric pressure. All the tubes represented are made of copper.

Table 7-1 gives the tube specifications for the finned tubes represented in Fig. 7-2.

Table 7-1 Finned tube specifications for Fig. 7-2

Tube	Outside diameter, mm	Root diameter, mm	Fin height, mm	Mean fin thickness, mm	Area/length, m^2/m	Area ratio
Gewa			No specifications given			
26 fpi	18.2	15.9	1.1			
Hitachi			No specifications given			
34.2 fpi	18.9	15.9	1.5	0.4	0.19	3.18
19 fpi	22.2	19.0	1.6		0.18	2.58
40.8 fpi	19.0	17.2	0.9	0.25		
26.8 fpi	18.9	16.3	1.3	0.36		

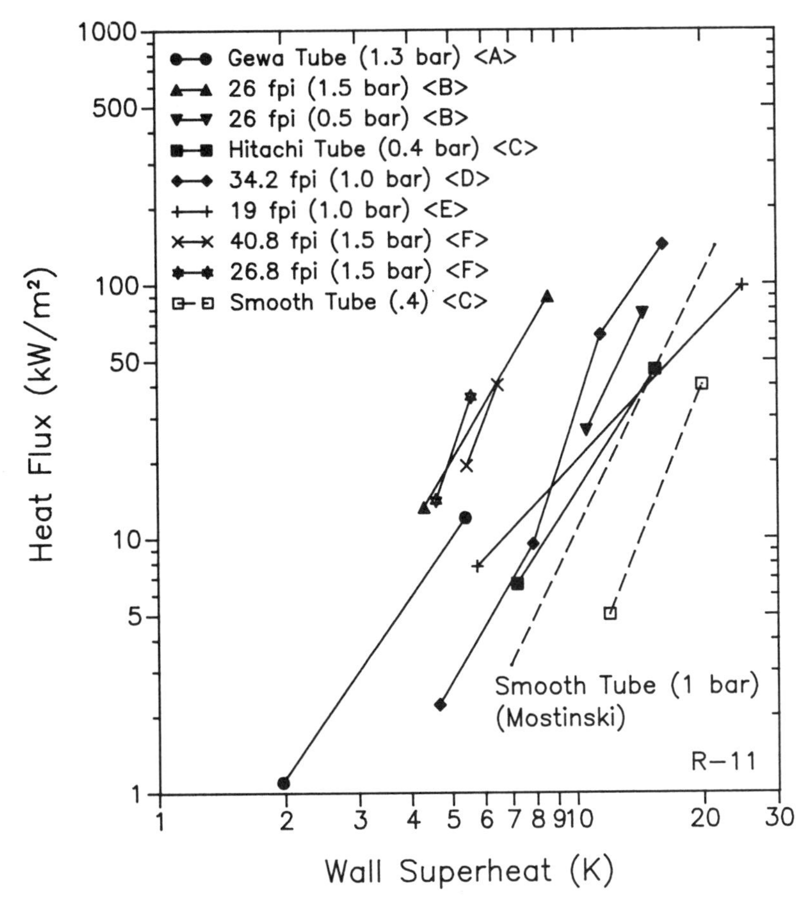

Figure 7-2 Boiling performances of finned tubes in R-11. References: (A) Gorenflo (1966), (B) Ever-Fin (Undated), (C) Hitachi (1984), (D) Hahne and Muller (1983), (E) Sauer, Davidson, and Chongrungreong (1980), and (F) Carnavos (1981).

As can be seen, specific information about the tube geometry has not been given in some reports. At pressures of 1.3 and 1.5 bar, there is no appreciable difference between the performances of the unidentified Gewa finned tube and the 26-, 26.8-, and 40.8-fpi tubes. In fact, the independent data sets for the 26- and 26.8-fpi tubes are nearly identical. At subatmospheric pressures of 0.5 and 0.4 bar, the 26-fpi tube is marginally better than the unspecified Hitachi finned tube. These two finned tubes performed substantially better than the smooth tube, as judged from the measured boiling data at 0.4 bar given by Hitachi (1984) for the latter, especially when performances are compared at the same wall superheat. The 1.0-bar data for the 34.2- and 19-fpi finned tubes are only slightly better than those predicted for a smooth tube by the Mostinski correlation. It should be noted that these two studies used test sections with wall thermocouples embedded in an insert inside the finned tube, which may explain the poor performances observed.

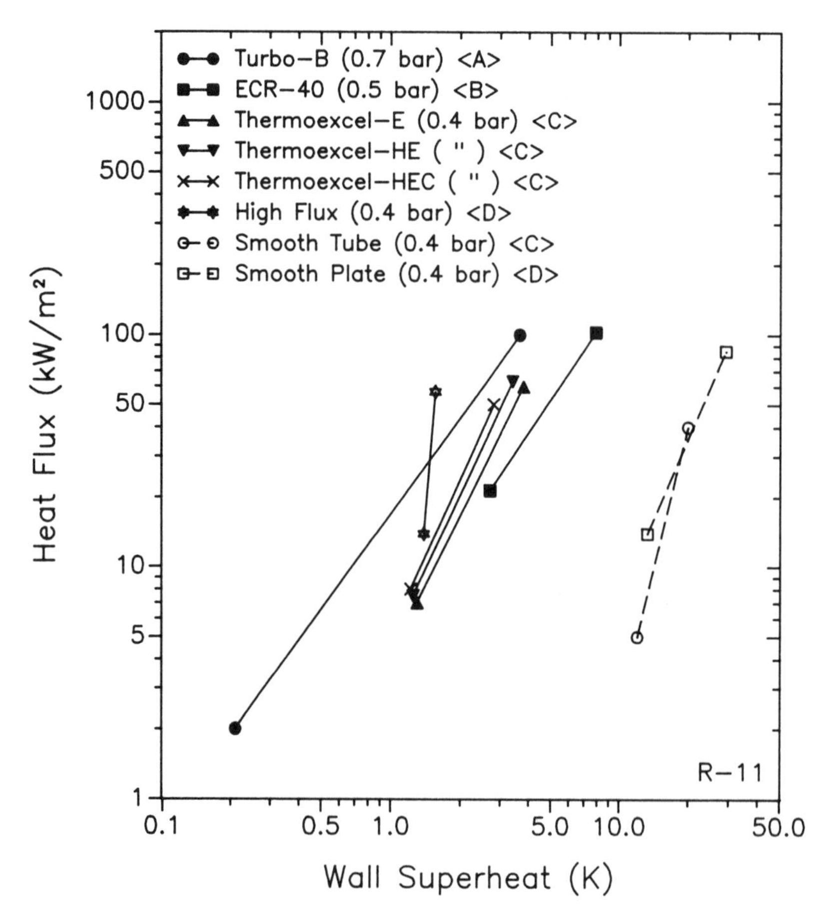

Figure 7-3 Subatmospheric boiling performances in R-11. References: (A) Turbo-B (1985), (B) Ever-Fin (Undated), (C) Hitachi (1984), and (D) Czikk et al. (1970).

The subatmospheric boiling performances of enhanced boiling tubes with R-11 are shown in Fig. 7-3. The Turbo-B boiling curve was calculated from the correlation supplied in the Turbo-B (1985) product bulletin for a 19.0-mm diameter tube at 10.5 psia. The Ever-Fin tube tested was specified as ECR-405028, which has an outside diameter of 18.2 mm and is made from 40-fpi low-finned tubing. Boiling curves for three versions of the Thermoexcel-E tube are shown, where E is the standard version, HE is the higher-performance version with fine projections formed just beneath the openings in the outer skin, and HEC is the corrugated version of the HE tube. The High Flux tube is the R-11 version with a porous layer thickness of 0.25 to 0.38 mm and a void fraction of about 60%.

At wall superheats less than 1.5 K, the Turbo-B tube performs best. Thermoexcel performance improves only marginally with the more sophisticated versions. The ECR-40 tube performs within experimental error of the Thermoexcel-E tubes. The two independent sets of smooth surface data are in good agreement. The enhanced boiling

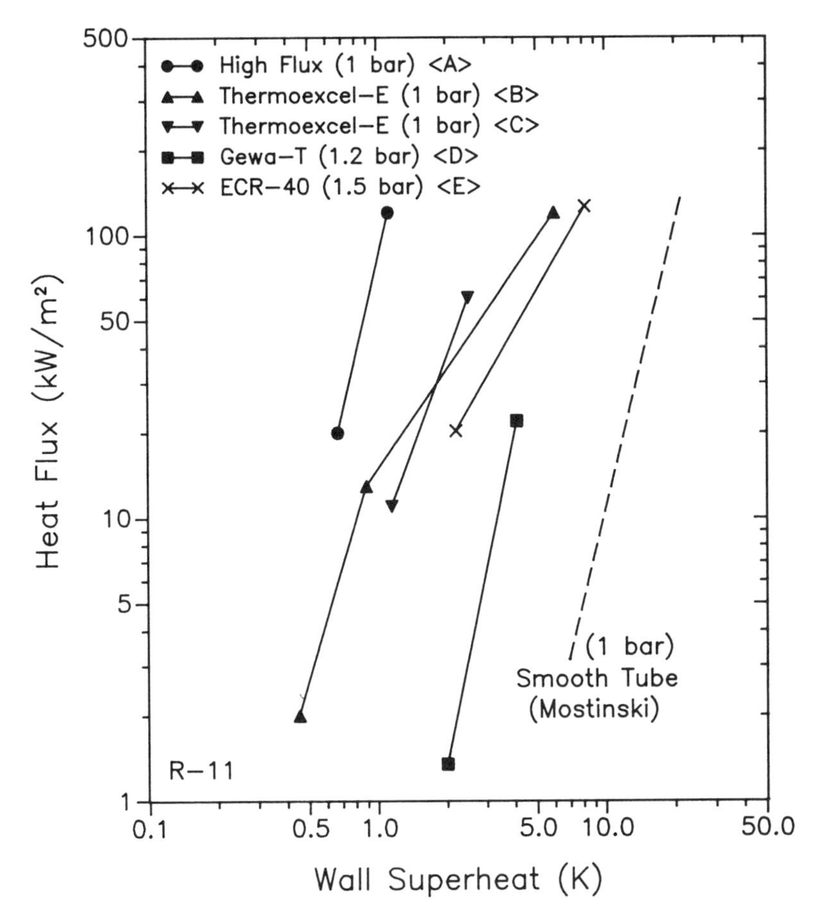

Figure 7-4 Performances in R-11 at atmospheric pressure and above. References: (A) O'Neill, Gottzmann, and Terbot (1972), (B) Nakayama et al. (1979, 1980), (C) Kuwahara, Nakayama, and Daikoku (1977), (D) Stephan and Mitrovic (1981), and (E) Ever-Fin (Undated).

tubes yield large augmentation ratios for these test conditions. A more realistic performance comparison for refrigerants is relative to integral low-finned tubes, which have been the industry standard for a long time. Comparing the boiling performances at a pressure of 0.4 bar in Fig. 7-3 to the finned tube boiling curves at the same pressure in Fig. 7-2, the complex-geometry enhanced boiling surfaces are in general about five times better than finned tubes operating at the same heat flux. At the same wall superheat, they are about ten times better.

Figure 7-4 depicts complex-geometry enhanced boiling tube performances for R-11 at 1 bar and above. The High Flux tube is the R-11 version. The Thermoexcel-E data are for a planar test surface designated as R(11)-1. This set agrees well with the Thermoexcel-E data set published earlier by Kuwahara, Nakayama, and Daikoku (1977), although their enhancement geometries may not be the same. The ECR-40 tube is the same one described for Fig. 7-3. The Gewa-T tube tested by Stephan and Mitrovic (1981) was 18.1 mm in outside diameter with a T-shaped fin height of 1.1 mm, a fin density of 18.8 fpi (741 fins per meter), and an outside wetted surface area 2.80 times that of a comparable 18.1-mm diameter smooth tube. The relative positions of the boiling curves for these surfaces are similar to those in Fig. 7-3, except that the Gewa-T tube performs more poorly than the others.

Refrigerant-12

Nucleate pool boiling data for R-12 at 3.1 bar are shown in Fig. 7-5 for copper tubes. The types of High Flux tube tested by Starner and Cromis (1977) and Antonelli and O'Neill (1981, 1986) were not specified. The Turbo-B (1985) boiling curve was calculated from the correlation provided in its product bulletin at 76 psia (5.2 bar) for a 19.0-mm outside diameter tube. The Thermoexcel-E tube had pore diameters of 0.1 mm with a channel density of 2,000 channels per meter (the Thermoexcel-HE tube has the same boiling performance as the E version shown). The 19-fpi (748 fins per meter) low-finned tube tested by Arai et al. (1977) was 18.0 mm in outside diameter with a fin height of 1.4 mm. No fin density was cited. The 19.5-fpi (770 fins per meter) low-finned tube of Myers and Katz (1953) was tested over the pressure range from 3.6 to 4.6 bar, and the line in Fig. 7-5 is the best representation of all their data. Their finned tube had an outside diameter of 18.8 mm, a fin height of 1.37 mm, and a wetted surface area 2.71 times that of a plain tube. The 26-fpi (1,026 fins per meter) tube tested by Starner and Cromis (1977) was 19.0 mm in diameter, but no test pressure was cited.

For Refrigerant-12, the High Flux tube performs better than the Turbo-B at wall superheats greater than 0.5 K. The two High Flux boiling curves agree quite well, although their slopes differ appreciably. The performances of the Thermoexcel-E tubes are essentially identical to the performance of the Turbo-B tube. The second boiling curve for Thermoexcel-E obtained by Arai et al. (1977) is in fairly good agreement with the Hitachi (1984) curve. The 19-fpi finned tube studied by Myers and Katz (1953) out-performs the 19.5-fpi finned tube tested by Arai et al. (1977). The smooth tube tested by Hitachi (1984) is modeled well by the Mostinski boiling curve. The slopes of nearly all the enhanced boiling tubes are similar, with only two exceptions. The special geometry-enhanced boiling tubes out-perform low-finned tubes by a factor of about 3 when compared at the same heat flux and by a factor of about 6 or more when compared at the same wall superheat of 2.0 K.

Refrigerant-22

Figure 7-6 depicts the enhanced boiling data available for R-22. The particular High Flux tube version was not specified by Antonelli and O'Neill (1981, 1986). The copper Gewa-T tube tested by Stephan and Mitrovic (1981) is the same one described above for R-12. The specific Thermoexcel-E and low-finned tubes were not cited by Hitachi

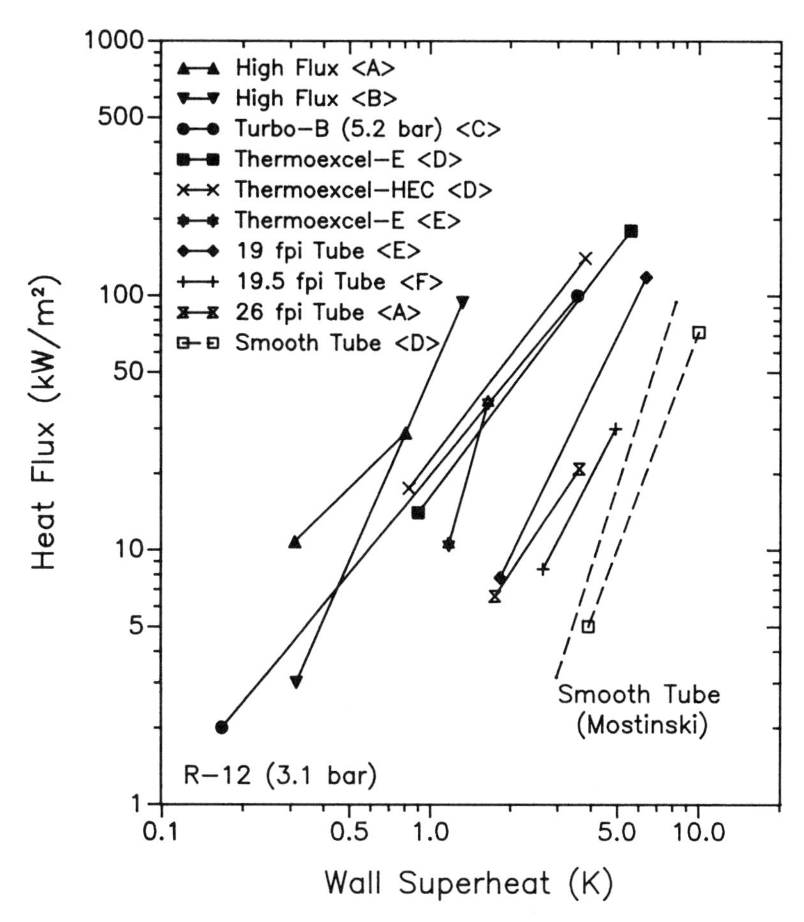

Figure 7-5 Boiling performances in R-12. References: (A) Starner and Cromis (1977), (B) Antonelli and O'Neill (1981, 1986), (C) Turbo-B (1985), (D) Hitachi (1984), (E) Arai et al. (1977), and (F) Myers and Katz (1953).

Cable (1984). The smooth tube boiling curves were calculated with the Mostinski correlation.

The High Flux tube performs about ten times better than the smooth tube when compared at the same heat flux, and at the same wall superheat the augmentation is by more than a factor of 100 (extrapolating the boiling curves a little to overlap at a wall superheat of 2.0 K). The Thermoexcel-E tube also performs well. The performance of the Gewa-T tube is only about two to four times better than that of the smooth tube at the same heat flux and four times better at the same wall superheat of 3.0 K. The effect of pressure on the Gewa-T tube's boiling curve in R-22 is observed to be slightly greater than the range from 2.5 to 5.0 bar. At 5 bar, the finned tube performs like the Gewa-T (additional data for the low-finned and Gewa-T tubes at high reduced pressures are shown in Fig. 6-17).

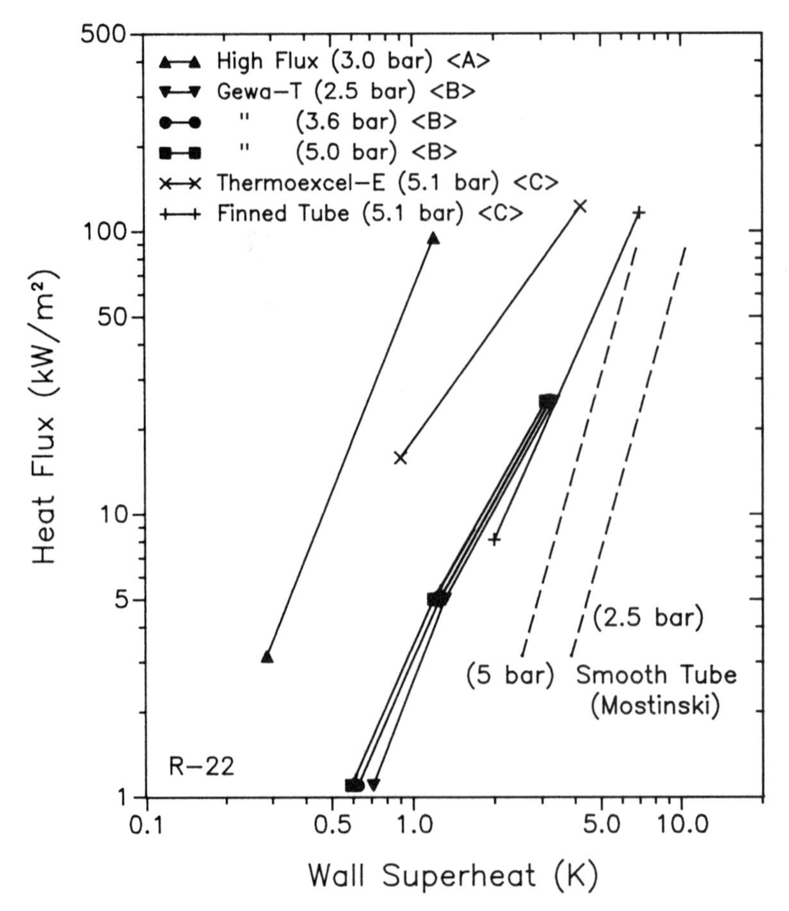

Figure 7-6 Boiling performances in R-22. References: (A) Antonelli and O'Neill (1981, 1986), (B) Stephan and Mitrovic (1981), and (C) Hitachi Cable (1984).

Refrigerant-113

Boiling performances for three enhanced boiling tubes are compared to boiling on a smooth tube in Fig. 7-7 for R-113. The High Flux boiling curve of O'Neill, Gottzmann, and Terbot (1972) is for a copper version. The High Flux tube studied by Bergles and Chyu (1981) is from the same stock as their test section described above for water (Sec. 7-1). Marto and Lepere (1981, 1982) tested a High Flux tube with essentially the same particle size range as Bergles and Chyu but with a porous layer thickness of only 0.08 mm rather than 0.38 mm. Remarkably, their boiling experiment gave exactly the same boiling curve as the Bergles and Chyu curve (only one is shown). The Thermoexcel-E tube tested by Hitachi Cable (1978) was their commercial version in copper. The copper Thermoexcel-E tube tested by Marto and Lepere (1981, 1982) had an outside diameter of 16.5 mm, an enhancement thickness of 0.19 mm, and an average cavity mouth opening of about 0.1 mm. The copper Gewa-T tube had an outside diameter of

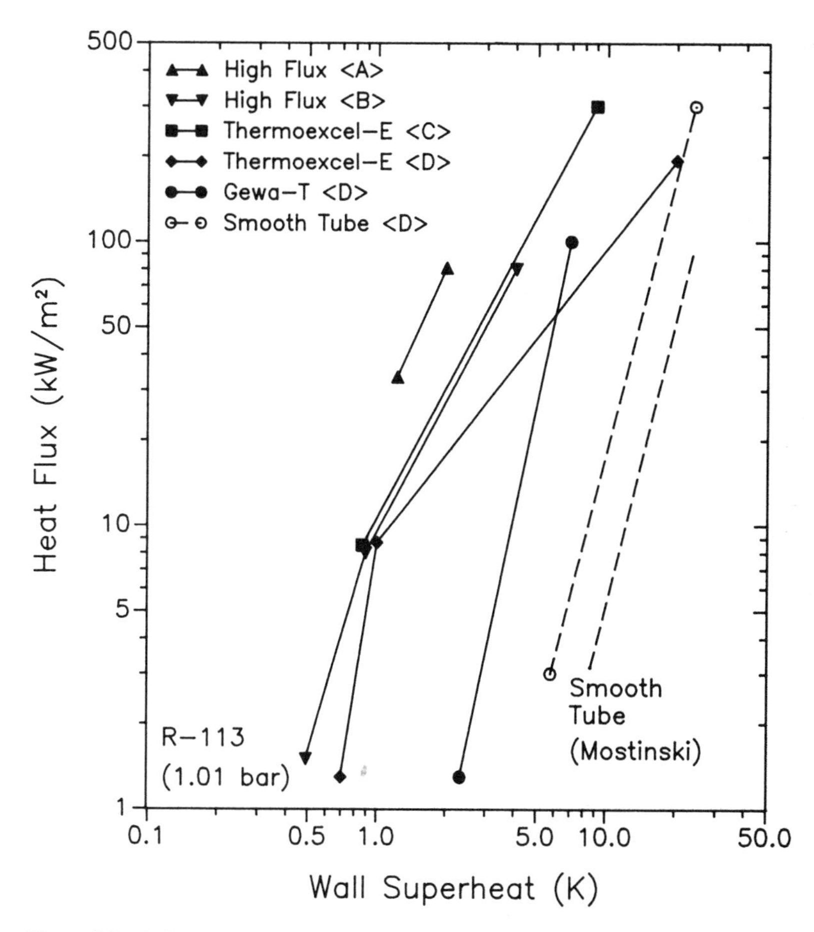

Figure 7-7 Boiling performances in R-113. References: (A) O'Neill, Gottzmann, and Terbot (1972), (B) Bergles and Chyu (1981), (C) Hitachi Cable (1978), and (D) Marto and Lepere (1981, 1982).

17.9 mm, a T-shaped fin height of 1.02 mm, 741 T-fins per meter (29.2 fpi), and a gap opening between the fin tips of 0.18 mm. The Marto and Lepere smooth tube boiling curve was obtained with a 15.8-mm diameter copper tube tested in its "as received" condition.

The two High Flux boiling curves shown in Fig. 7-7 have the same slope, but the industrial data <A> show better performance. The two independently obtained boiling curves for Thermoexcel-E also are in close agreement at low and medium heat fluxes. The High Flux and Thermoexcel-E tubes perform about ten times better than the smooth tubes when compared at the same heat flux and are several orders of magnitude better when compared at the wall superheat of 6.0 K. The Gewa-T boiling performance is about midway between that of the other enhanced boiling tubes and that of the smooth tube (apparently no data are available in the open literature for the newer version Gewa-TX). The Mostinski correlation is shown to predict wall superheats about 30% higher than the smooth tube experimental data, although their two slopes agree closely.

Refrigerant-114

Figure 7-8 depicts boiling results for R-114. Figure 7-8(*a*) compares the enhanced boiling data available for R-114 at several pressures greater than 1 bar. The copper Gewa-T tube test section of Stephan and Mitrovic (1981) is the same one described above for R-12. The 14.0-mm diameter finned tube studied by Hesse (1973) was made of nickel and had a fin density of 1,000 fins per meter (25.3 fpi). The fins had a rectangular cross-section 0.4 mm thick and 0.5 mm high. The mean surface roughness was measured to be 0.42 μm. The nickel smooth tube tested by Hesse (1973) was also 14.0 mm in diameter with a slightly coarser surface roughness (0.61 μm).

For R-114, the Gewa-T tube displays about a 30% improvement in performance with increasing pressure. Its level of augmentation compared to the smooth tube data, however, is less for this fluid than for the other fluids already reviewed, being about the same as that of the nickel finned tube. If the heat flux for the nickel finned tube

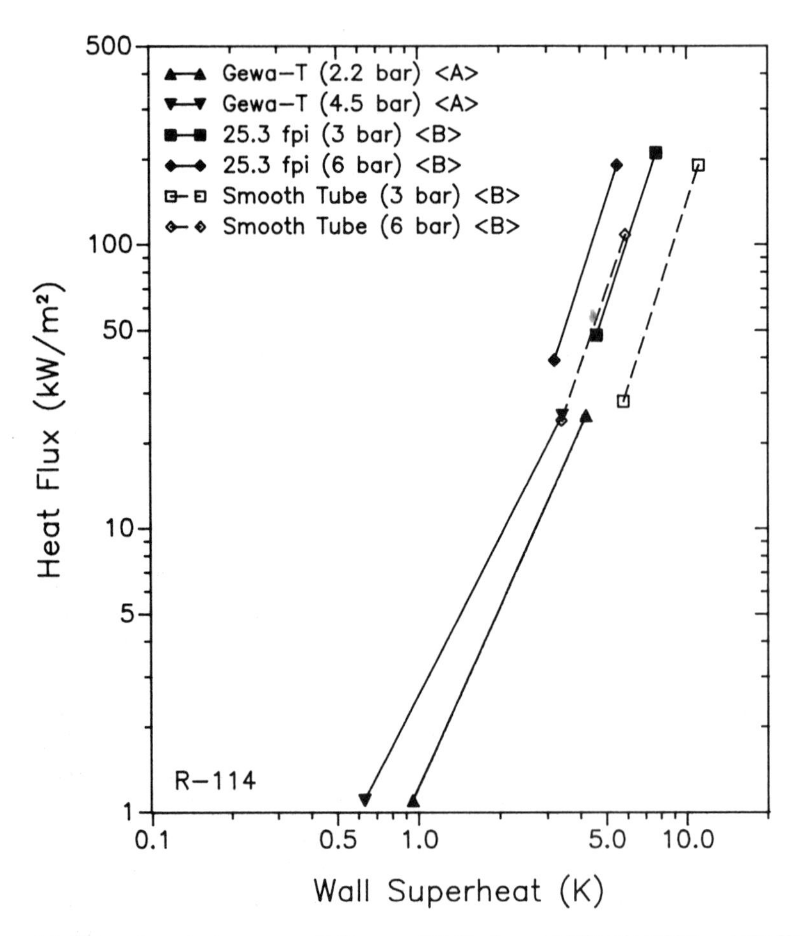

(*a*) Above atmospheric pressure. References: (A) Stephan and Mitrovic (1981) and (B) Hesse (1973).

Figure 7-8 Boiling performances in R-114.

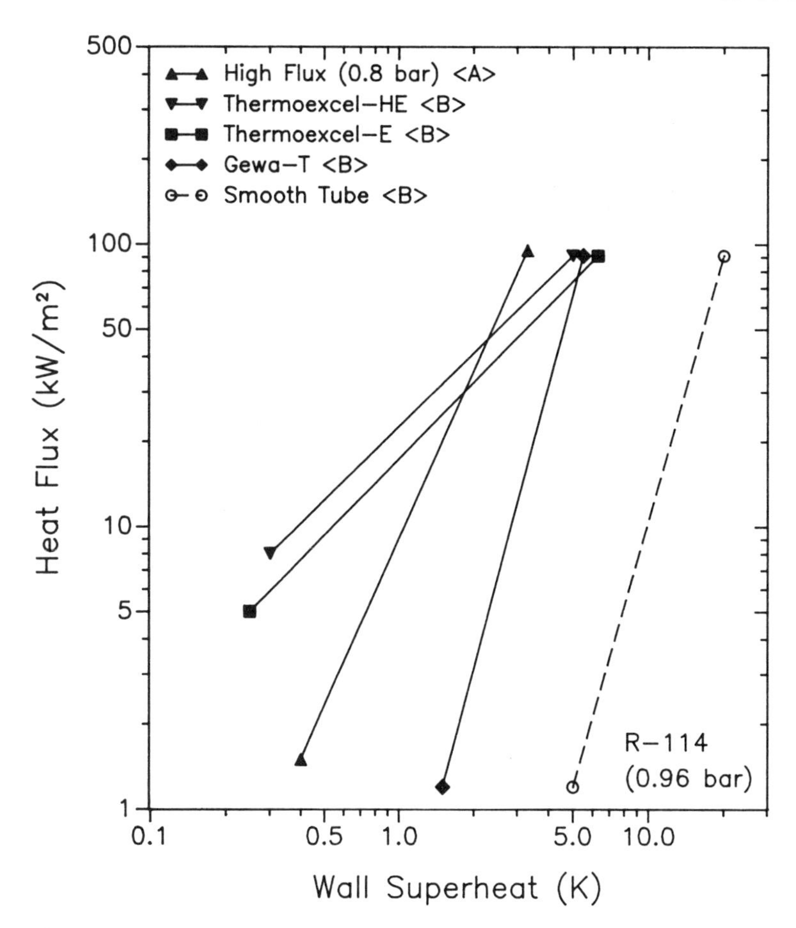

(b) Subatmospheric pressures. References: (A) Wanniarachchi, Marto, and Reilly (1986) and (B) Wanniarachchi, Sawyer, and Marto (1987).

were calculated on the basis of its total outside wetted surface area the boiling curve at 6 bar would fall exactly on the smooth tube boiling curve, whereas at 3 bar it would demonstrate a small enhancement.

The results of Wanniarachchi and co-workers (1986, 1987) are plotted in Fig. 7-8(b) at pressures less than 1 bar. The High Flux boiling curve was obtained at −2.2°C, and the others wre obtained at +2.2°C. The Gewa-T tube had 1,020 fins per meter (25.8 fpi) and 2.7 times the outside area of the plain tube tested. The outside diameters were reported to be 15.9 mm (although it is not clear whether all the tubes had this dimension), and all the tubes were made of copper. At a fixed heat flux of 30 kW/m^2, the enhancement ratios relative to the plain tube were reported as 9.1, 8.2, 6.8, and 4.4 for the High Flux, Thermoexcel-E, Thermoexcel-HE, and Gewa-T tubes, respectively.

7.3 CRYOGENS

Nitrogen

Nucleate pool boiling curves for liquid nitrogen are depicted in Fig. 7-9. The porous matrix geometry of the first High Flux surface listed was studied in detail by Czikk and O'Neill (1979). Results for their surface designated PS2 are shown because its performance was stated to match closely the standard performance of High Flux surfaces evaporating liquid nitrogen. This surface had a void fraction of 65.8% with a metal surface area per unit volume of porous matrix of 40,420 m^2/m^3. Its pore diameters ranged from 0.02 to 0.22 mm. The second High Flux boiling curve is for an unspecified version in Antonelli and O'Neill (1981, 1986). Both these High Flux surfaces were probably made with copper powder, although this was not stated. The copper Thermoex-

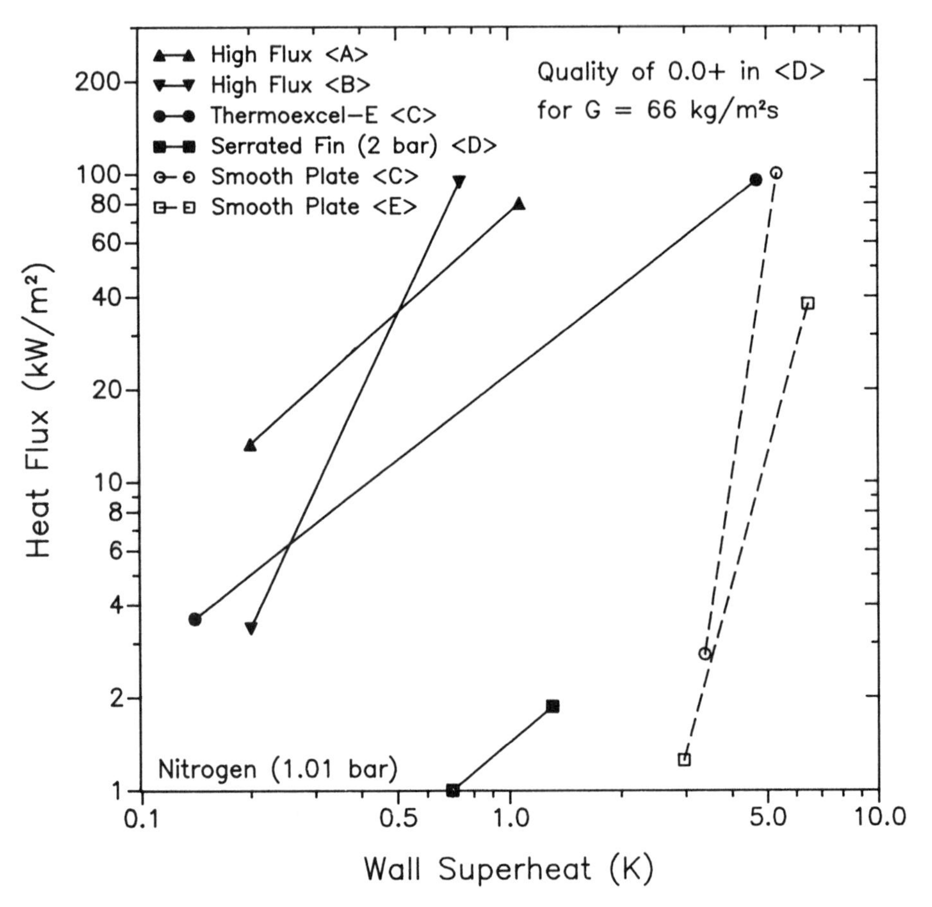

Figure 7-9 Boiling performances in liquid nitrogen. References: (A) Czikk and O'Neill (1979), (B) Antonelli and O'Neill (1981, 1986), (C) Nakayama et al. (1979, 1980), (D) Robertson (1979), (E) Thome and Bald (1978).

cel-E boiling curve shown is for the surface designated LN-3 in Nakayama et al. (1979, 1980). This planar test surface had pore diameters of 0.09 mm, 2,500 reentrant channels per meter (63.3 channels per inch), an enhancement height of 0.56 mm, and 4.05 times the wetted surface area of a plain, smooth surface. The smooth copper plate tested by Nakayama et al. (1979, 1980) was produced by lapping with grain diameters of 16 μm. The Thome and Bald (1978) copper test surface had a surface roughness of 76 μm.

The High Flux boiling curves intersect at a wall superheat of 0.5 K but have different slopes. This may be because the second one is the heat exchanger design curve, which the manufacturer derated to include any effects of fouling. The Thermoexcel-E boiling performance at low heat fluxes is equal to that of the High Flux tube and is about 20 times that of the smooth copper plates, which exceeds the 305% increase in its wetted surface area. At high heat fluxes, however, the Thermoexcel-E performance reverts to that of the smooth plates, which probably signifies that the reentrant channels have dried out and thus have been rendered ineffective.

Liquid oxygen is another cryogen of great importance in air separation plants. Enhanced boiling data for oxygen are apparently only available for the High Flux surface; liquid oxygen was the original fluid for which this surface was developed. The oxygen boiling design curve for the High Flux tube cited in Antonelli and O'Neill (1981, 1986) is the same as their nitrogen curve shown in Fig. 7-9 (Fig. 3-23 shows boiling inside a High Flux tube compared to boiling inside a plain tube). Oxygen's smooth surface boiling curve would be similar to that of nitrogen. Thus there is a substantial enhancement of boiling for oxygen.

For comparison with a competing technology in cryogenic services, some flow boiling data for a serrated plate-fin test channel obtained by Robertson (1979) are also shown in Fig. 7-9. The serrated finning was made of aluminum sheet 0.2 mm thick with 591 corrugations per meter (15 corrugations per inch). Each short serrated passageway was 6.35 mm high, 1.5 mm wide, and 1.7 mm long. The data are for a pressure of 2 bar and a vapor quality approaching zero. The heat transfer coefficients for this serrated fin are seen to be about three times better than those for nucleate pool boiling on a smooth plate at 1 bar when compared at the same heat flux. Therefore, these surfaces are also a type of enhanced boiling surface.

The serrated fin boiling performance under these conditions is poor compared to the High Flux and Thermoexcel-E surfaces. At a wall superheat of 0.7 K, for example, the High Flux tube transfers from 50 to 80 times more heat per unit surface area, whereas the Thermoexcel-E provides about 16 times as much. A serrated plate-fin heat exchanger has a higher surface area per unit volume than shell-and-tube heat exchangers, however, and this factor would have to be taken into account in a thermal design comparison. In addition, the plate-fin curve shown was for near-zero vapor quality conditions for comparison against nucleate pool boiling data; at larger vapor qualities, serrated plate-fins perform much better (boiling in plate-fin heat exchangers is reviewed in Chapter Thirteen).

Helium-I

Ogata and Nakayama (1982) tested eight different copper enhanced boiling geometries for augmenting the nucleate pool boiling of liquid Helium-I. They tested the surfaces under both clean and oxidized conditions. In Fig. 7-10 their results for two of these surfaces, the Thermoexcel-C surface (which is normally used for augmenting condensation and has a saw-toothed appearance) and a surface with rectangular cross-section fins (1.0 mm high, 1.0 mm thick, and 500 fins per meter), are shown for both clean and oxidized conditions together with the smooth plate results. Their data for increasing heat flux are represented here.

Comparing the clean surfaces, the Thermoexcel-C surface is marginally better than the other finned surface and about twice as effective as the smooth plate when evaluated at the same heat flux. Because the boiling curves are nearly parallel and have large slopes, however, the Thermoexcel-C surface is about six times better than the smooth plate at a fixed wall superheat of 0.14 K. Comparison of the oxidized surfaces yields similar conclusions.

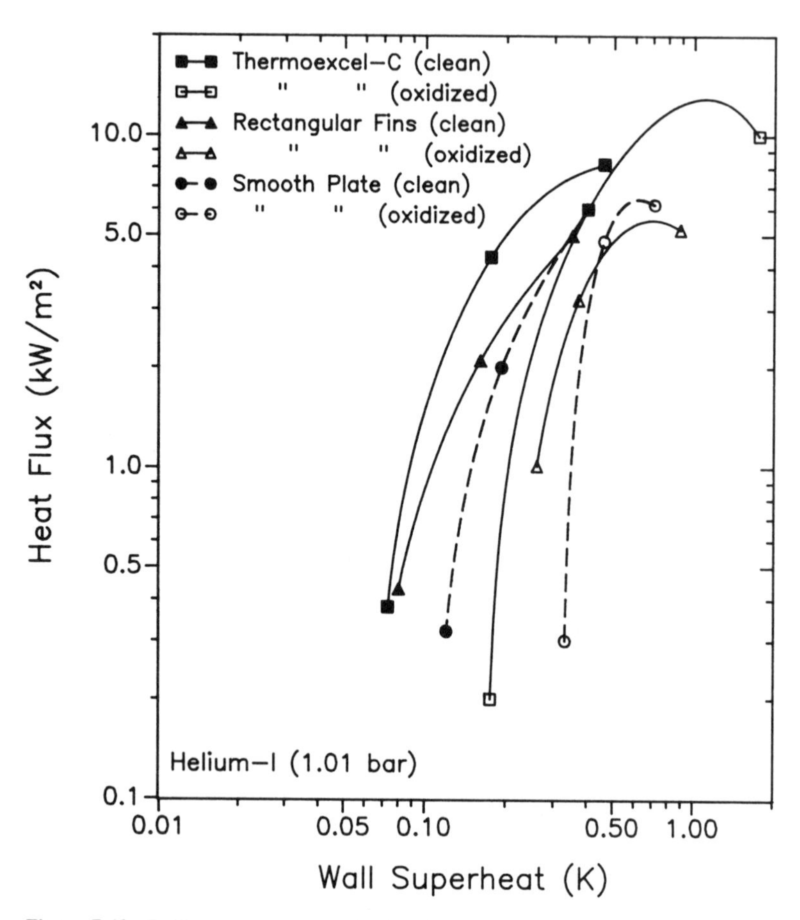

Figure 7-10 Boiling performances in Helium-I.

7-4 HYDROCARBONS

p-Xylene

Boiling enhancements in *p*-xylene are shown in Fig. 7-11 for copper tubes (performances in steel and copper alloys may differ appreciably because of their lower thermal conductivities). These results were obtained by Yilmaz, Hwalek, and Westwater (1980) and Yilmaz and Westwater (1981) by passing a 25-A electric current through the tubes and using each tube as an electric resistance thermometer. Heat was provided by condensing steam inside the tubes. The High Flux tube had a porous matrix 0.21 mm thick to give an outside diameter of 13.31 mm. About 46% of the copper particles were smaller than 44 μm, and the remaining ones were between 44 and 75 μm in diameter.

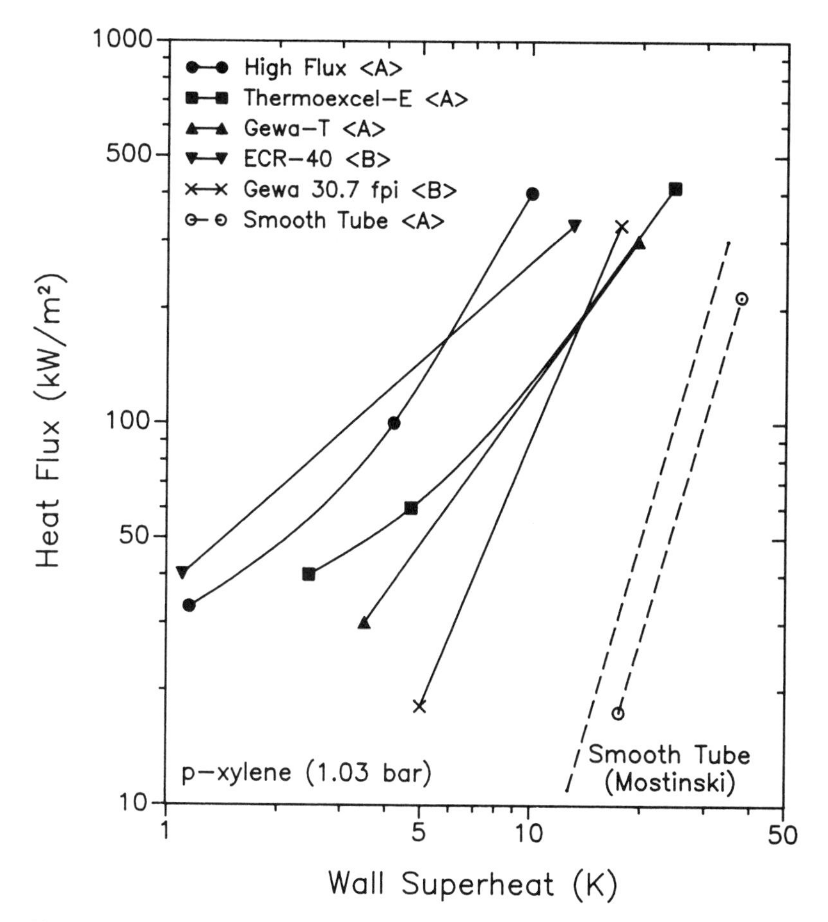

Figure 7-11 Boiling performances in *p*-xylene. References: (A) Yilmaz, Hwalek, and Westwater (1980) and (B) Yilmaz and Westwater (1981).

The Thermoexcel-E tube had an outside diameter of 13.16 mm with pores 0.1 mm in nominal diameter. The reentrant channels were 0.58 mm high, and there were 2,174 channels per meter of tubing (55 channels per inch). The Gewa-T tube had an outside diameter of 12.29 mm with 740 T-shaped fins per meter (18.7 fpi). The fins were 1.1 mm high and had a gap between their tips of 0.25 mm. The ECR-40 tube had an outside diameter of 12.0 mm, 1,575 circumferential channels per meter (39.9 channels per inch) with a gap width of 0.06 mm, and 1,333 axially oriented channels per meter (33.8 channels per inch). The Gewa finned tube was the same tube tested with water (see Sec. 7-1) and had an outside diameter of 12.57 mm with 1,213 fins per meter (30.7 fpi). The fins were 0.81 mm high with an outside wetted area ratio 2.36 times that of a 12.57-mm diameter plain tube. The smooth tube was made from 20-gauge seamless copper tubing.

The ECR-40 tube performed best over most of the heat flux range studied, although its boiling curve is within experimental error of the High Flux boiling curve. Their performances are followed by those of the Thermoexcel-E and Gewa-T tubes, in that order. The Gewa finned tube performed much better than the smooth tube and was comparable to the Thermoexcel-E and Gewa-T tubes at high heat fluxes. Also, the smooth tube data were predicted well by the Mostinski correlation. Augmentations provided by the ECR-40 and High Flux tubes are substantial when compared either at the same heat flux or at the same wall superheat as the smooth tube. Also, these two tubes out-performed the finned tube by a wide margin, which means that some finned tube bundles could be upgraded to augment reboiler capacity.

Benzene

Another commercially important hydrocarbon is benzene. Only one enhanced boiling tube (High Flux) has apparently been tested with this fluid, as is also true for other common hydrocarbons such as methane, ethane, propylene, and ethylene. The boiling performance for benzene is included here for completeness and also because some independent data sets are available for the High Flux tube. Figure 7-12 depicts these results. The particular High Flux tube tested by Antonelli and O'Neill (1981, 1986) was not specified. Ali and Thome (1984) tested the R-11 version with a porous layer thickness of 0.075 mm and an outside diameter of 18.7 mm. The porous coating was made of a copper alloy containing 1% iron. The smooth copper tube tested by Shakir (1986) was 22.2 mm in diameter, and its surface was prepared with 400 grade emery paper.

The boiling curves at 1.01 bar for the two High Flux tubes are close to one another. Boiling performance improves with an increase in pressure from 1 to 3 bar. The Mostinski correlation accurately fits the smooth tube boiling data at 1.01 bar. The augmentation in the boiling performance is about a factor of 10 when the High Flux tube is compared to the smooth tube at the same heat flux.

Propane

Figure 7-13 displays the comparison of boiling on a steel High Flux tube and an integral low-finned copper tube to boiling on a smooth copper tube. The thermal design curve

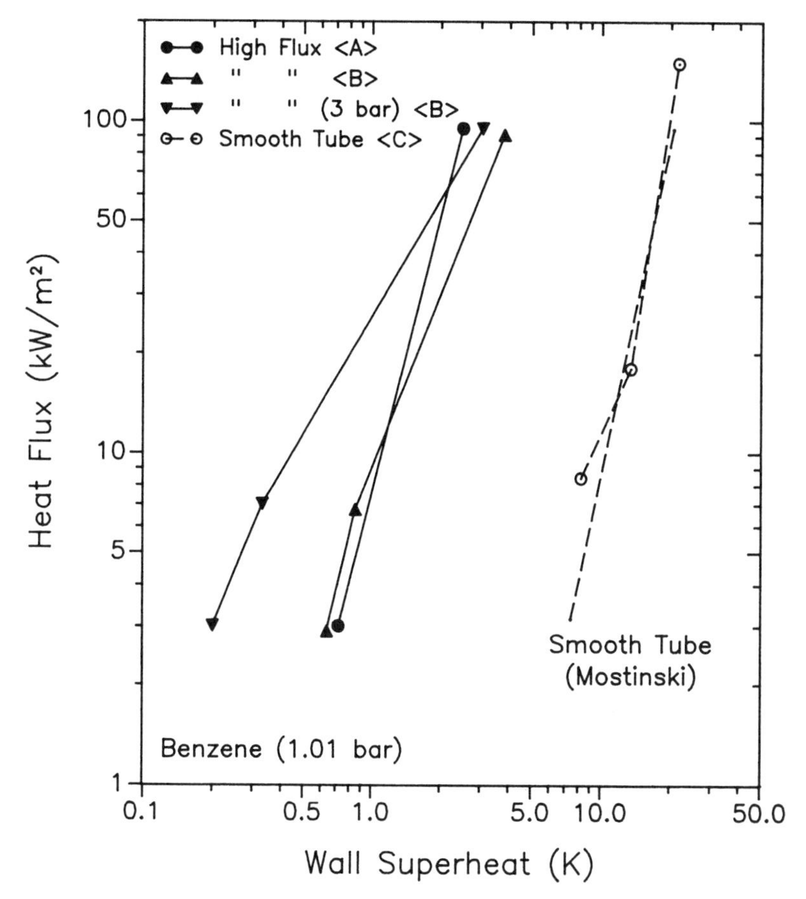

Figure 7-12 Boiling performances in benzene. References: (A) Antonelli and O'Neill (1981, 1986), (B) Ali and Thome (1984), and (C) Shakir (1986).

given by O'Neill, King, and Ragi (1980) is for the porous coating applied to either the inside or the outside of a tube and can also be used to model the boiling of propylene and ethylene over the pressure range from 1 to 5 bar. These investigators stated that the design curve already contains enough conservatism to include the effect of a 1% to 2% oil concentration in the liquid and also to account for the fact that the porous surface may not have the optimum pore size for the specific fluid at the particular operating pressure. The finned tube tested by Myers and Katz (1953) is the same one described above for refrigerant-12. Its boiling curve is a composite of data taken over the pressure range from 5.1 to 6.7 bar. The smooth copper tube was 19.0 mm in diameter.

The steel High Flux tube performs about twice as well as the copper finned tube at a fixed heat flux. The High Flux tube is about four times better than the finned tube when compared at a wall superheat of 1.7 K. If the finned tube were made of steel, the fin efficiency would substantially decrease its boiling performance. The boiling curve predicted by Mostinski for propane does not agree well with the experimental data.

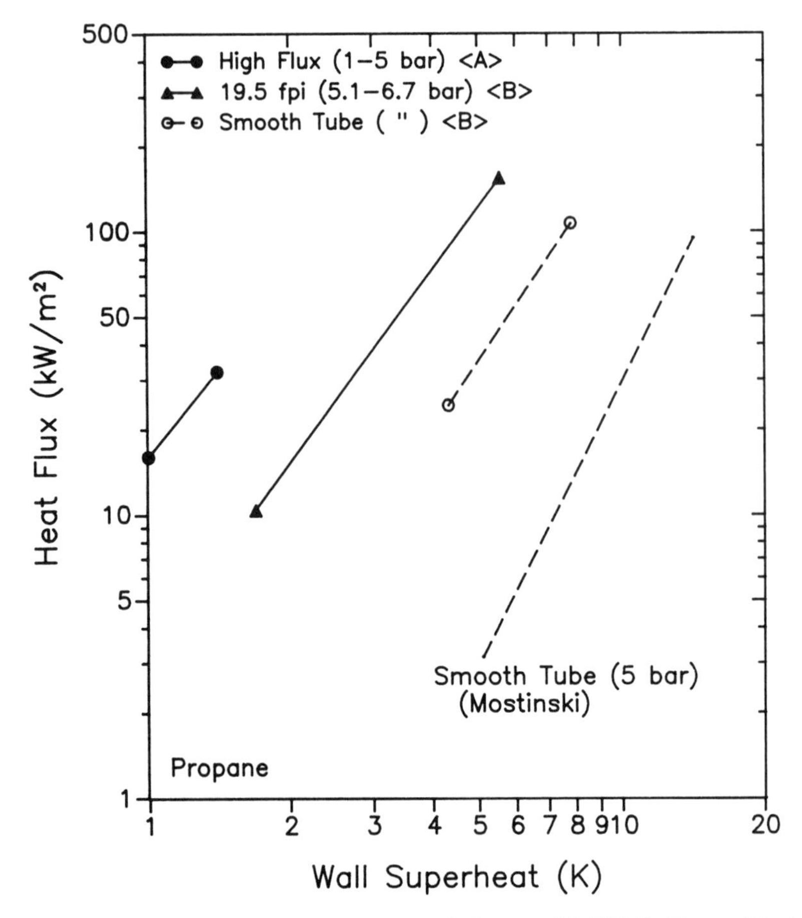

Figure 7-13 Boiling performances in propane. References: (A) O'Neill, King, and Ragi (1980) and (B) Myers and Katz (1953).

7-5 ALCOHOLS

Isopropanol

The boiling curves for copper enhanced boiling surfaces evaporating isopropyl alcohol were obtained in a study by Yilmaz and Westwater (1981). They are shown in Fig. 7-14. All the tube specifications are the same as in the study for *p*-xylene described above.

The ECR-40 and High Flux tubes have similar performances and out-perform the other tubes. The Thermoexcel-E, Gewa-T, and Gewa finned tube boiling curves are also close together, but, unexpectedly, the Gewa finned tube marginally out-performs the Gewa-T tube. The Mostinski correlation gives a conservative estimate of the boiling heat transfer coefficient compared to the experimental data, although its slope is close to that of the experimental curve. Compared to the Mostinski correlation, the ECR-40

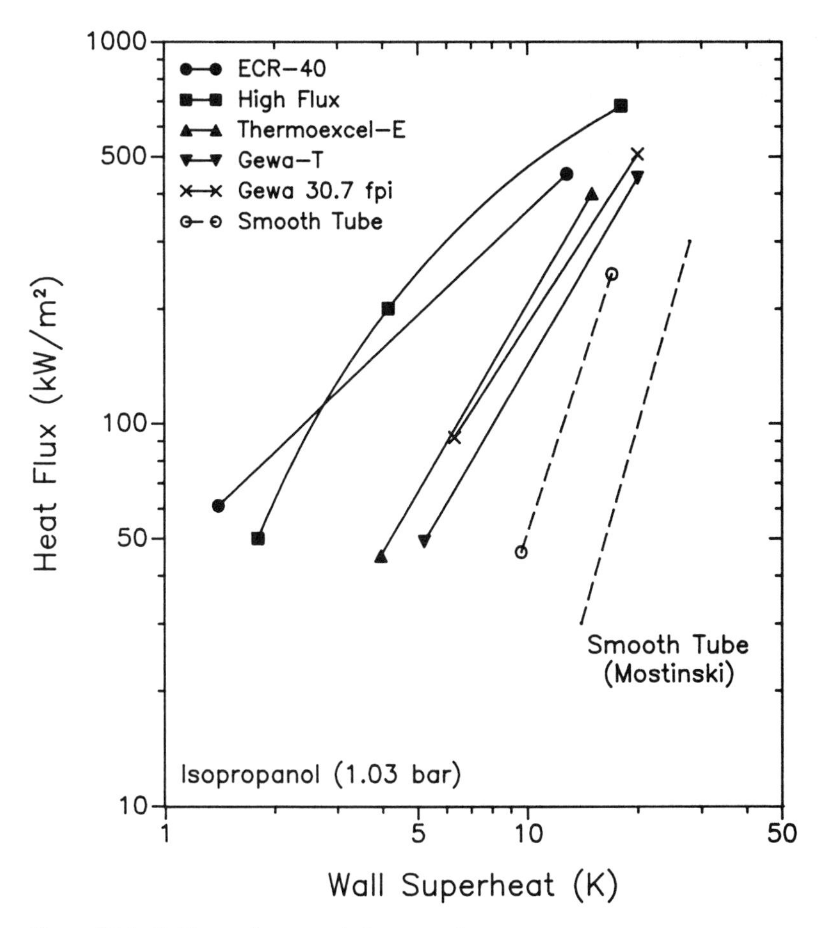

Figure 7-14 Boiling performances in isopropanol.

and the High Flux tubes provide a boiling enhancement of about a factor of 10 when compared at low and intermediate heat fluxes and are more than three times as effective as the low-finned tube. Because alcohols are often evaporated under vacuum, where plain tube boiling performance is relatively poor, these enhanced tubes could substantially upgrade the performance of reboilers in existing plants.

Ethanol

Another important alcohol is ethanol. Enhanced boiling data for the High Flux tube from two independent studies are shown in Fig. 7-15 along with low-finned tube data. The High Flux thermal design curve given in Antonelli and O'Neill (1981, 1986) is for an unspecified version. The High Flux tube tested by Ali and Thome (1984) for ethanol was the same R-11 version described above for benzene. The 18.8-fpi (741 fins per

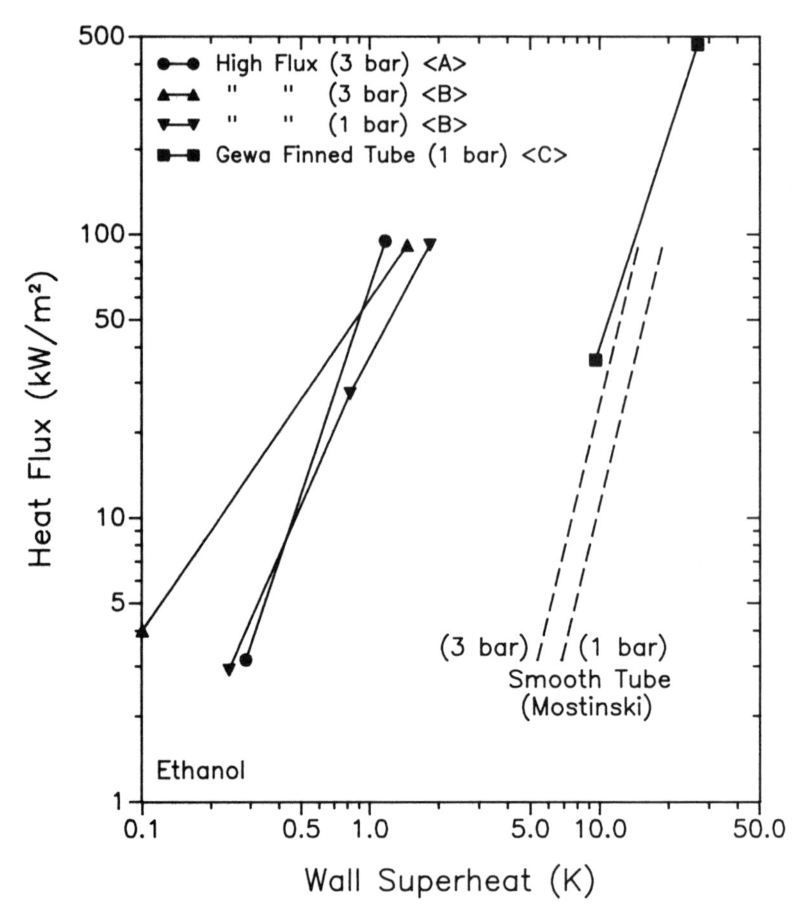

Figure 7-15 Boiling performances in ethanol. References: (A) Antonelli and O'Neill (1981, 1986), (B) Ali and Thome (1984), and (C) Bajorek (1988).

meter) low-finned tube tested by Bajorek (1988) was the same as that described above for water (see Sec. 7-1).

The 3-bar High Flux boiling curves from the two different studies are within experimental error of each other. The boiling performance of the High Flux tube increases with pressure over the small range tested. Boiling augmentation factor relative to the Mostinski correlation at 1 bar at a fixed heat flux is on the order of 20. Augmentation compared at a fixed wall superheat is also extremely high but not calculable because the boiling curves do not overlap. Similar results were obtained for the same High Flux tube in methanol by Shakir, Thome, and Lloyd (1985) and in *n*-propanol by Shakir (1986) and, for the low-finned tube, in methanol by Bajorek (1988). The High Flux tube out-performed the low-finned tube by a factor of about 10. The low-finned tube provides only marginal boiling augmentation at 1 bar when evaluated at a fixed heat flux. At a fixed wall superheat of 10 K, the low-finned tube provides about three times the plain tube performance predicted by the Mostinski correlation.

7-6 SOLVENTS (ACETONE)

Enhanced boiling of acetone has been studied by Uhlig and Thome (1985) for the High Flux tube and by Bajorek (1988) for a Gewa-K low-finned tube. Their results are compared to smooth tube data in Fig. 7-16. The same High Flux tube as described above for benzene was used here. The 18.8-fpi (741 fins per meter) low-finned tube was described earlier in Sec. 7-1. The smooth copper tube was 19.0 mm in diameter,

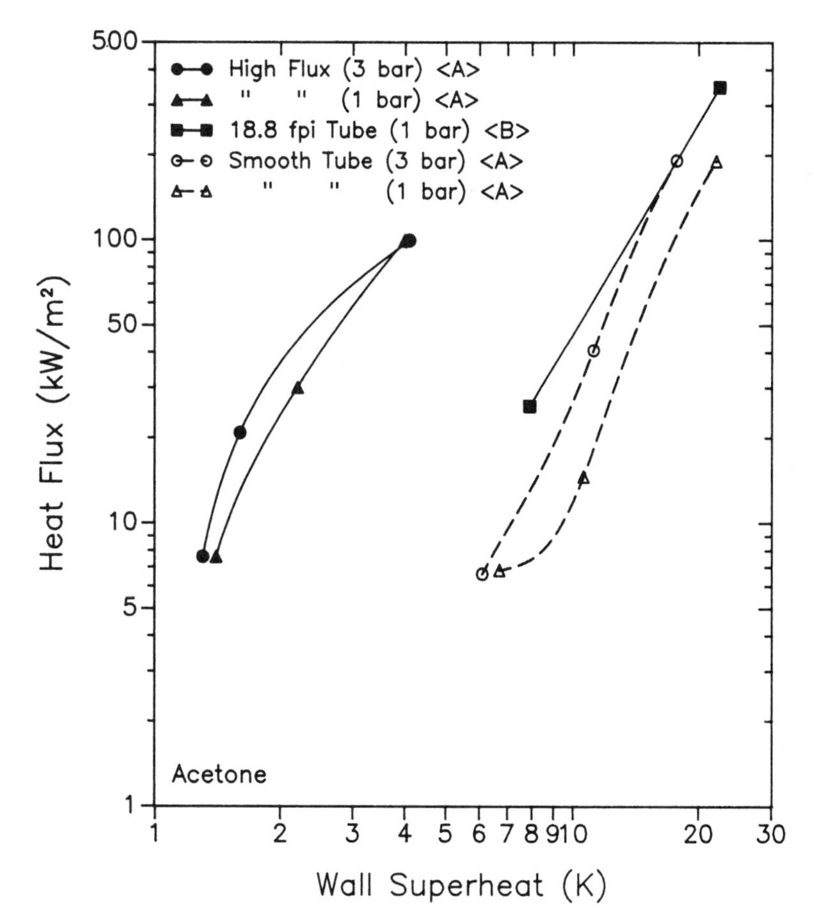

Figure 7-16 Boiling performances in acetone. References: (A) Uhlig and Thome (1985) and (B) Bajorek (1988).

and its surface was prepared by polishing with crocus paper to try to imitate the surface finish of commercial drawn tubing.

The High Flux augmentation factor at a fixed heat flux is about 6 relative to the smooth tube at 1 bar and about 5 at 3 bar. The High Flux tube again demonstrates a small improvement in performance with increasing pressure. Extrapolating the smooth tube boiling curve at 3 bar to a wall superheat of 4.0 K, the High Flux tube is shown to provide an enhancement in the boiling heat transfer coefficient of about a factor of 50. The High Flux tube substantially out-performs the low-finned tube. At 1 bar the low-finned tube demonstrates favorable performance relative to a plain tube at the same wall superheat.

7-7 DIELECTRICS (FLUORINERT FC-72)

Because enhanced boiling surfaces may eventually be utilized to cool high–power density electronic equipment, the dielectric fluid Fluorinert FC-72, a perfluorinated organic compound manufactured by 3M Company, has been tested in several studies to determine its enhanced boiling performance. Figure 7-17 depicts this fluid's performance on four different copper enhanced boiling surfaces. The Marto and Lepere (1981, 1982) tests were run with the same test sections as in their R-113 study described above. The Thermoexcel-E plate tested by Nakayama, Nakajima, and Hirasawa (1984) had pores 0.25 mm in diameter, an enhancement height of 0.4 mm, and reentrant channels 0.25 mm wide with a density of 1,818 channels per meter (46.1 channels per inch); this was their test surface designated P25. Their finned plate test surface had fins 0.3 mm thick and 0.5 mm high and a fin density of 1,667 fins per meter (42.2 fpi).

The boiling performances of the enhanced surfaces, with the exception of the finned plate, are all similar at low heat fluxes. At high heat fluxes the High Flux tube performs best, and the finned plate achieves a boiling performance equal to that of the other enhanced tubes. The Thermoexcel-E data obtained in the two independent studies match one another closely except at low heat fluxes. The boiling curve for the smooth tube is also seen to match that for the smooth plate reasonably well. At low heat fluxes, boiling augmentations are about four times that of the smooth plate. At a fixed wall superheat of 5.5 K the High Flux tube is able to dissipate about 30 times as much heat as the smooth surfaces, and the other enhanced boiling surfaces dissipate about 10 times as much.

7-8 SUMMARY

Figures 7-1 through 7-17 demonstrate that substantial heat transfer augmentation can be obtained for a wide variety of industrially important pure fluids. For a particular fluid, the best surface typically produces augmentation on the order of 10 times that of a smooth surface operating at the same heat flux; when compared at the same wall superheat, the augmentation may be 50 times greater. The relative position of the boiling curves for individual enhanced boiling tubes are also observed to vary from one

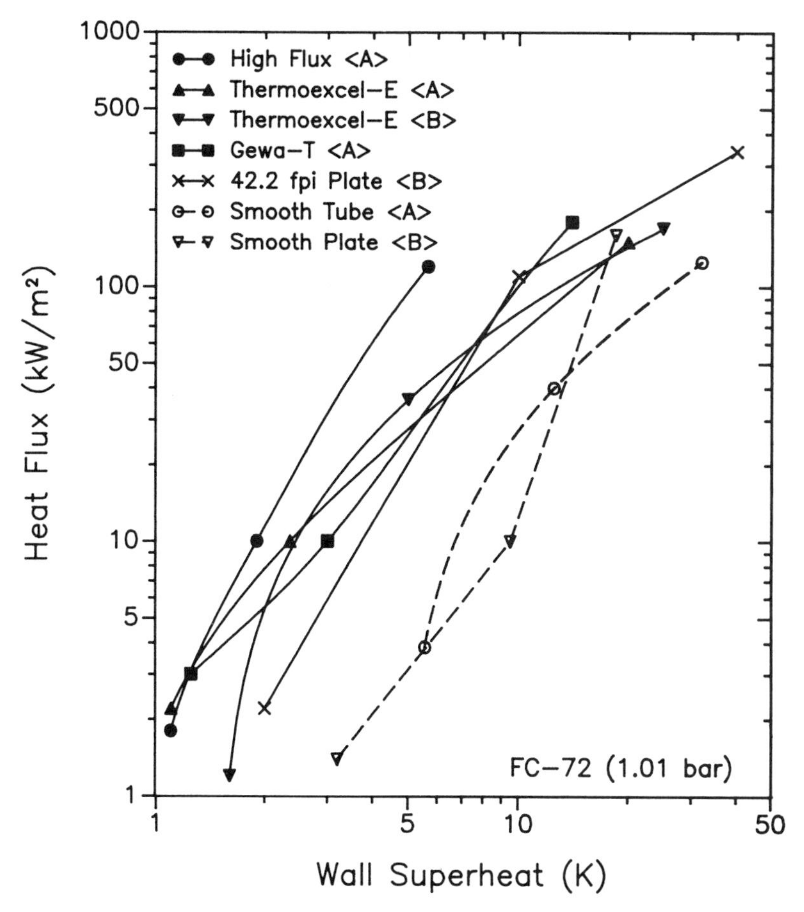

Figure 7-17 Boiling performances in Fluorinert FC-72. References: (A) Marto and Lepere (1981, 1982) and (B) Nakayama, Nakajima, and Hirasawa (1984).

fluid to another; thus extensive data are still required for the general application of these tubes and for developing a fundamentally based correlation for each.

7-9 NOMENCLATURE

A	area (m^2)
A^*	empirical constant
D	diameter (m)
F_P	pressure correction factor
L	tube length (m)
P_r	reduced pressure

q heat flux (W/m^2)

α_{nb} nucleate boiling heat transfer coefficient [W/(m^2·K)]

REFERENCES

Ali, S. M., and J. R. Thome. 1984. Boiling of ethanol-water and ethanol-benzene mixtures on an enhanced boiling surface. *Heat Transfer Eng.* 5(3–4):70–81.

Antonelli, R., and P. S. O'Neill. 1981. Design and application considerations for heat exchangers with enhanced boiling surfaces. Paper read at International Conference on Advances in Heat Exchangers, September, Dubrovnik, Yugoslavia.

———. 1986. Design and application considerations for heat exchangers with enhanced boiling surfaces. In *Heat exchanger sourcebook,* ed. J. W. Palen, 645–61. Washington, D.C.: Hemisphere.

Arai, N., T. Fukushima, A. Arai, T. Nakajima, K. Fujie, and Y. Nakayama. 1977. Heat transfer tubes enhancing boiling and condensation in heat exchangers of a refrigerating machine. *ASHRAE Trans.* 83 (part 2):58–70.

Bajorek, S. M. 1988. An experimental and theoretical investigation of multicomponent pool boiling on smooth and finned surfaces. Ph.D. diss., Michigan State University, East Lansing.

Bell, K. J., and A. C. Mueller. 1984. *Wolverine engineering data book II,* Sec. 5, p. 43. Decatur, Ala.: Wolverine Tube.

Bergles, A. E., and M. C. Chyu. 1981. Characteristics of nucleate pool boiling from porous metallic coatings. In *Advances in Enhanced Heat Transfer*, HTD Vol. 18, 61–71. New York: American Society of Mechanical Engineers.

Carnavos, T. C. 1981. An experimental study: Pool boiling R-11 with augmented tubes. In *Advances in Enhanced Heat Transfer*, HTD Vol. 18, 103–108. New York: American Society of Mechanical Engineers.

Collier, J. G. 1983. Pool Boiling. *Heat Exch. Des. Handb.* 2.7.2–6.

Czikk, A. M., and P. S. O'Neill. 1979. Correlation of nucleate boiling from porous metal films. In *Advances in Enhanced Heat Transfer*, 53–60. New York: American Society of Mechanical Engineers.

Ever-Fin. (Undated). *High performance heat transfer tube, product bulletin.* Tokyo: Furukawa Electric.

Gorenflo, D. 1966. Zum wärmeubergang bei der blasenverdampfung an rippenrohren. Ph.D diss., TH Karlsruhe, West Germany.

Gottzmann, C. F., P. S. O'Neill, and P. E. Minton. 1973. High efficiency heat exchangers. *Chem. Eng. Progr.* 69(7):69–75.

Hahne, E., and J. Muller. 1983. Boiling on a finned tube and a finned tube bundle. *Int. J. Heat Mass Transfer* 26:849–59.

Hesse, G. 1973. Heat transfer in nucleate boiling, maximum heat flux and transition boiling. *Int. J. Heat Mass Transfer* 16:1611–27.

Hitachi. 1984. Hitachi high-performance heat-transfer tubes (Cat. No. EA-500) Tokyo: Hitachi Cable, Ltd.

Hitachi Cable. 1978. *High Flux boiling and condensation tube, product bulletin.* Tokyo: Thermoexcel.

Kuwahara, H., W. Nakayama, and T. Daikoku. 1977. Boiling heat transfer from a surface with numerous tiny pores linked by small tunnels running below the surface. Paper read at 14th Symposium on Heat Transfer, Japan (paper B104).

Marto, P. J., and J. Lepere. 1981. Pool boiling heat transfer from enhanced surfaces to dielectric fluids. In *Advances in Enhanced Heat Transfer*, HTD Vol. 18, 93–102. New York: American Society of Mechanical Engineers.

———. 1982. Pool boiling heat transfer from enhanced surfaces to dielectric fluids. *J. Heat Transfer* 104:292–99.

Mostinski, I. L. 1963. Application of the rule of corresponding states for the calculation of heat transfer and critical heat flux. *Teploenergetika* 4:66.

Myers, J. E., and D. L. Katz. 1953. Boiling coefficients outside horizontal tubes. *Chem. Eng. Progr. Symp. Ser.* 49(5):107–14.

Nakayama, W., T. Kaikoku, H. Kuwahara, and T. Nakajima. 1979. Dynamic model of enhanced boiling heat transfer on porous surfaces. In *Advances in Enhanced Heat Transfer*, 31–43. New York: American Society of Mechanical Engineers.

———. 1980. Dynamic model of enhanced boiling heat transfer on porous surfaces, Part I: Experimental investigation. *J. Heat Transfer* 102:445–50.

Nakayama, W., T. Nakajima, and S. Hirasawa. 1984. *Heat sinks having enhanced boiling surfaces for cooling of microelectronic components* (ASME paper 84-WA/HT-89).

Ogata, H., and W. Nakayama. 1982. Heat transfer to boiling helium from machined and chemically treated copper surfaces. *Adv. Cryog. Eng.* 27:309–17.

O'Neill, P. S., C. F. Gottzmann, and J. W. Terbott. 1972. Novel heat exchanger increases cascade cycle efficiency for natural gas liquefaction. *Adv. Cryog. Eng.* 17:420–37.

O'Neill, P. S., R. C. King, and E. G. Ragi. 1980. Application of high performance evaporator tubing in refrigeration systems of large olefins plants. *AIChE Symp. Ser.* 76(199):289–303.

Robertson, J. M. 1979. Boiling heat transfer with liquid nitrogen in brazed aluminum plate-fin heat exchangers. *AIChE Symp. Ser.* 75(189):151–64.

Sauer, H. J. Jr., G. W. Davidson, and S. Chongrungreong. 1980. *Nucleate boiling of refrigerant-oil mixtures from finned tubing.* (ASME paper 80-HT-111).

Shakir, S. 1986. Boiling incipience and heat transfer on smooth and enhanced surfaces. Ph. D. diss., Michigan State University, East Lansing.

———. J. R. Thome, and J. R. Lloyd. 1985. Boiling of methanol-water mixtures on smooth and enhanced surfaces. In *Multiphase Flow and Heat Transfer,* ed. V. K. Dhir, J. C. Chen, and O. C. Jones. HTD vol. 47, 1–6. New York: American Society of Mechanical Engineers.

Starner, K. E., and R. A. Cromis. 1977. Energy savings using High Flux evaporator surface in centrifugal chillers. *ASHRAE J.* 19(12):24–27.

Stephan, K., and J. Mitrovic. 1981. Heat transfer in natural convective boiling of refrigerants and refrigerant-oil mixtures in bundles of T-shaped finned tubes. In *Advances in Enhanced Heat Transfer*, HTD Vol. 18, 131–46. New York: American Society of Mechanical Engineers.

Thome, J. R., and W. B. Bald. 1978. Nucleate pool boiling in cryogenic binary mixtures. *Proc. 7th Int. Cryog. Eng. Conf.* 523–30.

Turbo-B. 1985. *Turbo-B: An improved evaporator tube, product bulletin.* Decatur, Ala.: Wolverine Tube.

Uhlig, E., and J. R. Thome. 1985. Boiling of acetone-water mixtures on smooth and enhanced surfaces. In *Advances in Enhanced Heat Transfer*, HTD Vol. 43, 49–56. New York: American Society of Mechanical Engineers.

Wanniarachchi, A. S., P. J. Marto, and J. T. Reilly. 1986. The effect of oil contamination on the nucleate pool boiling performance of R-114 from a porous-coated surface. *ASHRAE Trans.* 92 (part 2).

Wanniarachchi, A. S., L. M. Sawyer, and P. J. Marto. 1987. Effect of oil on pool boiling performance of R-114 from enhanced surfaces. *Proc. 1987 ASME-JSME Therm. Eng. Joint Conf.* 1:531–37.

Yilmaz, S., J. J. Hwalek, and J. W. Westwater. 1980. *Pool boiling heat transfer performance for commercial enhanced tube surfaces* (ASME paper 80-HT-41).

Yilmaz, S., and J. W. Westwater. 1981. Effect of commercial enhanced surfaces on the boiling heat transfer curve. In *Advances in Enhanced Heat Transfer*, HTD Vol. 18, 73–91. New York: American Society of Mechanical Engineers.

CHAPTER

EIGHT

NUCLEATE POOL BOILING CORRELATIONS

8-1 INTRODUCTION

A number of nucleate pool boiling correlations have been developed specifically for enhanced boiling surfaces. Correlations are available for integral low-finned tubes, porous coatings, and several special extended surface geometries. In general, the accuracy and range of applicability of these correlations are still less than those of normal thermal design standards. This larger error band, however, will only have a marginal effect on the overall heat transfer coefficient when the heating fluid's thermal resistance is predominant. Some suggestions for the improvement of these boiling models and correlations are made.

8-2 LOW-FINNED TUBES

Many analytical studies of pool boiling on fins have been reported in the literature. Most of these studies have sought to determine either the fin efficiency or the optimum fin geometry for pool boiling on a single, isolated fin. The standard fin efficiency formulas listed in textbooks are for extended surfaces having uniform heat transfer coefficients. The heat transfer coefficient in pool boiling, however, varies as a function of the wall superheat and thus with distance from the base of a fin to its tip. Therefore, in calculating fin efficiency, the functional dependency of the boiling heat transfer coefficient on wall superheat must be accounted for.

Some of the earliest work on boiling fin efficiencies was carried out by Han and Leftowitz (1960) and Chen and Zyskowski (1963); both studies assumed a linear variation in the heat transfer coefficient along the length of the fin. Cumo, Lopez, and Pinchera (1965) improved the analysis by assuming that the heat flux along the fin's surface is proportional to the local wall superheat to the third power and then solving numerically for the fin efficiency. Haley and Westwater (1965) performed a similar numerical study that was based on the assumption that the local boiling heat transfer coefficient on the fin is not affected by the nonisothermal temperature of the fin and thus can be calculated from the experimental boiling curve obtained for a plain surface. Klein and Westwater (1971) later studied the effect of fin clearances on spine-shaped fins and concluded that the bubble departure diameter was a controlling factor on heat transfer performance on finned tubes. In addition, Cash, Klein, and Westwater (1971) analyzed pool boiling on a single, isolated fin to determine the optimum fin geometry; they obtained a shape resembling a turnip when they minimized the volume of metal in fins of circular cross-section. More recently, Jho et al. (1985) studied the boiling fin efficiencies for longitudinal and circumferential fins on tubes.

All the above analyses have severe limitations when compared to the actual boiling process on the outside of commercial low-finned tubes. First of all, most studies were for fins with high root temperatures, so that all regimes of nucleate boiling occurred on the fin, from film boiling at the fin root to the isolated bubble regime at the fin tip. These studies also assumed that a plain tube boiling curve was applicable without modification to model the boiling process, even though the surface finish of the finned tube may not be the same as that of the smooth test surface. More important, the effect of confining growing bubbles between fins was completely neglected. Therefore, application of fin efficiency alone to model boiling on a low-finned tube is useful only as a first approximation.

Fin Efficiencies for Low-Finned Tubes

Bell and Mueller (1984) prepared a diagram for calculating fin efficiencies for rectangular and triangular fins with nucleate pool boiling occurring over the entire wetted surface. Their diagram is shown in Fig. 8-1. A nucleate pool boiling curve of the form

$$q = b\Delta T^3 \tag{8-1}$$

was used to approximate the boiling heat transfer coefficient on the fin. The coefficient b can be calculated from the Mostinski (1963) correlation described in Chapters Two and Seven. The fin specifications required are the fin height H, the fin thickness at its base T, and the fin thermal conductivity. The wall superheat used in the calculation is that at the base of the fin.

As an example, the fin efficiency of the Gewa finned tube cited in Fig. 7-11 for p-xylene boiling at 1 bar is estimated as follows (assuming the tube to be cupro-nickel for a reboiler application). The fin height is 0.81 mm, and the fin thickness is assumed to be 0.4 mm. The thermal conductivity of cupro-nickel is 45 W/(m·K). At the low end of the Mostinski boiling curve in Fig. 7-11, the wall superheat is 12.74 K at a heat flux of 11,000 W/m^2. Evaluating the nondimensional expression on the abscissa in Fig. 8-1, one obtains 0.25 and reads a fin efficiency of 0.93 on the ordinate, assuming that the fin shape is approximately triangular.

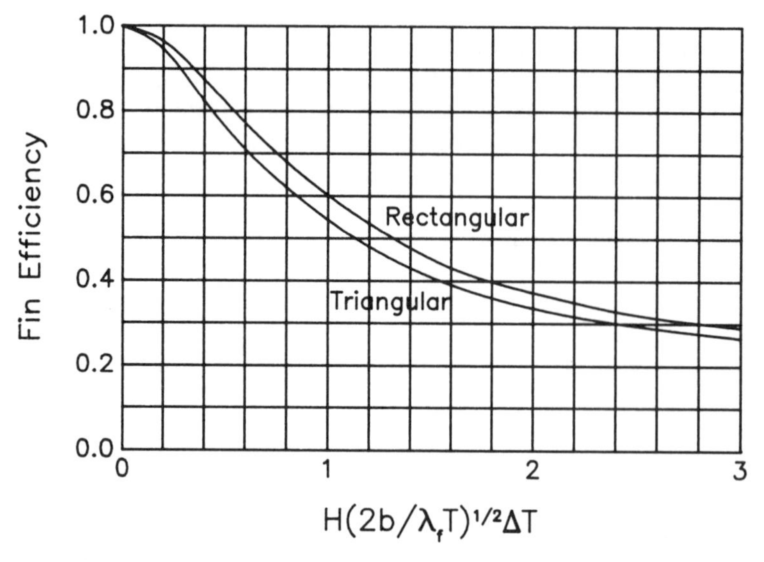

Figure 8-1 Fin efficiencies for nucleate pool boiling.

Finned Tube Correlation

Low-finned tubes augment nucleate pool boiling heat transfer in two different ways: (1) by increasing the wetted surface area per unit tube length, and (2) by modifying the boiling process. Palen and Yang (1983) recommend the following form of correlating expression to include these two factors for nucleate pool boiling on a single, horizontal low-finned tube:

$$\alpha_{\mathrm{f}} = F_{\mathrm{e}} F_{\mathrm{s}} F_{\mathrm{c}} \alpha_{\mathrm{nb}} + \alpha_{\mathrm{nc}} \tag{8-2}$$

In this expression the boiling heat transfer coefficient is composed of two contributions, nucleate pool boiling and natural convection (the latter is normally insignificant except at low heat fluxes). The factor F_{e} is the fin efficiency, which can be determined from Fig. 8-1. The fin efficiency in practice can vary from 1.0 for copper finned tubes at low heat fluxes to as little as 0.3 for metals with very low thermal conductivity at very high heat fluxes. From an economic standpoint, use of fins with efficiencies less than about 0.7 is probably not worthwhile.

The factor F_{s} is the surface correction factor that models the effect of the close proximity of the fins on the boiling process. According to Yilmaz and Palen (1984), the value of this factor varies from less than unity up to approximately 2.0 depending on the type of fin geometry, fluid, and heat flux. This factor has a correlating expression of the form

$$F_{\mathrm{s}} = C\left(\frac{q}{q_{\mathrm{r}}}\right)^{m1} P_{\mathrm{r}}^{\,m2} F_{\mathrm{c}}^{\,m3} \tag{8-3}$$

according to Palen, Taborek, and Yilmaz (1986). Here C is an empirical constant, m_1 and m_2 are negative exponents, and m_3 is a positive exponent (no numerical values were given). The heat flux q is divided by a reference heat flux q_r.

The factor F_c is the mixture correction factor for nucleate pool boiling on a plain tube (see Chapter Nine's discussion of nucleate pool boiling on smooth surfaces) in both Eqs. (8-2) and (8-3). It is less than unity for a mixture not at its azeotropic composition and equal to 1.0 for a pure fluid or azeotropic mixture. Because the exponent m_3 is positive, the mixture degradation for a low-finned tube may be greater or less than that for a plain tube. The nucleate pool boiling heat transfer coefficient in Eq. (8-2) is determined with a suitable correlation, such as the Mostinski correlation referred to earlier.

The empirical constants in Eq. (8-3) can be determined from nucleate pool boiling data obtained on copper finned tubes, which have fin efficiencies nearly equal to 1.0. They are determined by using the total wetted surface area because there is no way to differentiate between the respective contributions of the fins and the root area. Hence the surface correction factor F_s must be defined in such a way as not to apply the fin efficiency to the root area of the finned tube. The surface factor correlation does not include fluid physical properties and consequently is insensitive to their effects on the boiling process between adjacent fins, which would probably become most evident at high reduced pressures and under vacuum conditions. The above method provides a good framework for further improvements, however.

8-3 POROUS COATINGS

Original High Flux Correlation

Gottzmann, Wolf, and O'Neill (1971) developed the first nucleate pool boiling model to correlate enhanced boiling heat transfer on porous coated surfaces, in particular the High Flux surface. The order-of-magnitude increase in the boiling heat transfer coefficients of this surface relative to a smooth surface was surmised to be the result of the high density of relatively large interconnected reentrant cavities in its porous sintered layer. This special geometry was speculated to augment boiling heat transfer for the following reasons:

1. The porous matrix structure traps larger vapor nuclei than is possible in the pits and cavities of a smooth surface, which reduces the boiling nucleation superheat required to activate boiling sites.
2. The lower boiling nucleation superheat, the larger density of pores in the porous boiling surface, and the interconnection of these pores create a much larger boiling site density.
3. Thin film evaporation occurs within the passageways of the porous matrix over a much larger surface area than for microlayer evaporation under growing bubbles on a smooth surface.

These investigators proposed a boiling model that was based on two of these factors, the nucleation superheat and the thin film evaporation mechanism (they neglected the effect of the increased boiling site density on the external convection process). The total wall superheat between the heated porous surface and bulk saturated liquid was assumed to be the summation of the thermal resistances of these two factors. Thus the wall superheat could be written as

$$\Delta T = \Delta T_{\text{sat}} + \Delta T_{\text{film}} \tag{8-4}$$

where the nucleation superheat is given by

$$\Delta T_{\text{sat}} = \frac{2\sigma}{R(dP/dT)_{\text{sat}}} \tag{8-5}$$

[i.e., Eq. (2-1)]. The temperature drop across the thin evaporating liquid film was derived as

$$\Delta T_{\text{film}} = \frac{BqR^2}{\lambda_L} \tag{8-6}$$

where B is an empirical geometry factor to account for the film's thickness and surface area and q is the heat flux based on the outside nominal surface area of the porous layer. The origin of Eq. (8-6) is the heat transfer coefficient for heat conduction across a quiescent liquid layer of thickness t:

$$\alpha_{\text{film}} = \frac{\lambda_L}{t} \tag{8-7}$$

Rewriting this expression as a temperature difference, the temperature drop across the film becomes

$$\Delta T_{\text{film}} = \frac{qt}{\lambda_L} \tag{8-8}$$

To apply this expression to a porous layer, the film thickness was correlated as

$$t = BR^2 \tag{8-9}$$

where R is the average radius of the pores in the matrix and B is a geometric factor. The heat flux is defined with the nominal surface area rather than the unknown total wetted surface area of the porous matrix. Thus the wall superheat for boiling on a thin porous surface is given as

$$\Delta T = \frac{2\sigma}{R(dP/dT)_{\text{sat}}} + \frac{BqR^2}{\lambda_L} \tag{8-10}$$

The optimum pore radius R^* for a particular fluid, which yields the minimum wall superheat at a specified heat flux, was then obtained by differentiating Eq. (8-10) to obtain

$$R^* = \left[\frac{\sigma \lambda_L}{Bq\,(dP/dT)_{sat}} \right]^{1/3} \tag{8-11}$$

Substituting the optimum pore radius R^* into Eq. (8-10) leads to the following expression for the maximum boiling heat transfer coefficient, which is attained at R^*:

$$\alpha_{max} = \frac{1}{3B^{1/3}} \left[\frac{\lambda_L (dP/dT)^2_{sat}}{\sigma^2} \right]^{1/3} q^{2/3} \tag{8-12}$$

By using nucleate pool boiling results for ten different fluids on High Flux surfaces at subatmospheric and atmospheric pressures, the empirical constant B was determined by a least-squares fit of Eq. (8-12) to the data. The semiempirical correlation thus is given in final form as

$$\alpha_{max} = 0.0421 \left[\frac{\lambda_L (dP/dT)^2_{sat}}{\sigma^2} \right]^{1/3} q^{2/3} \tag{8-13}$$

where British units are used in the dimensional expression and the heat transfer coefficient is in British thermal units per hour per square foot per Fahrenheit degree.

Figure 8-2 depicts the investigators' comparison between Eq. (8-13) and their boiling data. Although most of the data correlate well, the error band is rather large. For instance, the expression overpredicts the ammonia and toluene data by about 100%. This error band, however, can be partially attributed to the inaccuracies in the measured heat transfer coefficients, which at wall superheats of 1.0 K and less become large. The method is particularly important for the novel theoretical model used to derive it, which most plain tube correlations still lack.

New High Flux Correlation

O'Neill, Gottzmann, and Terbot (1972) further improved the original porous layer boiling model to represent better the effects of the porous matrix geometry on the thin film evaporation process. [Because the development of the new improved model was not presented in detail in their publication, the reader is referred to Webb (1983) where it has been summarized]. Although these investigators retained the basic two-resistance model, many changes were made. Therefore, this newer version is also derived here to point out these differences, to note the assumptions, and to discuss some of the further improvements that could be made to this analysis.

Starting with the two-resistance assumption, the wall superheat is again defined from a summation of the temperature difference across the liquid film and the boiling nucleation superheat as

$$\Delta T = (T_w - T_{sat}) = (T_w - T_l) + (T_l - T_{sat}) \tag{8-14}$$

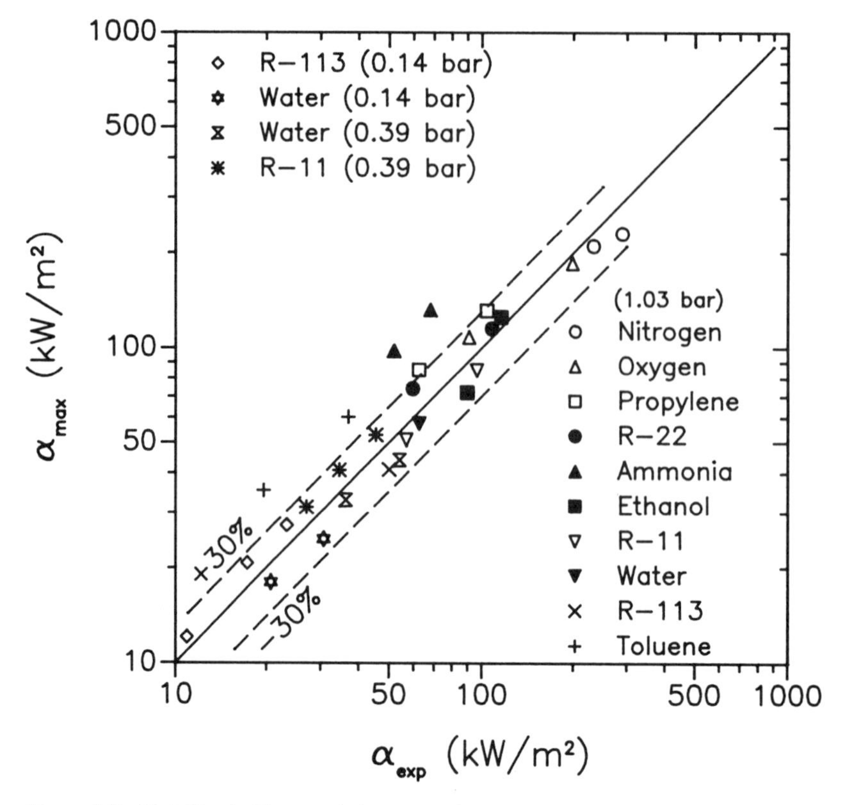

Figure 8-2 High Flux boiling correlation comparison.

where T_I is the interfacial temperature of the liquid film. The boiling nucleation superheat between the vapor-liquid interface and the bulk saturated liquid was again calculated with Eq. (8-5). The temperature drop across the thin liquid film was determined under the assumption of one-dimensional heat conduction, excluding any effect of convection in the liquid film. The expression for the heat flux then becomes

$$q = \left(\frac{S}{A}\right)\left(\frac{\lambda_L}{t}\right)(T_w - T_I) \tag{8-15}$$

where S is the surface area of the liquid film, A is the nominal surface area of the porous layer, and t is the thickness of the liquid film. The surface temperature of the particles in the matrix was assumed to be uniform and equal to that at the base. The following assumptions were then made to evaluate the area ratio (S/A):

1. the porous matrix is made of spherical particles of uniform size;
2. all the porous passageways are uniform in size and interconnected;
3. the pore radius is defined by the largest sphere that would fit within the void space between the closely packed spherical particles; and

4. each pore is an active evaporation site and a stable vapor trap.

The total volume of the matrix V is calculated as

$$V = At_p \tag{8-16}$$

where t_p is the thickness of the porous layer. Substituting for A in Eq. (8-15), the expression becomes

$$T_w - T_l = \left(\frac{q}{\lambda_L}\right)\left(\frac{tV}{t_p S}\right) \tag{8-17}$$

where the term (tV/t_pS) is the geometric factor for the liquid film in the matrix. The value of this term is dependent on the packing arrangement and is calculable once it has been specified; see Webb (1983) for two different packings. This geometry factor is then transformed into a new geometric factor B defined as

$$B = \frac{tV}{t_p S R^2} \tag{8-18}$$

so that, with the substitution of B, Eq. (8-17) becomes equivalent to Eq. (8-6) and the original expression of Gottzmann, Wolf, and O'Neill (1971) is obtainable, Eq. (8-10), except that now B is directly related to the geometry of the porous layer. For an in-line square packing arrangement as shown in Fig. 6-15, Webb (1983) calculated the product (Bt_p) to be 1.04 when the ratio of the pore radius R to the particle radius R_p is equal to 0.414. This ratio gives a porosity of 47.6%. With these values, the wall superheat correlation for porous coated surfaces becomes

$$T_w - T_{sat} = \frac{0.045 q d_p^2}{\lambda_L t_p} + \frac{9.66\sigma}{d_p (dP/dT)_{sat}} \tag{8-19}$$

where d_p is the particle diameter.

Equation (8-19) is only partially successful in predicting High Flux boiling data. For instance, the expression was shown to predict the wall superheat with errors as high as 56% for water and 74% for methanol in Shakir, Thome, and Lloyd (1985). Webb (1983) also demonstrated that it was inaccurate for predicting the boiling performances of R-11 and R-113 on various porous enhanced boiling surfaces. In all fairness, these boiling curves are not easy to predict because their slopes change substantially from one fluid to another.

There are several fundamental improvements that could be made to the model. First, the contribution of convection to the enhanced boiling process should be included. In its present form, the model assumes that all heat is transported as latent heat. Instead, Nakayama et al. (1979) have shown that most of the heat leaving the Thermoexcel-E surface is in the form of sensible heat at heat fluxes greater than 3 kW/m^2 in R-11. Therefore, the convective heat transfer mechanism inside the porous matrix (described in Chapter Six) and convection on the external surface produced by the large boiling site density should be included. The model also assumes that the evaporating liquid film geometric factor B is not a function of heat flux or fluid physical properties. Instead, Arshad and Thome (1983) observed that the thickness for the liquid film in various reentrant channels decreased with increasing heat flux. In addition, the correlation assumes the slope of the boiling curve to be fixed, although data indicate the contrary. This is also true for nucleate pool boiling curves for smooth surfaces, but the effect is magnified at low wall superheats. Therefore, future development of a more accurate correlation may have to include the effect of fluid physical properties on the slope of the boiling curve.

Czikk and O'Neill (1979) developed another correlation for the High Flux tube. This correlation is difficult to apply, however, because it requires extensive statistical information about the geometry of the porous matrix.

Nishikawa and Ito Empirical Correlation

A completely empirical correlation for boiling on sintered porous layers has been developed by Nishikawa and Ito (1982). They obtained extensive boiling data on porous layers made from either copper or bronze particles of uniform size for various particle sizes and layer thicknesses. They tested R-11, R-113, and benzene at 1.01 bar. Using a linear multiple regression analysis on a selected list of physical properties and several characteristic dimensions of the porous layer (particle diameter, porous layer thickness, and porosity), they developed the following enhanced boiling correlation:

$$\frac{qt_p}{\lambda_m(T_w - T_{sat})} = 0.001\left(\frac{\sigma^2 \Delta h_v}{q^2 t_p^2}\right)^{0.0284}\left(\frac{t_p}{d_p}\right)^{0.56}\left(\frac{qd_p}{\varepsilon \Delta h_v \mu_v}\right)^{0.593}\left(\frac{\lambda_L}{\lambda_m}\right)^{-0.708}\left(\frac{\rho_L}{\rho_v}\right)^{1.67} \tag{8-20}$$

where the thermal conductivity of the porous layer is calculated as

$$\lambda_m = \lambda_L + (1 - \varepsilon)\lambda_p \tag{8-21}$$

The term to the left of the equal sign in Eq. (8-20) is recognizable as a Nusselt number whose characteristic dimension is the porous layer thickness t_p. The third dimensionless group to the right of the equal sign is a vapor-phase Reynolds number, whose characteristic dimension is the particle diameter d_p. Therefore, this correlation empirically includes the effect of convection on the boiling process. The range of application of the correlation is as follows:

$0.1 < d_p < 1.0$ (in millimeters)
$1.6 < t_p/d_p < 20$
$0.38 < \varepsilon < 0.71$
$61 < \lambda_p < 372$ (in watts per meter per Kelvin)

The Nishikawa and Ito (1982) correlation was able to predict their own data to within about 30%. In a spot check by Webb (1983) with three independent data sets for R-11 and R-113, the errors were 51%, 8%, and 1%. The 51% error was for a comparison against the High Flux R-11 data of Milton (1968). The correlation is not necessarily applicable to the High Flux tube, however, because the particle diameters are not of uniform size. The correlation was only developed with boiling data for three different fluids at atmospheric pressure, so that caution should be exercised when applying it to other fluids or other pressures. This correlation's predictions are also sensitive to the thermal conductivity of the metallic particles. For copper the thermal conductivity drops rapidly with small impurities, and the effect of brazing powder in the matrix may be substantial.

8-4 MODIFIED EXTERNALLY FINNED TUBES

Correlation of Nakayama and Colleagues

An enhanced boiling model for the Thermoexcel-E and other similar surfaces with relatively large numbers of interconnected cavities and small openings to the outside pool of liquid was developed by Nakayama et al. (1979). They observed pulsating liquid menisci in the corners of a glass-walled channel that was constructed to be similar to that of the Thermoexcel-E tube's channel geometry (see Sec. 5-2 for a detailed description of their visualization experiment). On the basis of their observations, they hypothesized that the following sequence of events occurs during the enhanced boiling process:

1. *Phase I (pressure build-up)*—Evaporation of liquid in the passageways causes the internal vapor pressure to build up until the vapor nuclei located at the mouths of the pores attain a hemispherical shape, protruding outward into the external liquid pool.
2. *Phase II (pressure reduction)*—At some of the pores the vapor nuclei become active and grow into bubbles. Initially, the bubbles grow as a result of the pressure differential across the interface, which reduces the pressure in the passageways as the vapor flows into the bubbles. They then continue to grow as a result of the inertial forces that they have imparted to the surrounding liquid.
3. *Phase III (liquid intake)*—During the inertial stage of bubble growth the pressure inside the passageways is reduced to less than that of the external liquid pool, and liquid is drawn into the passageways at the inactive pores. At the end of this phase the bubbles depart, all pores are closed by liquid menisci, and the liquid spreads along the angled corners of the passageways by capillary forces. The cycle then repeats from this point.

The heat transfer model proposed to model this boiling process is as follows: (1) convective heat transfer occurs on the external surface, induced by the agitation of the liquid by the growing and departing bubbles; and (2) thin film evaporation occurs inside the enhanced surface's channels at the liquid films trapped in the corners. The heat leaving the surface is therefore the sum of the sensible heat transported by the external convection process and the latent heat transported in the vapor bubbles leaving the channels to the external pool of liquid. Evaporation taking place on the outside surface (i.e., the further growth of the bubbles leaving the pores in the external thermal boundary layer) is apparently assumed to be of negligible importance. The heat balance is then given by the expression

$$q = q_{ec} + q_{lat} \tag{8-22}$$

where q_{ec} is the external convective heat flux and q_{lat} is the latent heat flux based on the outside projected surface area. The external convective heat flux was correlated with the following expression:

$$q_{ec} = \left(\frac{\Delta T}{C}\right)^{1/y} \left(\frac{N}{A}\right)^{-x/y} \tag{8-23}$$

where C, x, and y are empirically determined constants and N/A is the boiling site density. The latent heat flux is calculated with the following equation:

$$q_{lat} = \frac{(m_{LI} + m_{LII})\Delta h_v}{(\theta_I + \theta_{II})A} \tag{8-24}$$

where the masses of liquid evaporated during the time periods θ_I and θ_{II} of phases I and II are m_{LI} and m_{LII}, respectively. Nakayama and co-workers then performed extensive analyses to derive expressions for these parameters.

Their final correlating expressions were complex, and the interested reader is referred to the original report. Practical application to thermal design is not feasible because the empirical constants were different for each fluid tested (nitrogen, R-11, and water). Thus it would be easier to use the experimental nucleate pool boiling curves themselves rather than to determine the constants. The model also assumed that liquid films only form in the corners, which is contrary to the visual observations of Arshad and Thome (1983) for this type of surface geometry. Nevertheless, the model and analysis of Nakayama and co-workers provide extensive insight into the enhanced boiling process.

Xin and Chao Gewa-T Correlation

An enhanced boiling model for Gewa-T types of surfaces was developed by Xin and Chao (1985). They approached the problem in a much different manner from that of either the High Flux two-resistance model or the liquid pumping model of Nakayama et al. (1979). Their basic premise was that the spreading of the thin evaporating liquid film is controlled by the dynamics of a countercurrent two-phase flow between the channel and the outside liquid pool, similar to that which occurs for boiling in a narrow slot. Their model for a planar (not tubular) Gewa-T type of geometry with T-shaped fins facing upward is illustrated in Fig. 8-3. They assumed that the vapor and liquid flows are steady state rather than cyclic. Thus vapor flows upward out of the channel through the center of the slit's opening, and liquid flows simultaneously downward through the slit and along the fin wall into the channel. Heat was assumed to be

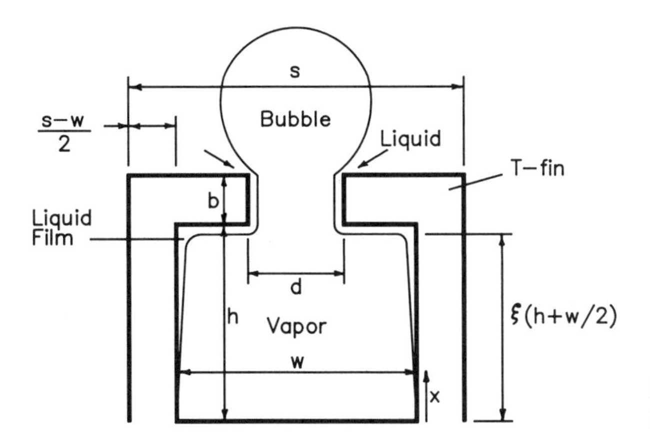

Figure 8-3 Countercurrent two-phase flow model.

transferred as latent heat from inside the channel and by convection from the external surface to the liquid pool.

The channel formed by the T-shaped fins was considered to have a well-defined geometry with a channel width w, a slit width d, a fin pitch s, and a fin wall height h. The fin thickness at the slit is b. The temperatures of the channel walls were assumed to be isothermal and equal to the temperature at the root of the fins. The vapor in the channel was assumed to be at the saturation temperature corresponding to the external bulk pressure; thus the boiling nucleation superheat was ignored because the slit opening is relatively large. A falling liquid film was assumed to form on the walls of the channel, and the slit opening was such that the channel was neither flooded by liquid nor vapor bound. The bottom of the channel was apparently assumed to be unwetted, and the capillary forces acting to thicken the film in the upper corners of the T were neglected.

The investigators began their analysis by applying the Wallis (1969) relationship for one-dimensional countercurrent flow of a vapor and a liquid film, so that the nondimensional film thickness t^* was correlated as

$$t^* = c\mathrm{Re}_L^m \tag{8-25}$$

and t^* is defined as

$$t^* \equiv t\left[\frac{g(\rho_L - \rho_v)\rho_L}{\mu_L^2}\right]^{1/3} = c_1 t \tag{8-26}$$

The liquid film Reynolds number is defined as

$$\mathrm{Re}_L \equiv \frac{4\Gamma}{\mu_L} \tag{8-27}$$

where Γ is the mass flux of liquid per unit length of channel and c, c_1, *and* m are empirical constants. The depth of penetration of the liquid film on the channel wall is approximated with the following expression:

$$h' = \xi\left(h + \frac{w}{2}\right) \tag{8-28}$$

where ξ is a revisory factor to account for the effect of vapor shear and surface tension forces on the liquid film. This expression in effect assumes that $b = (d/2)$ and that the bottom of the channel is dry and inactive. The latter assumption is contrary to the completely wetted channel walls observed by Arshad and Thome (1983).

At any height x from the bottom of the channel, the expression for the local liquid mass flux can be obtained from Eqs. (8-25) to (8-27) to be

$$\Gamma_x = \left(\frac{c_1}{c}\right)^{1/m} t_x^{1/m}\left(\frac{\mu_L}{4}\right) \tag{8-29}$$

The local heat flux for one-dimensional steady-state heat conduction through the film is

$$q_x = \left(\frac{\lambda_L}{t}\right)\Delta T \tag{8-30}$$

This expression is equivalent to Eq. (8-7). Performing a heat balance over the element dx between the heat conducted through the liquid to the liquid-vapor interface and the latent heat transported away gives

$$\Delta h_v \frac{d\Gamma_x}{dt} = \left(\frac{\lambda_L}{t}\right)\Delta T\ dx \tag{8-31}$$

and introducing Eq. (8-29) yields

$$\left[\Delta h_v\left(\frac{c_1}{c}\right)^{1/m}\left(\frac{\mu_L}{4m}\right)\right] t^{1/m}\ dt = \lambda_L \Delta T\ dx \tag{8-32}$$

Integrating this expression from $t = 0$ at $x = 0$ to $t = t_x$ at x gives the following expression for the local film thickness:

$$t_x = \left[\left(\frac{m+1}{m}\right)\left(\frac{\lambda_L \Delta T}{B_1 \Delta h_v}\right) x\right]^{\left(\frac{m}{m+1}\right)} \tag{8-33}$$

where

$$B_1 = \left(\frac{c_1}{c}\right)^{1/m}\left(\frac{\mu_L}{4m}\right) \tag{8-34}$$

The latent heat transferred per unit length of channel can now be obtained by integrating the local latent heat flux q_x over the entire height of the fin:

$$Q_{lat} = 2\int_0^{h'} q_x\, dx = 2\int_0^{h'} \frac{\Delta T \lambda_L}{t_x}\, dx \tag{8-35}$$

which yields

$$Q_{lat} = \frac{2\Delta T \lambda_L h' (1+m)}{t_{h'}} \tag{8-36}$$

Solving for $t_{h'}$ in Eq. (8-36) together with Eqs. (8-25) to (8-27) gives the liquid film thickness at a height h' as

$$t_{h'} = \frac{c}{c_1}\left[\frac{4\Gamma_{h'}}{\mu_L}\right]^m \tag{8-37}$$

where $\Gamma_{h'}$ denotes the mass flow per unit length at height h'. Because liquid can only enter the channel from the top slit opening and because no liquid is assumed to leave the channel, the liquid mass flow entering on the two channel walls at h' must be equal to that which is evaporated:

$$2\Gamma_{h'} = \frac{Q_{lat}}{\Delta h_v} = \frac{q_{lat}\, s}{\Delta h_v} \tag{8-38}$$

where q_{lat} is the latent heat flux per unit projected area. Substituting Eq. (8-38) into Eq. (8-37) gives

$$\left(\frac{\Delta T \lambda_L}{q_{lat}\, s}\right) t^* \xi = C'\left(\frac{2 q_{lat}\, s}{\Delta h_v \mu_L}\right)^m \tag{8-39}$$

where the dimensionless liquid film thickness t^* is

$$t^* = c_1\left(h + \frac{w}{2}\right) = \left[\frac{g(\rho_L - \rho_v)\rho_L}{\mu_L^2}\right]^{1/3}\left(h + \frac{w}{2}\right) \tag{8-40}$$

Rewritten with the Archimedes number Ar_d, t^* becomes

$$t^* = Ar_d^{1/3}\left(\frac{2h + w}{2d}\right) \tag{8-41}$$

The empirical constant C' is given as

$$C' = \frac{c}{2\,(m+1)} \tag{8-42}$$

and the Achimedes number is defined as

$$\mathrm{Ar}_d \equiv \frac{gd^3}{\nu_{\mathrm{L}}^2}\left(\frac{\rho_{\mathrm{L}} - \rho_{\mathrm{v}}}{\rho_{\mathrm{L}}}\right) \tag{8-43}$$

In the model, the influences of surface tension and vapor shear stresses on the liquid film have been included in the factor ξ. The shear stress imparted on the liquid film by the countercurrent flow of liquid and vapor influences the depth of penetration of the liquid film down the channel wall and also its thickness. The interfacial shear stress is directly proportional to the vapor velocity and hence is also inversely proportional to the heat flux. Therefore, the higher the heat flux, the lower the liquid penetration into the channel. This relationship coincides well with experimental results that show the Gewa-T boiling curve approaching that of a smooth surface at high heat fluxes, apparently because its channel walls dry out. The influence of surface tension instead is to draw liquid into the corners of the T fins, which tends to increase the liquid film's thickness. The interaction of these two forces was represented by the ratio

$$\xi \propto \left(\frac{\sigma/d}{\tau_{\mathrm{I}}}\right)^n \tag{8-44}$$

The interfacial shear stress is calculated from the following standard expression:

$$\tau_{\mathrm{I}} = \frac{1}{2}\,c_f \rho_{\mathrm{v}}\, u_{\mathrm{v}} \tag{8-45}$$

where c_f is the skin friction coefficient and u_{v} is the vapor velocity. The average vapor velocity in the slit of the channel can be determined from an energy balance to be

$$u_{\mathrm{v}} = \frac{q_{\mathrm{lat}}\, s}{\rho_{\mathrm{v}}\,\Delta h_{\mathrm{v}}\, d} \tag{8-46}$$

Thus the expression for the interfacial shear stress becomes

$$\tau_{\mathrm{I}} = \frac{1}{2}\,c_f \rho_{\mathrm{v}} \left(\frac{q_{\mathrm{lat}}\, s}{\rho_{\mathrm{v}}\,\Delta h_{\mathrm{v}}\, d}\right)^2 \tag{8-47}$$

Now, substituting Eqs. (8-44) and (8-47) into Eq. (8-39) and introducing a new empirical constant C'' gives the following expression:

$$\left(\frac{\Delta T \lambda_L}{q_{lat}\, s}\right) t^* = C'' \left(\frac{2 q_{lat}\, s}{\Delta h_v\, \mu_L}\right)^m \left(\frac{q_{lat}^2\, s^2}{\sigma \rho_v\, \Delta h_v^2\, d}\right)^n \tag{8-48}$$

The last term to the right is a special expression for the Weber number:

$$\mathrm{We} \equiv \frac{q_{lat}^2\, s^2}{\sigma \rho_v\, \Delta h_v^2 d} \tag{8-49}$$

At this point, an expression relationing the total heat flux from the enhanced boiling surface q to the latent heat flux q_{lat} is required. Xin and Chao (1985) noted that various studies have found the relationship to be close to linear. In addition, because the sensible heat flux is the difference between q and q_{lat}, the liquid Prandtl number affects the relationship. On the basis of these concepts, the following empirical expression was utilized:

$$q \propto q_{lat}\, \mathrm{Pr}_L^e \tag{8-50}$$

Substituting this relationship into Eq. (8-48), the heat transfer correlating expression is obtained:

$$\mathrm{Nu} = C_1'' \left(\frac{2h+w}{2d}\right) \mathrm{Ar}_d^{1/3}\, \mathrm{Re}_L^{m_1}\ \mathrm{We}^{n_1}\ \mathrm{Pr}_L^{g_1} \tag{8-51}$$

where the empirical constants are determined with experimental data. The Nusselt number is defined as

$$\mathrm{Nu} \equiv \frac{qs}{\Delta T\, \lambda_L} = \frac{\alpha\, s}{\lambda_L} \tag{8-52}$$

with the fin pitch as the characteristic dimension. The Reynolds number can now be defined in the following convenient form:

$$\mathrm{Re}_L \equiv \frac{2\, qs}{\Delta h_v \mu_L} \tag{8-53}$$

Experimental data were obtained at atmospheric pressure for ten different copper Gewa-T type surfaces with ethanol and water and for six Gewa-T types surfaces for R-113. Their structural dimensions were varied as follows: the slit width d was varied incrementally from 0.09 to 0.24 mm; two channel widths w were used, 0.6 and 0.8 mm; two fin pitches s were tested, 0.8 and 1.2 mm; and two channel heights h were studied, 0.5 and 1.0 mm. The optimum test surface was that with a slit width of 0.11 mm, a channel width of 0.6 mm, a fin pitch of 0.8 mm, and a channel height of 0.5 mm, giving the best overall results for all three test fluids. Using regression analysis on Eq. (8-51), the investigators obtained the following enhanced boiling correlation for Gewa-T type surfaces:

$$\mathrm{Nu} = 3.76\left(\frac{2h+w}{2d}\right)\mathrm{Ar}_d^{1/3}\,\mathrm{Re}_\mathrm{L}^{-0.15}\,\mathrm{We}^{0.29}\,\mathrm{Pr}_\mathrm{L}^{0.76} \tag{8-54}$$

This expression correlated with the experimental data to within about 30%. The ranges of the dimensionless groups in the data were as follows:

$0.016 < \mathrm{Re}_\mathrm{L} < 7.03$
$0.0000014 < \mathrm{We} < 0.17$
$1.76 < \mathrm{Pr}_\mathrm{L} < 7.86.$

A check of this correlation has been performed here with independently obtained boiling data for Gewa-T tubes (not Gewa-T type plates). The channels of actual Gewa-T tubes are not rectangular in cross-section as in their model (Fig. 8-3) but instead have rounded corners that produce a rather circular channel cross-section as shown in cutaway photographs in Marto and Lepere (1981) and Marto and Hernandez (1983). This presents a problem for selecting the dimensions to use for h and w.

Table 8-1 depicts the comparison of predictions from Eq. (8-54) to representative boiling data for the fluids R-11, R-12, R-22, R-113, isopropyl alcohol, and p-xylene over the pressure range 0.7 to 5.0 bar and heat fluxes from 1 to 400 kW/m^2. The channel dimensions for the Stephan and Mitrovic (1981) Gewa-T tube were taken as follows:

Table 8-1 Comparison of Xin and Chao (1985) Gewa-T correlation with selected data

Fluid	Pressure, bar	q,kW/m^2	ΔT_{exp},K	ΔT_{pred},K	Error, %
R-11*	0.7	2.0	3.1	0.8	−74
	0.7	20.0	3.8	3.1	−18
R-12*	3.1	1.0	0.7	0.8	+13
	3.1	20.0	2.9	4.6	+58
R-22*	5.0	2.0	3.3	1.3	−61
	5.0	20.0	4.7	2.5	+88
R-113†	1.0	10.0	3.4	2.0	−41
	1.0	100.0	7.0	9.3	+33
R-113‡	1.0	10.0	4.0	2.5	−37
	1.0	100.0	9.4	9.3	−1
Isopropanol§	1.0	50.0	5.2	4.1	−21
	1.0	100.0	7.8	6.1	−22
	1.0	400.0	16.0	12.8	−20
p-Xylene#	1.0	60.0	5.8	5.7	−2
	1.0	400.0	22.5	16.8	−25

*Stephan and Mitrovic (1981)
†Marto and Lepere (1981)
‡Marto and Hernandez (1983)
§Yilmaz and Westwater (1981)
#Yilmaz, Hwalek, and Westwater (1980).

$s = 1.35$ mm, $d = 0.25$ mm, $w = 0.95$ mm, and $h = 0.70$ mm, where the widest width of the channel was used for w and h was obtained from the measured height of the T fin minus its thickness (b in Fig. 8-3). In Marto and Lepere (1981) the dimensions were $s = 1.35$ mm, $d = 0.18$ mm, $w = 0.95$ mm, and $h = 0.60$ mm. The Marto and Hernandez (1983), Yilmaz and Westwater (1981), and Yilmaz, Hwalek, and Westwater (1980) dimensions were the same as those of Stephan and Mitrovic (1981). The Xin and Chao (1985) correlation was found to perform most accurately for the nonrefrigerants.

The Xin and Chao (1985) enhanced boiling model represents an advance over the two-resistance and liquid pumping models described above (Secs. 8-2 and 8-3) because it models both the thickness and the surface area of the thin evaporating liquid film. No effort was apparently made, however, to determine whether the predicted film thickness at the slit was actually less than half the width of the slit opening d of their test surfaces.

Xin and Chao Thermoexcel-E Correlation

The above model for the Gewa-T geometry was extended to the Thermoexcel-E geometry by making several modifications. The dimensions of the Thermoexcel-E surface are characterized by the channel width w, the channel height h, the fin pitch s, the pore diameter D, and the pore spacing along the channel r. The heat transfer mechanism was assumed to be similar to that of the T-shaped finned surface, except that it was recognized that the thickness of the liquid film also varies along the length of the channel because liquid can enter only at the pores. For simplification of the model, the average thickness of the liquid film at an arbitrary channel height was utilized to correct for the pore spacing effect. The countercurrent two-phase flow model described above was then applied to the Thermoexcel-E surface by modifying the revisory factor in Eq. (8-28).

For the Thermoexcel-E channel, the average velocity of vapor flow at the pore was obtained by a heat balance similar to that used to derive Eq. (8-46). This gives u_v as

$$u_v = \frac{4\, q_{lat}\, sr}{\pi D^2\, \Delta h_v \rho_v} \tag{8-55}$$

with the pore diameter D as the characteristic dimension, the revisory factor was expressed as

$$\xi \propto \left(\frac{\sigma / D}{\tau_I} \right)^k \tag{8-56}$$

or

$$\xi \propto \left(\frac{D^3\, \Delta h_v\, \rho_v \sigma}{q_{lat}^2\, s^2\, r^2} \right)^k \tag{8-57}$$

Then, following the same derivation as for the Gewa-T type surface, a correlating expression similar to Eq. (8-48) was obtained. The empirical constants were determined with water and R-11 data at 1.0 bar taken by Nakayama et al. (1979) for flat Thermoexcel-E surfaces. Curiously, the liquid nitrogen data in the same study were not included. The resulting Thermoexcel-E correlating expression is

$$\mathrm{Nu} = 6.02\left(\frac{2h + w}{2D}\right)\mathrm{Ar}_D^{1/3}\,\mathrm{Re}_\mathrm{L}^{0.51}\,\mathrm{We}_D^{-0.08}\,\mathrm{Pr}_\mathrm{L}^{-0.34} \tag{8-58}$$

where the Nusselt and Reynolds numbers are defined by Eqs. (8-52) and (8-53), respectively, and the slit width d in the Archimedes number is replaced with the pore diameter D to give

$$\mathrm{Ar}_D \equiv \frac{gD^3}{\nu_\mathrm{L}^2}\left(\frac{\rho_\mathrm{L} - \rho_\mathrm{v}}{\rho_\mathrm{L}}\right) \tag{8-59}$$

The slit width d in the Weber number is replaced with the ratio D^3/r^2 to define We_D as

$$\mathrm{We}_D \equiv \frac{q^2 s^2 r^2}{D^3\,\Delta h_\mathrm{v}^2\,\rho_\mathrm{v}\,\sigma} \tag{8-60}$$

Most of the water and R-11 data were able to be predicted to within 20% by the correlation. It was not possible to compare this correlation against the other Thermoexcel-E data sets shown in Chapter Seven because the values of the pore spacing r are not available. Commercial versions of the Thermoexcel-E tube do not have channels with rectangular cross-sections according to the schematic representation in Yilmaz and Westwater (1981), nor are the pores circular in shape (they are approximately triangular). The correlation could still be applied, however, with the use of the wetted perimeter of the channel instead of $(2h + w)$ and the nominal diameter of the triangular pores for D.

Comparison of Eqs. (8-54) and (8-58) shows that the dependencies of the Nusselt number on the Reynolds, Weber, and Prandtl numbers are completely different. For instance, the Nusselt number changes from being directly proportional to the Prandtl number in the first correlation to being inversely proportional in the second. Also, it is unlikely that the liquid and vapor could flow countercurrently contemporaneously through the small pores, as assumed by the model.

8-5 CONCLUSIONS

Correlations have been developed for several enhanced boiling geometries. Much work can still be done to improve the modeling of the enhanced boiling process, however. It appears that the trend is to develop a different model and correlating expression for each separate type of enhanced boiling geometry.

8-6 NOMENCLATURE

A	area (m^2)
Ar_d	Archimedes number based on d
Ar_D	Archimedes number based on D
b	boiling curve coefficient in Eq. (8-1)
b	T-fin thickness in Fig. 8-3 (m)
B	geometric factor
B_1	liquid film factor
c	empirical constant
c_f	skin friction coefficient
c_1	empirical constant
C	empirical constant in Eq. (8-3)
C	empirical constant in Eq. (8-23)
C'	empirical constant
C''	empirical constant
C_1''	empirical constant
d	slit width (m)
d_p	particle diameter (m)
D	pore diameter (m)
e	empirical exponent
F_c	mixture boiling correction factor
F_e	fin efficiency
F_s	nucleate boiling surface correction factor
g	gravitational acceleration (m/s^2)
g_1	empirical exponent
h	height of T-fin wall (m)
h'	height of liquid film penetration (m)
Δh_v	latent heat of vaporization (J/kg)
H	height of fin (m)
k	empirical constant
m	empirical constant
m_{LI}	mass of evaporating liquid during phase I (kg)
m_{LII}	mass of evaporating liquid during phase II (kg)
m_1, m_2, m_3	empirical constants in Eq. (8-3)
n, n_1	empirical constants
N/A	boiling site density (site/m^2)
Nu	Nusselt number
P_r	reduced pressure
Pr_L	liquid Prandtl number
$(dP/dT)_{sat}$	slope of vapor pressure curve [$N/(m^2 \cdot K)$]
q	heat flux (W/m^2)
q_{ec}	external convective heat flux (W/m^2)
q_{lat}	latent heat flux (W/m^2)
q_r	reference heat flux (W/m^2)

q_x	heat flux at position x (W/m^2)
Q_{lat}	latent heat flow (W)
r	pore spacing (m)
R	radius of vapor nucleus or pore (m)
R^*	optimum pore radius (m)
R_p	particle radius (m)
Re_L	liquid Prandtl number
s	fin pitch (m)
S	surface area of liquid film (m^2)
t	liquid film thickness (m)
t^*	dimensionless liquid film thickness
$t_{h'}$	liquid film thickness at h' (m)
t_p	porous layer thickness (m)
t_x	liquid film thickness at position x (m)
T	thickness of fin (m) in Fig. 8-1.
T_I	interfacial temperature (K)
T_w	wall temperature (K)
ΔT	wall superheat (K)
ΔT_{exp}	experimentally measured wall superheat (K)
ΔT_{film}	film temperature difference (K)
ΔT_{pred}	predicted wall superheat (K)
ΔT_{sat}	boiling nucleation superheat (K)
u_v	vapor velocity (m/s)
V	volume of porous matrix (m^3)
w	width of channel (m)
We	Weber number
We_D	Weber number based on D
x	distance from bottom of liquid layer (m)
x,y	empirical constants in Eq. (8-23)
α	heat transfer coefficient [W/(m^2·K)]
α_f	fin boiling heat transfer coefficient [W/(m^2·K)]
α_{film}	liquid film heat transfer coefficient [W/(m^2·K)]
α_{max}	maximum enhanced boiling heat transfer coefficient [W/(m^2·K)]
α_{nb}	nucleate pool boiling heat transfer coefficient [W/(m^2·K)]
α_{nc}	natural convective heat transfer coefficient [W/(m^2·K)]
Γ	mass flow rate per unit length [kg/(m·s)]
$\Gamma_{h'}$	mass flow rate per unit length at h' [kg/(m·s)]
Γ_x	mass flow rate per unit length at position x [kg/(m·s)]
ε	porosity of porous matrix
θ_I	time period of phase I (s)
θ_{II}	time period of phase II (s)
λ_f	thermal conductivity of fin [W/(m·K)]
λ_L	liquid thermal conductivity [W/(m·K)]
λ_m	thermal conductivity of porous matrix [W/(m·K)]
λ_p	thermal conductivity of particles [W/(m·K)]

μ_L	dynamic viscosity of liquid (N·s/m)
μ_v	dynamic viscosity of vapor (N·s/m)
ν_L	kinematic viscosity of liquid (m^2/s)
ξ	revisory factor
ρ_L	liquid density (kg/m^3)
ρ_v	vapor density (kg/m^3)
σ	surface tension (N/m)
τ_I	interfacial shear stress (N/m^2)

REFERENCES

Arshad, J., and J. R. Thome. 1983. Enhanced boiling surfaces: Heat transfer mechanism and mixture boiling. *Proc. ASME-JSME Therm. Eng. Joint Conf.* 1:191–97.

Bell, K. J., and A. C. Mueller. 1984. *Wolverine engineering data book II,* Sec. 5, 43. Decatur, Ala.: Wolverine Tube.

Cash, D. N., G. J. Klein, and J. W. Westwater. 1971. Approximate optimum fin design for boiling heat transfer. *J. Heat Transfer* 93:19–24.

Chen, S., and G. L. Zyskowski. 1963. *Steady-state heat conduction in a straight fin with variable film coefficient* (ASME paper 63-HT-12).

Cumo, M., S. Lopez, and G. C. Pinchera. 1965. Numerical calculation of extended surface efficiencies. *Chem. Eng. Progr. Symp. Serv.* 61(59):225.

Czikk, A. M., and P. S. O'Neill. 1979. Correlation of nucleate boiling from porous metal films. In *Advances in Enhanced Heat Transfer*, 53–60. New York: American Society of Mechanical Engineers.

Gottzmann, C. F., J. B. Wolf, and P. S. O'Neill. 1971. Theory and application of high performance boiling surfaces to components of absorption cycle air conditioners. *Proc. Conf. Natl. Gas Res. Technol.* Session V, Paper 3, February 28.

Haley, K. W., and J. W. Westwater. 1965. Heat transfer from a fin to a boiling liquid. *Chem. Eng. Sci.* 20:711.

Han, L. S., and S. G. Leftowitz. 1960. *Constant cross-section fin efficiencies for non-uniform surface heat transfer coefficients* (ASME paper 60-WA-41).

Jho, S. G., R. Simada, S. Kumagai, T. Takeyama, and M. Izumi. 1985. Theoretical study on boiling heat transfer from finned tube bundles. *Heat Transfer Jap. Res.* 14(4): 44–59.

Klein, G. J., and J. W. Westwater. 1971. Heat transfer from multiple splines to boiling liquids. *AIChE J.* 17:1050–56.

Marto, P. J., and B. Hernandez. 1983. Nucleate pool boiling characteristics of a Gewa-T surface in Freon-113. *AIChE Symp. Ser.* 79(225):1–10.

Marto, P. J., and J. Lepere. 1981. Pool boiling heat transfer from enhanced surfaces to dielectric fluids. In *Advances in Enhanced Heat Transfer*, HTD Vol. 18, 93–102. New York: American Society of Mechanical Engineers.

Milton, R. M. 1968. Heat exchanger system. U. S. Patent 3,384,154, May 21.

Mostinski, I. L. 1963. Application of the rule of corresponding states for the calculation of heat transfer and critical heat flux. *Teploenergetika* 4:66.

Nakayama, W., T. Daikoku, H. Kuwahara, and T. Nakajima. 1979. Dynamic model of enhanced boiling heat transfer on porous surfaces. In *Advances in Enhanced Heat Transfer*, 31–43. New York: American Society of Mechanical Engineers.

Nishikawa, K., and T. Ito. 1982. Augmentation of nucleate boiling heat transfer by prepared surfaces. In *Heat transfer in energy problems,* ed. T. Mizushina and W. J. Yang, 111–18. Washington, D.C.: Hemisphere.

O'Neill, P. S., C. F. Gottzmann, and J. W. Terbot. 1972. Novel heat exchanger increases cascade cycle efficiency for natural gas liquefaction. *Adv Cryog. Eng.* 17:420–37.

Palen, J. W., J. Taborek, and S. Yilmaz. 1986. Comments to the application of enhanced boiling surfaces in tube bundles. In *Heat exchanger sourcebook,* ed. J. W. Palen, 663–73. Washington, D.C.: Hemisphere.

Palen, J. W., and C. C. Yang. 1983. Circulation boiling model for analysis of kettle and internal reboiler performance. In *Heat exchangers for two-phase applications,* HTD vol. 27, 55–61. New York: American Society of Mechanical Engineers.

Shakir, S., J. R. Thome, and J. R. Lloyd. 1985. Boiling of methanol-water mixtures on smooth and enhanced surfaces. In *Multiphase flow and heat transfer,* ed. V. K. Dhir, J. C. Chen, and O. C. Jones. HTD vol. 47, 1–6. New York: American Society of Mechanical Engineers.

Stephan, K., and J. Mitrovic. 1981. Heat transfer in natural convective boiling of refrigerants and refrigerant-oil mixtures in bundles of T-shaped finned tubes. In *Advances in Enhanced Heat Transfer*, HTD Vol. 18, 131–46. New York: American Society of Mechanical Engineers.

Wallis, G. B. 1969. *One-Dimensional Two-Phase Flow*. New York: McGraw-Hill.

Webb, R. L. 1983. Nucleate boiling on porous coated surfaces. *Heat Transfer Eng*. 4(3–4):71–82.

Xin, M., and Y. Chao. 1985. Analysis and experiment of boiling heat transfer on T-shaped finned surfaces. Paper read at 23rd National Heat Transfer Conference (Enhanced heat transfer equipment session), August 4–7, Denver.

Yilmaz, S., J. J. Hwalek, and J. W. Westwater. 1980. *Pool boiling heat transfer performance for commercial enhanced tube surfaces* (ASME paper 80-HT-41).

Yilmaz, S., and J. W. Palen. 1984. *Performance of finned tube reboilers in hydrocarbon service* (ASME paper 84-HT-91).

Yilmaz, S., and J. W. Westwater. 1981. Effect of commercial enhanced surfaces on the boiling heat transfer curve. In *Advances in Enhanced Heat Transfer*, HTD Vol. 18, 73–91. New York: American Society of Mechanical Engineers.

CHAPTER

NINE

ENHANCED BOILING OF MIXTURES

An important area of application of enhanced boiling surfaces is the evaporation of binary and multicomponent mixtures on the shell side of horizontal reboilers and on the tube side of vertical thermosyphon reboilers. Also, the application of enhanced boiling tubes to refrigeration systems can involve evaporation of refrigerant-oil mixtures caused by leakage of the compressor lubricant into the refrigeration fluid. Enhanced boiling in mixtures differs significantly from that in pure fluids because of the effect of mass diffusion on the evaporation and nucleation processes. Therefore, the subject is surveyed separately in this chapter.

9-1 VAPOR-LIQUID PHASE EQUILIBRIA

A working knowledge of the principles of vapor-liquid phase equilibrium is a prerequisite to the understanding of mixture boiling. Only a brief review of the subject can be presented here, however, to cover the more basic ideas.

Phase equilibria for binary mixtures are represented on phase diagrams. Figure 9-1 depicts a phase equilibrium diagram at a constant pressure for an ideal binary mixture system. The saturation temperature T_{sat} is plotted along the ordinate, and the vapor and liquid mole fractions of the more volatile component (also sometimes referred to as the lighter component) are plotted along the abscissa. The more volatile component is defined as the fluid that has the lower boiling point temperature. The other fluid is referred to as the less volatile component. The dew point line represents the locus of the dew point temperatures, which is the temperature at which a superheated mixture of vapor will first begin to condense on cooling. The bubble point line defines the locus

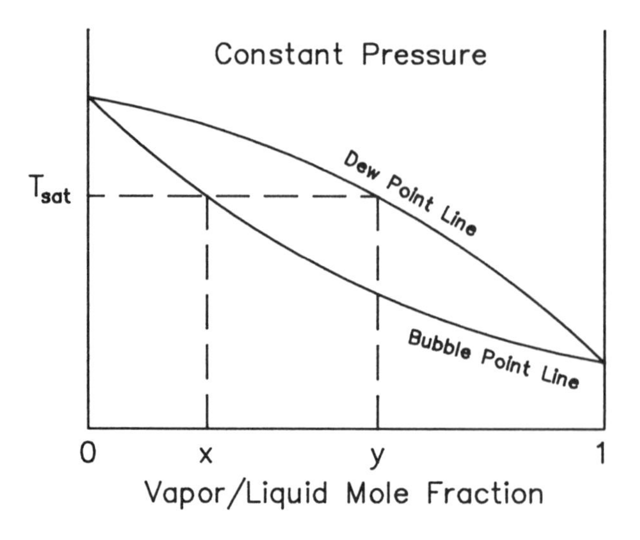

Figure 9-1 Phase equilibrium diagram.

of the bubble point temperatures, which is the temperature at which the first vapor will be formed on heating a subcooled liquid mixture. The equilibrium vapor mole fraction of the more volatile component y is always greater than the corresponding liquid mole fraction x, except at the two extremes.

Consider now a subcooled liquid mixture of composition x enclosed by a frictionless piston in a cylindrical vessel. On heating in a quasi-equilibrium isobaric process, the subcooled liquid is brought to its bubble point temperature. Additional heating causes the initial vapor phase to form with a composition equal to y, which represents the equilibrium vapor mole fraction. As heating proceeds, the liquid composition x of the more volatile component decreases as its molecules preferentially enter the vapor phase. Hence the bubble point temperature rises correspondingly along the bubble point line, and the vapor composition y in the vapor phase is also reduced. Eventually, if heating continues, the last drop of liquid will evaporate, with the vapor's equilibrium composition corresponding to that of the original subcooled liquid, and the bubble point temperature will have reached its maximum value. In contrast, a pure fluid at a given pressure evaporates at a fixed saturation temperature. This example serves to illustrate the important thermodynamic differences between mixture boiling and the boiling of a pure fluid, namely that the local saturation temperature and composition at a given pressure are not fixed when a mixture evaporates but depend in part on the heat and mass transfer processes themselves.

Some vapor-liquid mixture systems form an azeotrope at an intermediate composition. Figure 9-2 illustrates such a phase equilibrium diagram for a binary mixture system. At the azeotrope the liquid and vapor compositions are identical, and hence the fluid behaves like a pure component. The boiling range is also depicted in Fig. 9-2. The boiling range is here defined as the temperature difference between the dew point and the bubble point temperatures at the same composition. For an azeotropic binary mixture system, the boiling range goes through a maximum on both sides of the azeotrope.

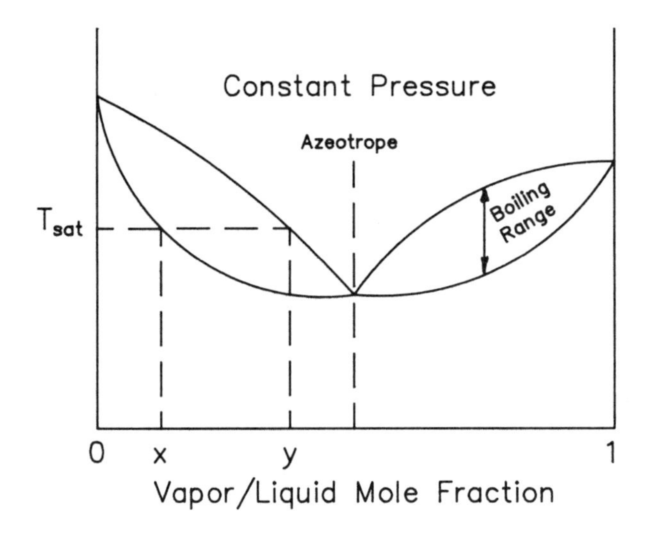

Figure 9-2 Phase diagram of an azeotropic mixture system.

The methods for predicting phase equilibria of vapor-liquid mixture systems are beyond the scope of the present review. The interested reader is referred to Prausnitz (1969) or Fredenslund, Gmehling, and Rasmussen (1979) for details. For phase equilibrium data for specific mixture systems, the reader is directed to Kirschbaum (1948), Chu (1950), Hala et al. (1968), and the DACHEMA data in Gmehling et al. (1984).

9-2 MIXTURE PHYSICAL PROPERTIES

Mixture physical properties have a large influence on the processes of heat and mass transfer in the boiling of mixtures. Sometimes only a small change in liquid composition can result in a large change in properties, especially surface tension, latent heat of vaporization, and viscosity. Because boiling nucleation, bubble growth and departure, and other boiling phenomena are particularly sensitive to changes in these properties, methods for accurately predicting the variations in physical properties with composition and temperature are essential to the study of enhanced boiling of mixtures. These predictive methods, however, are beyond the scope of this review. Further information can be sought in Reid, Prausnitz, and Poling (1987).

As an example of large nonlinear variations in properties with composition, some properties for the aqueous mixture system of acetone-water at 1.01 bar are shown in Fig. 9-3 (taken from Uhlig and Thome [1985]). These properties are for equilibrium conditions, so that they are not only a function of composition but also include the effect of the variation in the bubble point temperature with composition at this pressure. The solid curves represent the predicted variations, and the dashed lines depict idealized linear molar interpolations between the pure component values.

The differential latent heat of vaporization is shown in Fig. 9-3 (*a*), which is defined as the vapor enthalpy at the equilibrium composition y (corresponding to the liquid composition x) minus the liquid enthalpy at composition x. This form of latent

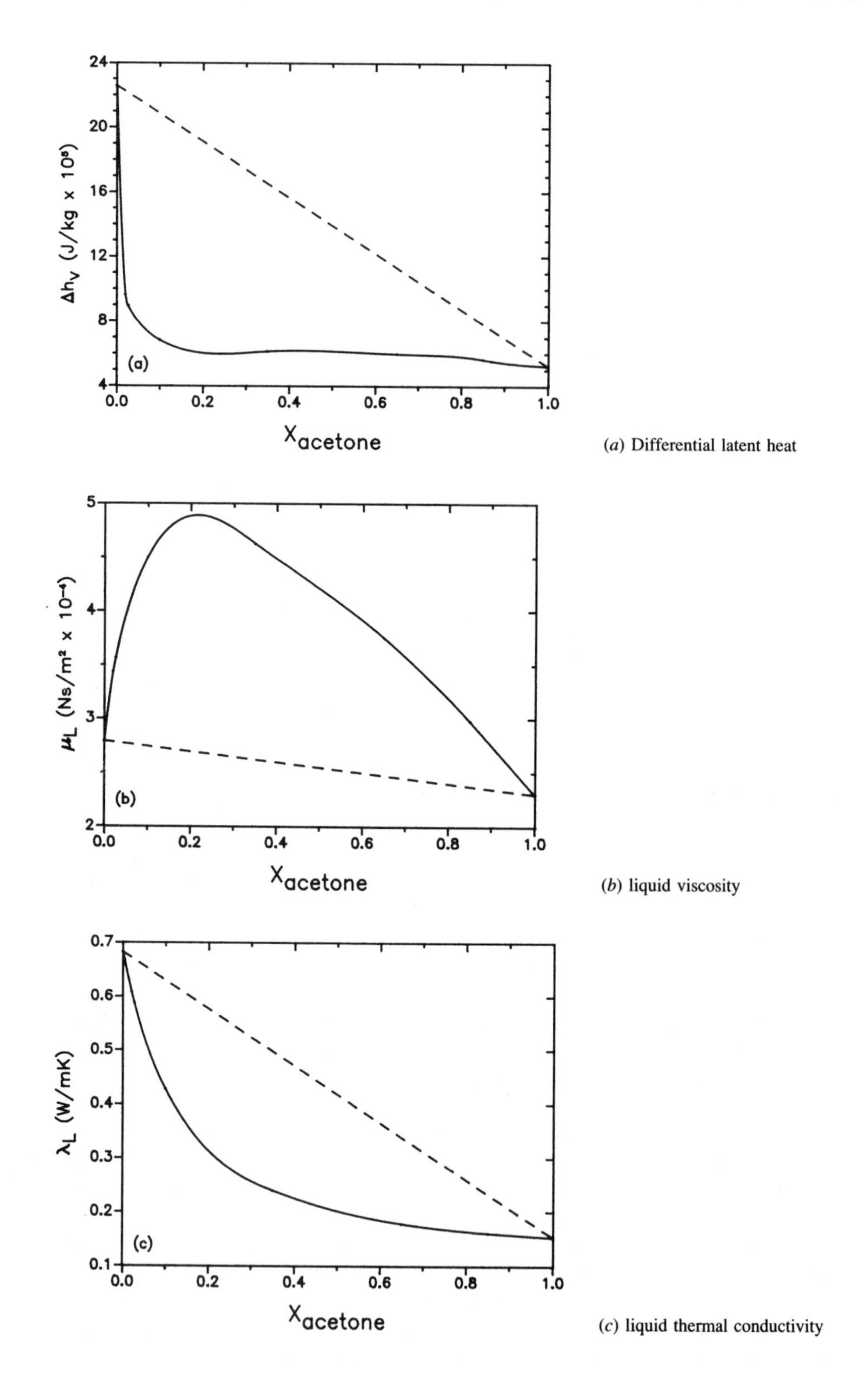

(*a*) Differential latent heat

(*b*) liquid viscosity

(*c*) liquid thermal conductivity

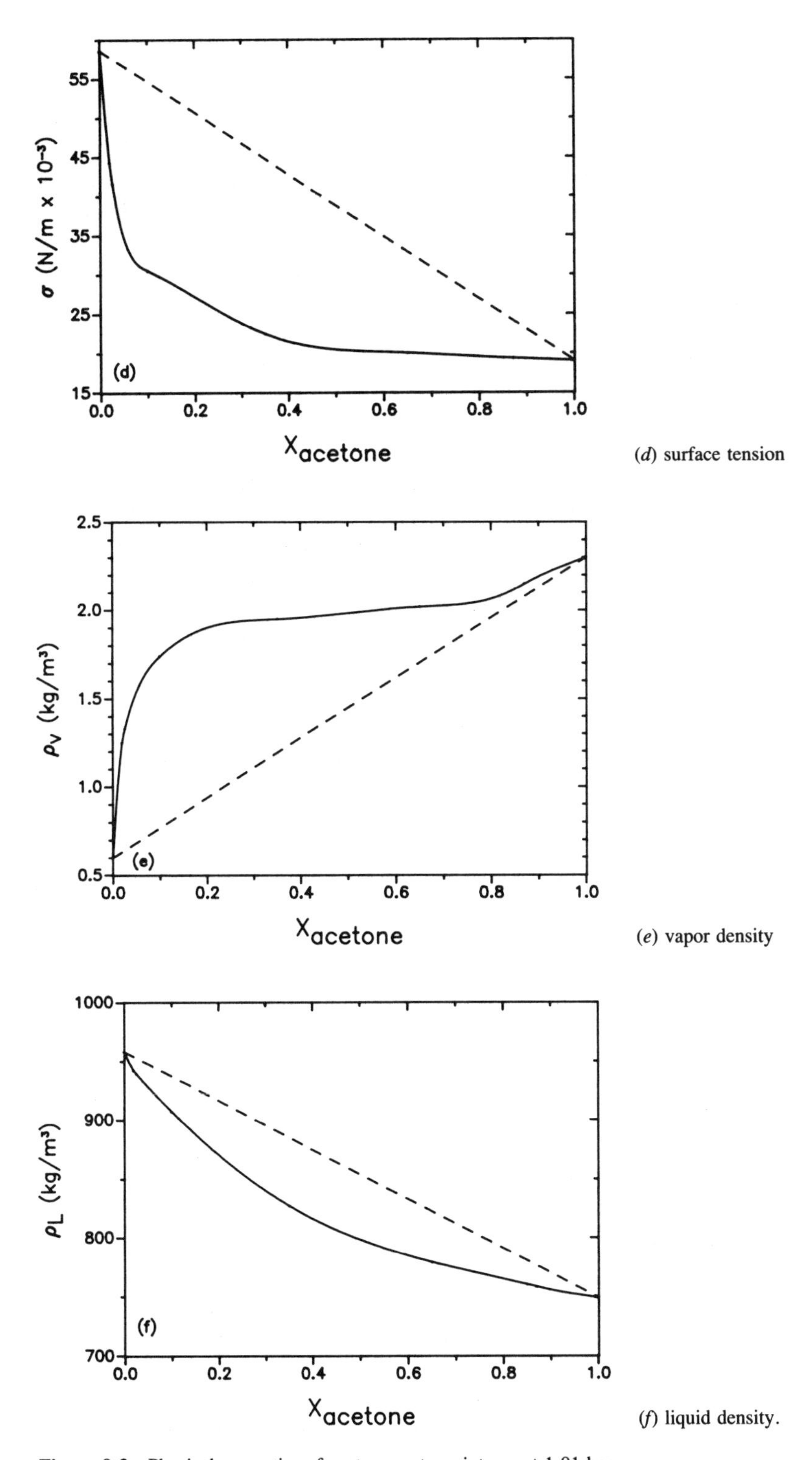

(*d*) surface tension

(*e*) vapor density

(*f*) liquid density.

Figure 9-3 Physical properties of acetone-water mixtures at 1.01 bar.

heat of a mixture is utilized rather than the integral latent heat, which is the enthalpy difference evaluated at identical vapor and liquid compositions. The differential values are used in predicting mixture boiling heat transfer coefficients because the evaporation process at a growing bubble is thought to follow phase equilibrium, and hence the vapor formed corresponds to the equilibrium composition. The disparity between differential and integral values can be substantial when one pure component has a much larger latent heat than the other, as in the acetone-water system. The latent heat is observed to decrease substantially with a small addition of acetone to water.

The vapor density of a mixture [Fig. 9-3 (*e*)] is also calculated with the equilibrium vapor composition (not the liquid composition). Hence it is nonlinear when plotted against the liquid mole fraction.

Turning to the other fluid properties in Fig. 9-3, one sees that the liquid dynamic viscosity [Fig. 9-3(*b*)] passes through a large maximum at a liquid mole fraction of about 0.1. The liquid thermal conductivity [Fig. 9-3(*c*)] deviates negatively from a linear molar interpolation between the pure component values. The surface tension [Fig. 9-3(*d*)] drops drastically on addition of a small amount of acetone to water. Only the liquid density [Fig. 9-3(*f*)] varies fairly linearly with composition.

A physical property important to boiling nucleation in mixtures is the contact angle β. Figure 9-4 depicts the contact angle as a function of liquid mole fraction for a binary mixture system of methanol-water measured under various test conditions. The values of Shakir and Thome (1986) were obtained at 25° C against nitrogen gas on brass and copper surfaces. The values of Ponter and Peier (1978) are for equilibrium conditions at 1.01 bar on graphite surfaces and for total mass flux conditions at the same pressure. The equilibrium values on graphite are consistently lower than those obtained in the presence of nitrogen gas on brass and copper. The total reflux values are lower still. Thus care must be exercised when choosing the test conditions most appropriate to the circumstances when contact angles are used to interpret enhanced boiling nucleation results for mixtures.

The nonlinear variations in physical and transport properties with composition shown in Fig. 9-3 are typical of aqueous mixture systems but are also representative of those that can occur for other types of mixture systems. Therefore, the prediction of the mixture properties to use in a low-finned tube boiling correlation, for instance, should be done with care, as is the case for predicting the phase equilibria of mixtures.

9-3 BOILING NUCLEATION

Boiling nucleation in pure fluids on enhanced boiling surfaces is discussed in Chapter Five, and nucleation in reentrant cavities is reviewed in Chapter Three. Therefore, only the additional effects particular to mixtures are surveyed here. Thome, Shakir, and Mercier (1982) and Shakir and Thome (1986) experimentally studied the problem of boiling nucleation of mixtures under pool boiling conditions on conventional, smooth surfaces. Shock (1977) and Toral, Kenning, and Shock (1982) reported experimental results for the onset of nucleate boiling of mixtures evaporating inside plain, smooth tubes. Much of this work is summarized in Thome and Shock (1984), and interested

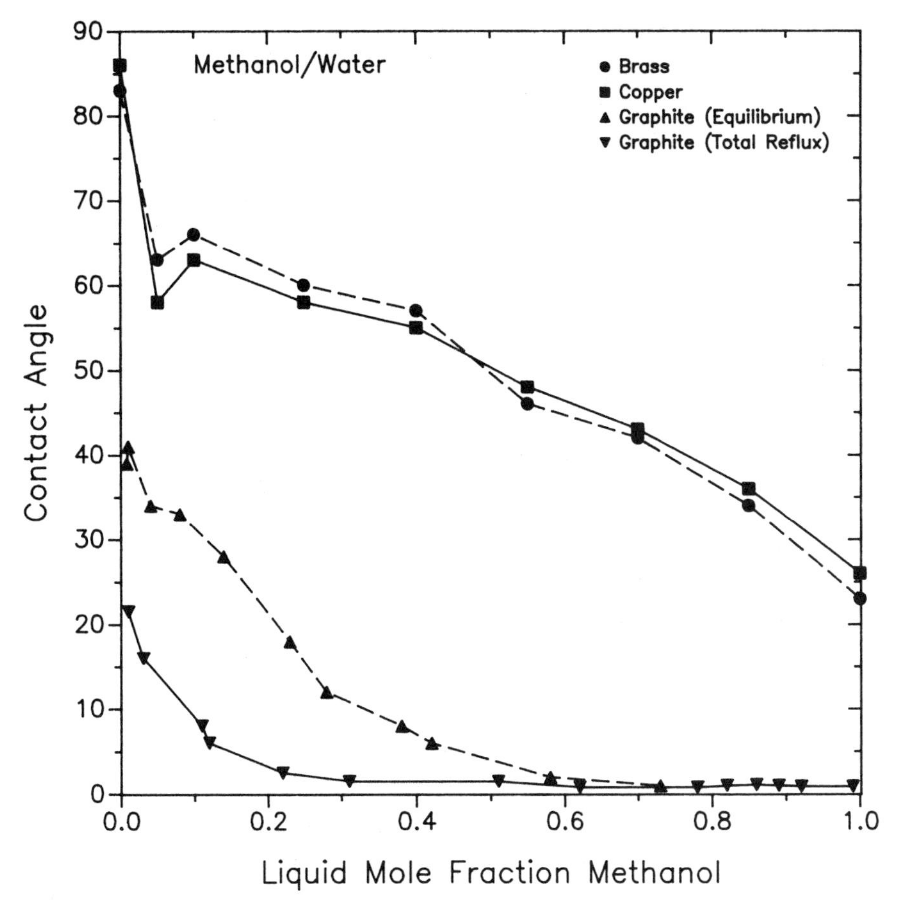

Figure 9-4 Contact angles of methanol-water mixtures.

readers are referred there for information about boiling nucleation in mixtures on conventional surfaces.

Ali and Thome (1984) were apparently the first to investigate systematically the mixture effect on boiling nucleation for enhanced surfaces as a function of composition. Figure 9-5 shows their data for the variation in the wall superheat required to initiate boiling on a commercial, copper High Flux tube for ethanol-benzene mixtures at 1.0 bar compared to data for several smooth surfaces. The test conditions were as follows: (1) the surface was preboiled at a large heat flux; (2) the surface was then allowed to cool to the bulk liquid temperature, which was at saturation conditions at 1.0 bar; and (3) the heat flux was then reapplied (by means of a cartridge heater) in several fixed steps until boiling initiated. This procedure was repeated identically for each mixture composition tested to obtain only the mixture effect on boiling nucleation and to eliminate any effects of subcooling and other factors. The activation superheat of the High Flux tube was found to be a function of composition, deviating positively from

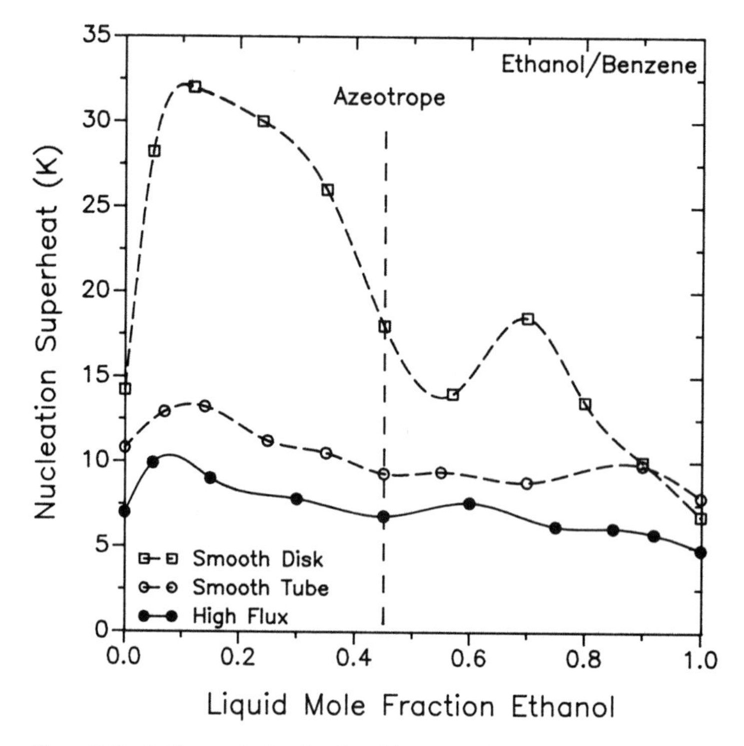

Figure 9-5 Boiling nucleation in ethanol-benzene mixtures.

a linear interpolation between the pure fluid values and the azeotrope value. The High Flux tube had the advantage of nucleating at lower superheats than the smooth surfaces, however. Yet the results demonstrated that nucleation superheats for pure fluids cannot be relied on to characterize values for their mixtures.

For the ethanol-benzene mixture system at 1.0 bar, the physical properties pertinent to nucleation—the surface tension, the slope of the vapor pressure–temperature curve, and the contact angle—vary nearly linearly with composition. Thus evaluating the boiling nucleation superheat expression, Eq. (2-1) does not predict the maximum observed experimentally at either side of the azeotrope, nor do the wetting characteristics suggest that one should occur. Shakir and Thome (1986) attributed the positive deviation in the nucleation superheat for the plain and enhanced surfaces to be the result of supersaturation and mass diffusion effects on trapped vapor nuclei. This process is described below.

When a vapor bubble grows in a liquid mixture, the more volatile component preferentially evaporates to provide the extra more volative component in the vapor, $(y - x)$, required to maintain phase equilibrium. Thus the more volatile component in the surrounding liquid is partially depleted and diffuses to the bubble interface through the diffusion shell formed. Therefore, the local liquid mole fraction of the more volatile component at the interface is reduced together with its local equilibrium vapor mole fraction, as depicted in Fig. 9-6 for the ethanol-benzene system at 1.0 bar. The liquid

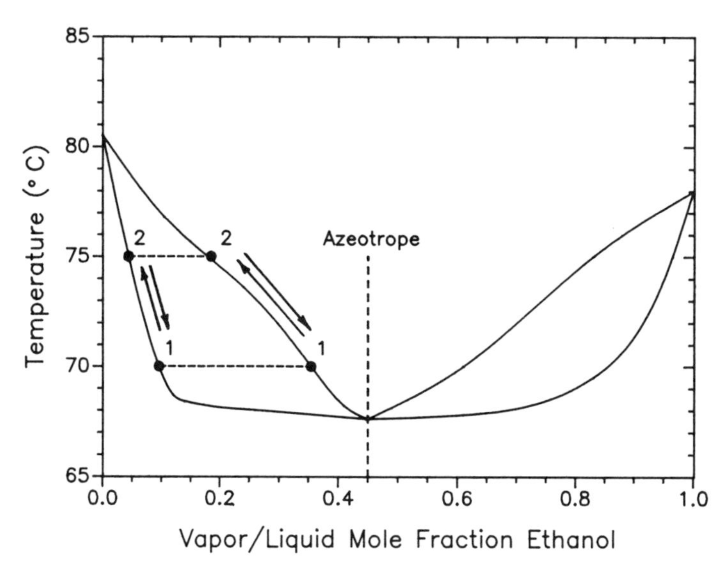

Figure 9-6 Ethanol-benzene phase diagram showing the effect of supersaturation.

mole fraction decreases from its bulk value (1) toward its minimum local value (2) during the bubble growth process. Then, when the heat to the surface is shut off, the liquid composition at the surface returns to the bulk value (1). Consequently, the vapor nucleus remaining in the cavity is supersaturated with the less volatile component (with respect to its equilibrium composition), and this component partially condenses to bring the system back to equilibrium. The volume and radius of the vapor nucleus are thus reduced, which increases the nucleation superheat required to activate the nucleus during subsequent heating. This process does not occur in pure fluids or at the azeotrope but is similar in many respects to the effect of subcooling on a vapor nucleus in a pure fluid.

Shakir and Thome (1986) also measured boiling activation and deactivation wall superheats for the mixture systems ethanol-water, methanol-water, and *n*-propanol–water for the same High Flux and smooth copper tubes. The mixture activation superheats for the High Flux tube were larger than those at its deactivation of boiling, and its activation and deactivation superheats were always less than those for the smooth tube. The supersaturation process described above, however, cannot explain some of the trends observed in the aqueous mixture activation superheats, and thus other factors must also affect the nucleation process.

9-4 BOILING HEAT TRANSFER IN MIXTURES

Pool boiling heat transfer of mixtures on conventional, smooth surfaces has been studied extensively over the years. The subject has been thoroughly reviewed by Thome and

Shock (1984). Therefore, only a brief summary of mixture effects on conventional boiling is presented here to provide some background before discussing the enhanced boiling of mixtures. Mixture boiling results for enhanced surfaces are then surveyed and other factors affecting the enhanced boiling of mixtures discussed (nucleate pool boiling of refrigerant-oil mixtures on enhanced boiling surfaces is reviewed in Sec. 9-5).

Nucleate Pool Boiling on Smooth Surfaces

The degradation in mixture boiling heat transfer coefficients relative to their pure component values, as illustrated in Fig. 9-7 for methanol-water mixtures boiling on a smooth copper tube, has been explained as being the result of several different effects. Van Wijk, Vos, and Van Stralen (1956) presented the first physical explanation for this behavior. They pointed out that, because the equilibrium vapor mole fraction of the more volatile component y is greater than that of x, an additional amount of the more volatile component, equal to $(y - x)$, must be evaporated at a bubble's interface to maintain phase equilibrium. This causes the more volatile component to be partially depleted from the surrounding liquid. For the more volatile component to diffuse to the bubble interface, its composition at the bubble interface must be less than that in the

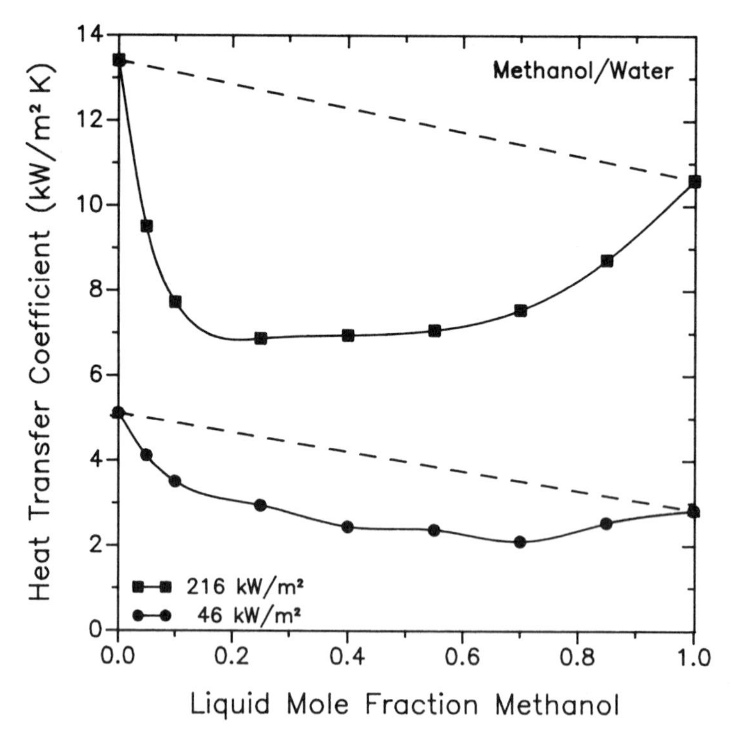

Figure 9-7 Boiling of methanol-water mixtures on a smooth tube at 1.01 bar (from Shakir, Thome, and Lloyd [1985]).

bulk liquid. The reduction in x at the bubble interface results in a rise in the local bubble point temperature at the interface. Consequently, the temperature difference driving the evaporation process is reduced by an amount equal to the rise in the bubble point temperature, which in turn adversely affects the heat transfer process. Thus, to transfer heat at the same rate, the heated wall's surface temperature must be raised to compensate for the reduction in the effective temperature difference caused by the mass diffusion process. The mixture boiling heat transfer coefficient, which is defined as the heat flux divided by the temperature difference between the heated wall and the bulk bubble point temperature of the mixture (not the local value at the bubble interface), is thus reduced.

Sternling and Tichacek (1961), Grigor'ev (1962), and Stephan and Korner (1969) postulated that the lower heat transfer coefficients were the result of the adverse effect of mass diffusion on boiling nucleation, which would lower the boiling site density and hence the boiling heat transfer coefficient. This point has been validated experimentally by Hui and Thome (1985) with a photographic study on boiling site densities in ethanol-water and ethanol-benzene mixtures. In addition, by evaluating several pure fluid nucleate pool boiling correlations with mixture physical properties for the acetone-water mixture system, Stephan and Preusser (1979) demonstrated that part of the reduction in the mixture boiling heat transfer coefficient is simply the result of the nonlinear variations in the physical properties of the mixture with composition. Thome (1981) has shown that bubble departure diameters and frequencies are affected by the mass diffusion process and demonstrated analytically that this reduces the mixture boiling heat transfer coefficient. Thome (1982) also showed that the heat transfer mechanisms of evaporation and cyclic thermal boundary layer stripping are diminished by the mass diffusion process and are hence partially responsible for the decreases that he observed in the boiling heat transfer coefficients in liquid nitrogen–argon mixtures. Thome (1983), in addition, has pointed out that the maximum rise in the local bubble point temperature resulting from the mass diffusion process is limited by the mixture's boiling range.

Thome (1986*a*) and Thome and Shakir (1987) modified the Schlünder (1982) correlation to predict nucleate pool boiling heat transfer coefficients of binary and multicomponent mixtures for systems with low to moderate boiling ranges. The method incorporates the effects of both composition and heat flux on the mass diffusion effect. The modified version is given as

$$\frac{\alpha}{\alpha_{\mathrm{I}}} = \left(1 + \left\{\frac{\alpha_{\mathrm{I}}}{q}\Delta T_{\mathrm{bp}}\left[1 - \exp\left(\frac{-B_0 q}{\rho_{\mathrm{L}}\Delta h_{\mathrm{v}}\beta_{\mathrm{L}}}\right)\right]\right\}\right)^{-1} \tag{9-1}$$

where the ideal heat transfer coefficient is calculated from a suitable single component correlation with the mixture physical properties or is calculated as

$$\alpha_{\mathrm{I}} = \frac{q}{\Delta T_{\mathrm{I}}} \tag{9-2}$$

The ideal wall superheat in Eq. (9-2) is determined from a linear molar mixing law, given as

$$\Delta T_1 = \sum_{i=1}^{n} x_i \Delta T_i \tag{9-3}$$

for a nonazeotropic mixture system. For an azeotropic system, the mixing law is applied between the pure component and the azeotrope compositions. If a correlation is utilized to determine the ideal heat transfer coefficient, then the effects of nonlinear variations in the physical properties with composition are included in the calculation.

Nucleate Pool Boiling on Enhanced Boiling Surfaces

Enhanced boiling of liquid mixtures differs substantially from boiling of mixtures on plain, smooth surfaces. From a practical standpoint, this means that correlations like the one described above are not adequate for predicting the variation in enhanced boiling heat transfer coefficients with composition. Enhanced boiling data available for various enhancements are described below.

High Flux tube. Antonelli and O'Neill (1981) reported nine mixture boiling design curves for various types of High Flux tubes with the coating applied to either the inside or the outside of the tube wall. Figure 9-8 depicts these curves for the fluids described

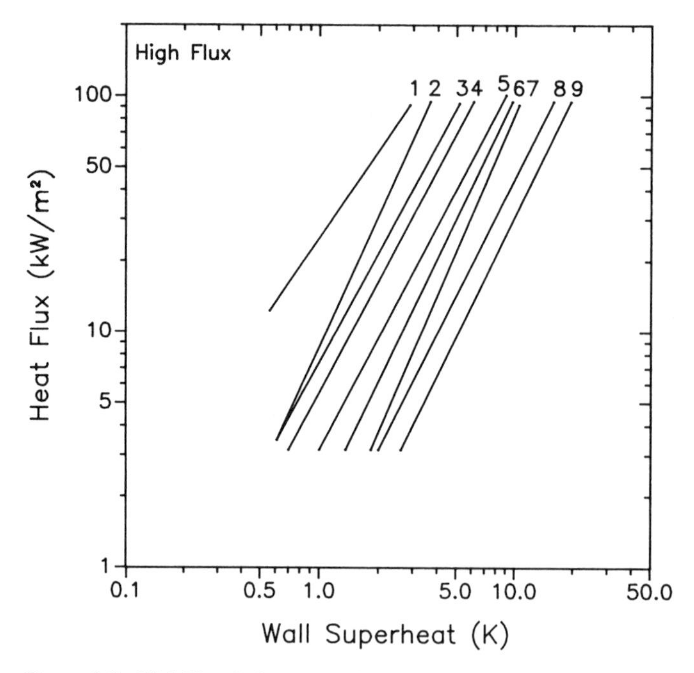

Figure 9-8 High Flux design curves (legend in Table 9-1).

in Table 9-1, which already have allowances for normal fouling from oil, magnetite, pipe rust, and the like. Czikk, O'Neill, and Gottzmann (1981) also presented some specific data in tabular form for eight mixture systems, including two ternary systems, at various pressures for seven different High Flux tubes made from various materials (e.g., copper, cupro-nickel alloy, and steel). These studies do not, however, present the corresponding plain tube boiling data for comparison purposes or the functional dependency of the High Flux tube's boiling performance with composition, which makes it difficult to draw specific conclusions about the enhanced boiling process of mixtures. The wide variation between boiling curves 1 and 9 demonstrates the need for a fundamentally based correlation to predict the boiling performances when specific boiling data are not available.

Thome and co-workers have also extensively studied the enhanced boiling of mixtures on High Flux tubing. Ali and Thome (1984) tested a copper High Flux tube with a 1% iron alloy with ethanol-water and ethanol-benzene mixtures at 1.01 and 3.03 bar over the entire composition range. In addition, Uhlig and Thome (1985) obtained data for acetone-water mixtures at 1.01 and 3.03 bar; Shakir, Thome, and Lloyd (1985) investigated methanol-water mixtures at 1.01 bar; and Shakir (1987) boiled ethanol-water and *n*-propanol–water mixtures at 1.01 bar, all for the same High Flux tubing.

Figure 9-9 depicts the Ali and Thome (1984) results for ethanol-water at 1.01 bar on a 18.55-mm tube with a 0.15-mm thick porous coating. Smooth tube data from Shakir (1987) for a smooth copper tube 22.2 mm in diameter (roughened with a 400 grade emery paper) are also shown for comparison purposes. The initial boiling curve for water was repeatable within experimental error after taking all the other mixture boiling curves, demonstrating that aging of the surface did not occur. Although the boiling heat transfer coefficient for the smooth tube was always less than or equal to a linear interpolation between the water value and the azeotrope value, the enhanced boiling heat transfer coefficient was unexpectedly found to be higher than the ideal value at some concentrations. The maxima and minima were later reconfirmed by Shakir (1987) in tests with a different piece of the same commercial tubing. Boiling augmentation in mixtures was therefore noted to be a strong function of composition, varying from about 3 times greater than the smooth tube values at low ethanol mole fractions to about 14 times greater for pure ethanol. Consequently, the boiling augmen-

Table 9-1 Mixtures represented in Fig. 9-8

Mixture	Pressure, bar
1. 15% Ethylene glycol–85% water	6.1
2. Propane-propylene splitter bottoms	10.1
3. Ethane-ethylene splitter bottoms	20.3
4. 60% Ethylene glycol–40% water	1.0
5. Demethanizer bottoms	20.3 to 30.4
6. Mixed xylenes–ethyl benzene–toluene	1.0
7. Mixed xylenes–30% C_{9+} aromatics	1.0
8. Refinery debutanizer bottoms	1.0
9. Naphtha stabilizer bottoms	15.2

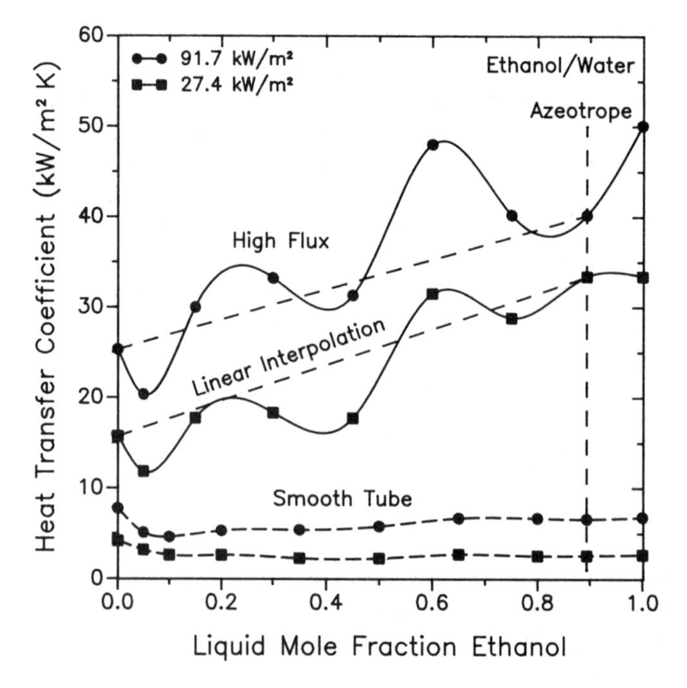

Figure 9-9 Boiling of ethanol-water mixtures at a fixed heat flux.

tation of a mixture may be greater than that of one of its pure components (water in this case).

When an enhanced boiling tube is being considered for the retubing of an existing reboiler for a fixed log mean temperature difference, the augmentation at the same wall superheat should be used for comparison of boiling performance. Because the wall superheats for the High Flux tube are small relative to those of a smooth tube, comparing the heat transfer coefficients at the same wall superheat compares the top end of the enhanced boiling curve to the bottom end of the smooth tube curve. To have an overlap in wall superheats, the Shakir (1987) data were used in constructing the comparison in Fig. 9-10, where the heat transfer coefficients are plotted as a function of liquid mole fraction for a fixed wall superheat of 4.0 K. The smooth tube still performs worse than or equal to its linear mixing law value. The High Flux tube performance, however, ranges from about 10% less than its mixing law value at low ethanol mole fractions to about 40% higher at 0.65 mole fraction ethanol. Thus the enhanced boiling tube is seen to have a strong, positive mixture effect at some compositions. Boiling augmentations in Fig. 9-10 vary in the range 300% to 3,200%.

A comparison of boiling heat transfer coefficients for the acetone-water mixture system obtained by Uhlig and Thome (1985), the physical properties of which are shown in Fig. 9-3, is depicted in Fig. 9-11 at two pressures. For this mixture system, the High Flux tube has a sharp drop in performance with the addition of small concentrations of acetone; this drop is greater than the degradation in the smooth tube values. At

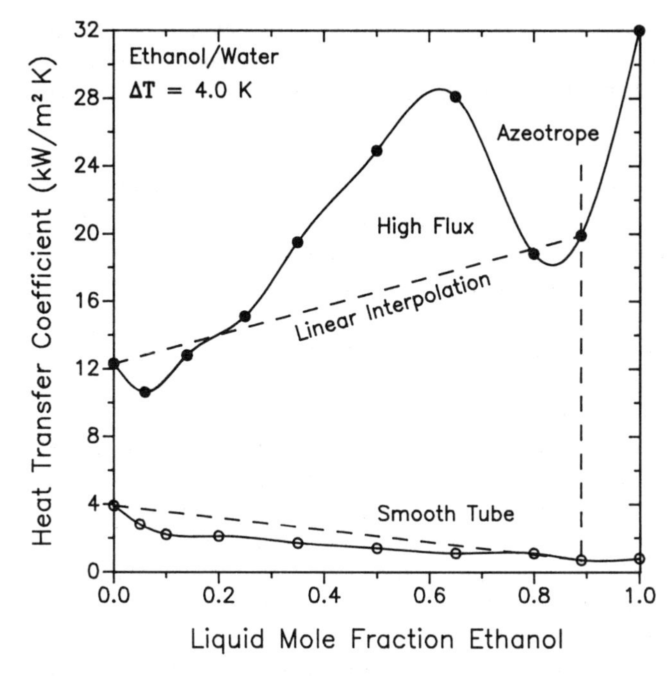

Figure 9-10 Boiling of ethanol-water mixtures at a fixed wall superheat.

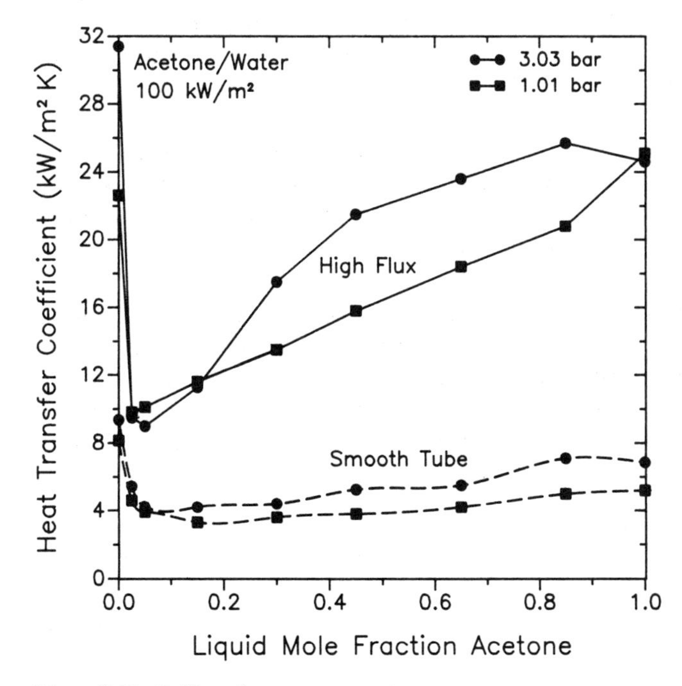

Figure 9-11 Boiling of acetone-water mixtures.

high concentrations, however, the High Flux performance increases substantially, and the smooth tube values remain low. Thus these data sets provide a good indication that the mixture effects on enhanced boiling are different from those on conventional surfaces.

Gewa-TX tube. Some mixture boiling data are available for the Gewa-TX tube of Wieland-Werke (see Chapter Four for a physical description of this tube). Thome (1986*b*, 1989) tested a copper version with several different hydrocarbon mixtures having wide boiling ranges to investigate the application of this tube to reboilers. The test section specifications are given in Table 9-2.

Figure 9-12 (*a*) shows the nucleate pool boiling data obtained for a binary mixture of 95% *n*-pentane–5% 1-tetradecene (by weight percent) boiling on the Gewa-TX tube. The boiling range for this mixture was estimated to be 93 K. The plain tube curve depicted was predicted with the Mostinski (1963) correlation, Eqs. (2-8) and (2-9) without correcting for the mixture effect because the mixture demonstrated little degradation in its boiling heat transfer coefficient on a plain tube compared to boiling in pure pentane in a study by Sardesai, Palen, and Thome (1986). In a comparison of performances at the same heat flux, the augmentation ranges from about a factor of 8 at low heat fluxes to about a factor of 4 at high ones. At a fixed wall superheat of 7.5 K, the boiling augmentation is about a factor of 80.

Figure 9-12 (*b*) depicts additional Gewa-TX boiling curves obtained for a five-component mixture (20% *n*-pentane, 20% *n*-heptane, 20% cyclohexane, 35% *p*-xylene, and 5% 1-tetradecene, by weight percent) at 2.07 and 6.90 bar. This mixture has boiling ranges of 68 K and 74 K, respectively, at these two pressures. The plain tube boiling curves were again estimated from the Mostinski correlation but were corrected for the mixture boiling effect with Eqs. (9-1) to (9-3). The boiling performance for the Gewa-TX tube improved with increasing pressure, notwithstanding the increase in the boiling range. The plain tube performance was predicted to diminish slightly with pressure at high heat fluxes as the result of its larger boiling range. The boiling augmentation was similar to that obtained for the binary mixture.

Low-finned tube. Bajorek (1988) obtained nucleate pool boiling data for a copper Gewa-K tube of Wieland-Werke. This tube had a diameter of 19 mm over the fins with 741 fins per meter (18.8 fins per inch [fpi]). The fins were essentially trapezoidal in cross-section, although their tips were rounded. The fins were 1.55 mm high and 0.46 and 0.30 mm thick at the base and tip, respectively. The following aqueous mixtures

Table 9-2 Gewa-TX tube specifications

Outside diameter	18.65mm
Root diameter	16.93mm
Inside tube diameter	12.73mm
Fin Height	0.86mm
Fin pitch	1.35mm
Gap size	0.20mm
Heated length	127.0mm

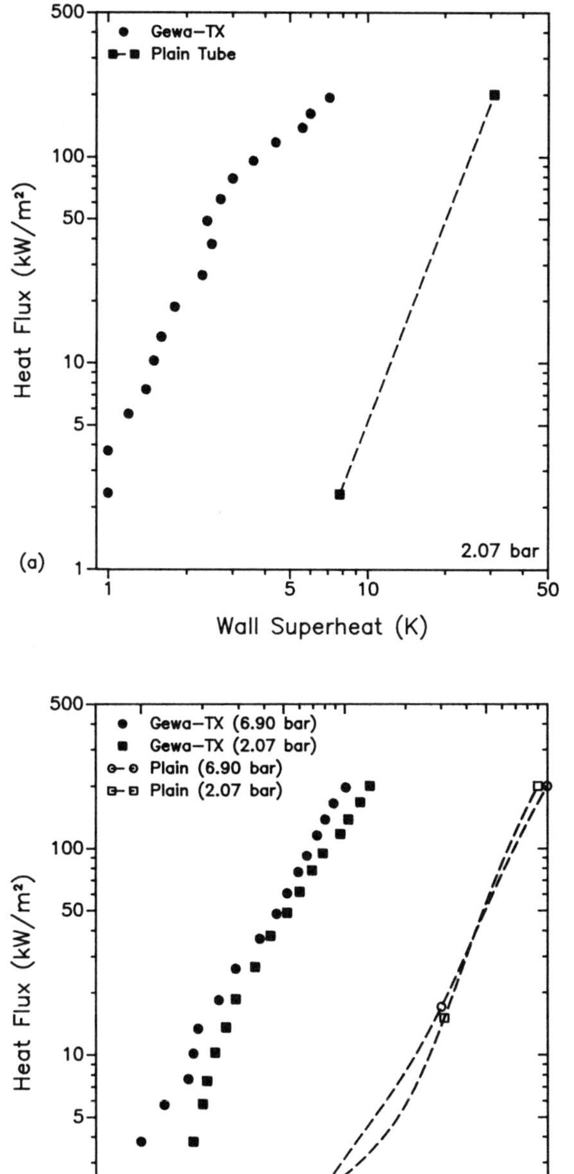

(*a*) 95% *n*-Pentane–5% 1-tetradecene mixture

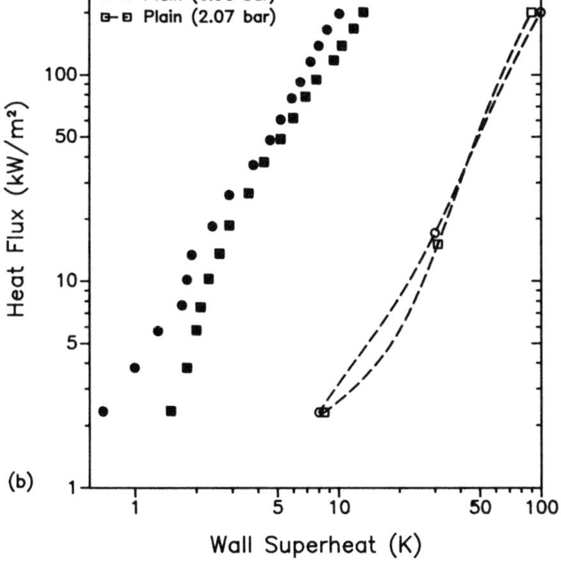

(*b*) five-component mixture.

Figure 9-12 Boiling on a Gewa-TX tube.

were tested at 1.01 bar: methanol-water, acetone-water, ethanol-water, and acetone-methanol-water.

Figure 9-13 shows the finned tube heat transfer coefficient as a function of composition for the acetone-water mixture. The heat transfer coefficients were calculated with the use of the nominal outside surface area, not the total wetted area. The degradation in the heat transfer coefficient relative to a linear interpolation between the pure fluid values is sizable. Compared to the 1.01-bar acetone-water data in Fig. 9-11, the finned tube's values at 97 kW/m^2 are higher than those of the plain tube. The finned tube performs like the High Flux tube at small acetone mole fractions but more poorly at medium and high concentrations. The minimum for the finned tube appears to be to the right of the minima for the other two tubes; this probably results from no data points being taken at 0.05 and 0.10 mole fractions. The plain tube mixture boiling correlation (Eqs. [9-1] to [9-3]) predicted the finned tube mixture data fairly well.

Summary. In summary, these experimental studies have demonstrated that the well-documented boiling augmentation for pure fluids is also obtainable with mixtures, even those with very wide boiling ranges. In addition, this work has shown that both porous type and integral extended surface type enhanced boiling surfaces perform well with mixtures. The extensive experimental results available for the High Flux tube, however, illustrate that the mixture boiling behavior of an enhanced boiling tube can be much different from that for a plain, smooth tube, as indicated by their dissimilar variations with composition shown in Figs. 9-9 and 9-10.

Enhanced Boiling Mixture Effects

For investigating the effect of mixtures on boiling, the mixture wall superheat can be broken down into two parts as in the following expression:

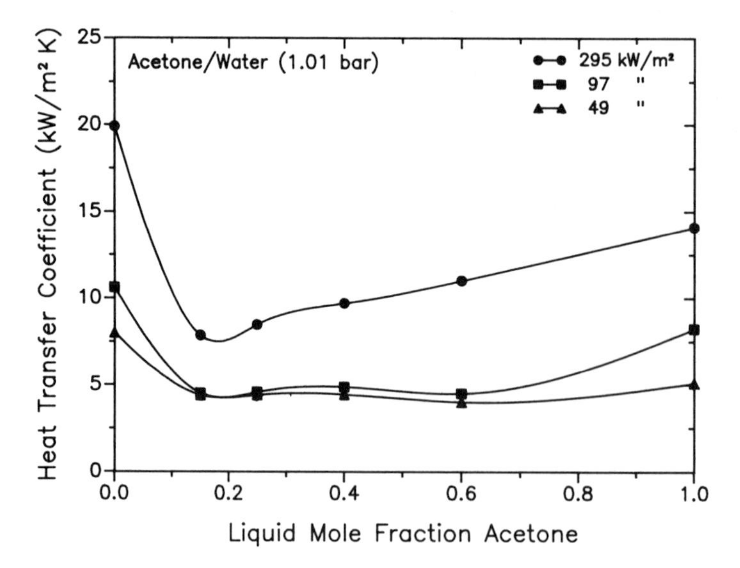

Figure 9-13 Boiling of acetone-water mixtures on a low-finned tube.

$$\Delta T = \Delta T_{\mathrm{I}} + \Delta\theta \tag{9-4}$$

where the ideal wall superheat is as defined in Eq. (9-3) and represents the pure fluid effects. $\Delta\theta$ is the wall temperature rise attributable to the mixture effects and will be called the mixture effect superheat. Thus the mixture effects of enhanced boiling can be compared to those for conventional boiling by determining their respective mixture effect superheat values at the same heat flux and composition.

Figure 9-14 (from Thome [1987]) shows such a comparison for acetone-water mixtures boiling on High Flux and smooth tubes at 1.01 bar. The mixture effect superheats for the High Flux tube are much lower than those for the plain tube. This was contrary to prior expectations, those being that the more intensive boiling on an enhanced boiling tube would increase the preferential evaporation of the more volatile component and thus result in a higher rise in the local boiling point temperature. Instead, the data demonstrated exactly the opposite effect. This trend also occurred for the High Flux tube with other aqueous mixture systems: ethanol-water, methanol-water, and *n*-propanol-water. In addition, for the low-finned tube above, the mixture effect superheats were less than those of the plain tube over most of the composition range but substantially higher than those of the High Flux tube. A possible explanation of these trends is that there is a heat transfer mechanism that is augmented in these mixtures to offset the mass diffusion effect. Yet Ali and Thome (1984) showed that the mixture effect superheats for ethanol-benzene mixtures on High Flux tubes were identical to those for plain tubes.

One explanation for the lower mixture effect superheats was proposed by Ali and Thome (1984) for their ethanol-water data. The evaporating thin film inside the enhancement matrix is replenished by fresh bulk liquid entering from inactive pores,

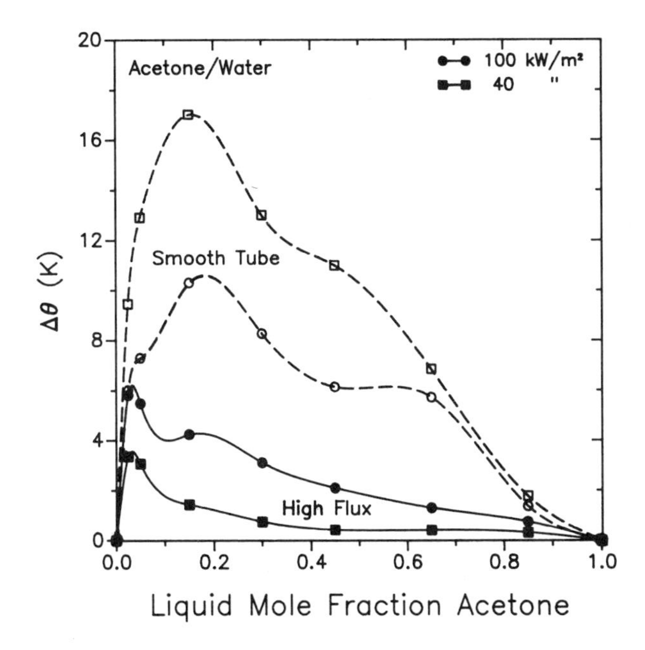

Figure 9-14 Mixture effect superheats for acetone-water mixtures at 1.01 bar.

perhaps such that a substantial concentration gradient in the liquid film does not form; instead, no replenishment is possible for the liquid microlayer trapped underneath a growing bubble on a plain surface. This does not explain the apparently anomalous behavior of the ethanol-benzene mixtures, however.

The enhanced mixture boiling effect has been studied in more detail by Thome (1987). He based his investigation on a qualitative analogy between the boiling process inside the enhancement matrix and a mixture boiling inside a smooth tube. For the latter situation, Bennett and Chen (1980) developed a superposition model that sums the convective and nucleate boiling contributions to heat transfer while correcting separately for the mixture effects on these two processes. Their expression is

$$\alpha = \alpha_L F'C + \alpha_p SC' \tag{9-5}$$

where the first term on the right hand side is the convective contribution and the second term is the nucleate boiling contribution. α_L, α_p, and S are the liquid-only convective heat transfer coefficient, the pool boiling heat transfer coefficient, and the boiling suppression factor, respectively, as in the original Chen (1963) flow boiling correlation for pure fluids. The mixture boiling factor C' accounts for the adverse effect of preferential evaporation of the more volatile component on the boiling process, similar to the situation described by Eq. (9-1). The convection factor C models the decrease in the liquid convection contribution to heat transfer due to the smaller temperature difference across the evaporating liquid film, which results from the increase in the bubble point temperature caused by the preferential evaporation of the more volatile component. F' is a modified version of the original Chen (1963) two-phase multiplier that includes the augmentation of the convection process provided by the mass diffusion. This effect was incorporated in the two-phase multiplier by inclusion of the liquid Prandtl number.

Assuming the liquid convection and annular flow boiling processes in a tube to be qualitatively analogous to the convection and thin film evaporation processes inside an enhancement matrix, Thome (1987) observed that the positive mass diffusion effect on F' could partially counterbalance the negative mass diffusion effect of C' on evaporation. For instance, Fig. 9-15 depicts the variation in F' for the various mixture systems that he studied at 1.01 bar. The maxima in the two-phase multiplier at intermediate compositions for the aqueous mixtures substantially augments convection and hence counterbalances the negative effect of mass diffusion on evaporation. Thus the smaller mixture effect superheats, such as those shown in Fig. 9-14, are the result of the maximum in the liquid Prandtl number in F'. F' varies nearly linearly with composition for ethanol-benzene mixtures; therefore, little augmentation occurs through the convection contribution to heat transfer, and the mixture effect superheats should be the same as those for conventional boiling, as was observed experimentally.

In summary, the smaller degradation in enhanced boiling heat transfer coefficients for aqueous mixtures has been shown to be caused by the augmentation of the convection process. Only mixture systems with positive maxima in their liquid Prandtl numbers can be expected to demonstrate this behavior, however. Because the convection process is characteristic of the boiling enhancement geometry, the mixture boiling effect is

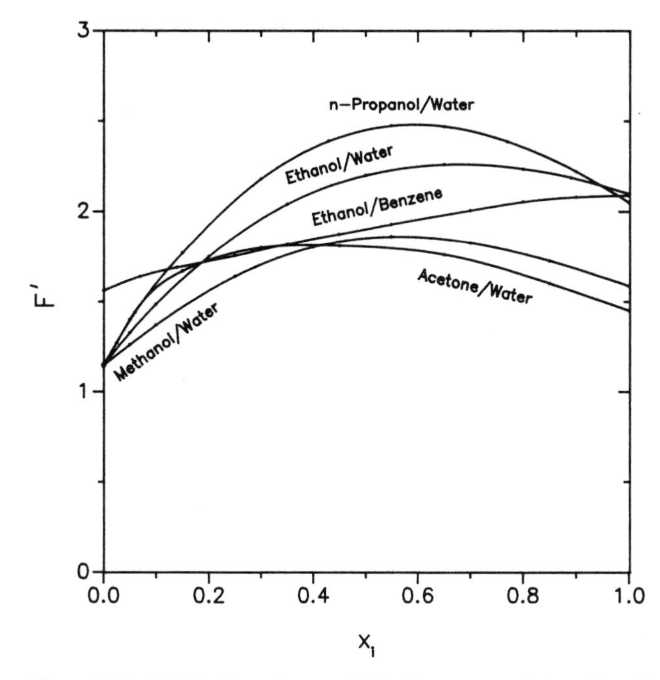

Figure 9-15 Modified two-phase multiplier for aqueous mixtures (i = ethanol, methanol, n-propanol, and acetone).

different for each type of enhanced boiling tube, as indicated by the data for the High Flux and low-finned tubes.

9-5 BOILING OF REFRIGERANT-OIL MIXTURES

There are many applications of enhanced boiling tubes that involve contamination of the evaporating fluid by lubricating oils. Because these lubricants are soluble, over a period of time they are leached from the bearings and into the working fluid, building up their concentration until the unit must be shut down to recharge it with new refrigerant. Therefore, it is important to understand the effect of small concentrations of oil on the performance of enhanced boiling tubes. The present state of the art in this area is still inadequate not only for enhanced boiling tubes but also for conventional boiling. As yet, no comprehensive model is available to describe and predict the effect of oil on a refrigerant's boiling characteristics. Thus the following summary is limited to a review of the experimental work on refrigerant-oil mixtures on smooth and enhanced tubes (the effect of oil on intube enhanced boiling of refrigerants is described in Chapter Ten, Sec. 10-1).

Conventional Surfaces

Nucleate pool boiling studies by Stephan (1963), Dougherty and Sauer (1975), Sauer, Gibson, and Chongrungreong (1978), and Chongrungreong and Sauer (1979) have established that increasing percentages of oil in R-11, R-12, and R-22 produced a decrease in the boiling heat transfer coefficient of a single tube or flat plates. A decrease with oil concentrations up to about 3% by weight is generally small enough to be ignored because it is within the experimental error of the measurements. At higher concentrations, however, the decrease is substantial. For instance, at an oil concentration of 10% the reduction in the nucleate pool boiling heat transfer coefficient is about 50% or more compared to that for boiling in the pure refrigerant under the same conditions.

Further details of refrigerant-oil mixture boiling on conventional surfaces and in refrigeration equipment can be found in Chaddock (1976) and Schlager, Pate, and Bergles (1987).

Enhanced Boiling Surfaces

A number of experimental studies on the effect of oil on the performance of enhanced boiling tubes have been reported in the literature. These are described below for various types of enhancement.

Finned tubes. Sauer, Davidson, and Chongrungreong (1980) investigated the effect of two different oils on nucleate pool boiling of R-11 on an integral low-finned tube. The finned tube test section was made from commercial copper finned tubing with the following characteristics: an outside diameter of 22 mm, a root diameter of 19 mm, a fin density of 19 fpi (750 fins per meter), and an outside wetted surface 2.76 times that of an equivalent smooth tube. The oils tested were of naphthene base, with viscosities at 37.8°C of 33.1 and 111.1 centistokes (cs), respectively.

Figure 9-16 depicts the effect of oil concentration (in weight percent) on the boiling heat transfer coefficient for the 33-cs oil. The two pure fluid boiling curves obtained from the investigation show that some variation was due to experimental uncertainties. The boiling heat transfer coefficient decreased with the addition of the lighter oil at all heat fluxes. In tests with the higher-viscosity oil, however, the heat transfer coefficient first increased with respect to one of the two pure refrigerant boiling curves and then decreased in a manner similar to that of the lower-viscosity oil. This type of behavior has also been observed for boiling on plain tubes by Stephan (1963). The increase was not substantial, however, and a monotonic decrease for all oil concentrations would have been observed if the curves had been compared to the top pure refrigerant boiling curve in Fig. 9-16.

Sauer, Davidson, and Chongrungreong (1980) also studied boiling nucleation on the same finned tube as a function of oil concentration. Table 9-3 shows their results for the two different oils tested. The general trend in the nucleation superheat is to increase with oil concentration. The increase becomes large at high concentrations and would have an adverse effect on evaporator startup. The rise in the nucleation superheat

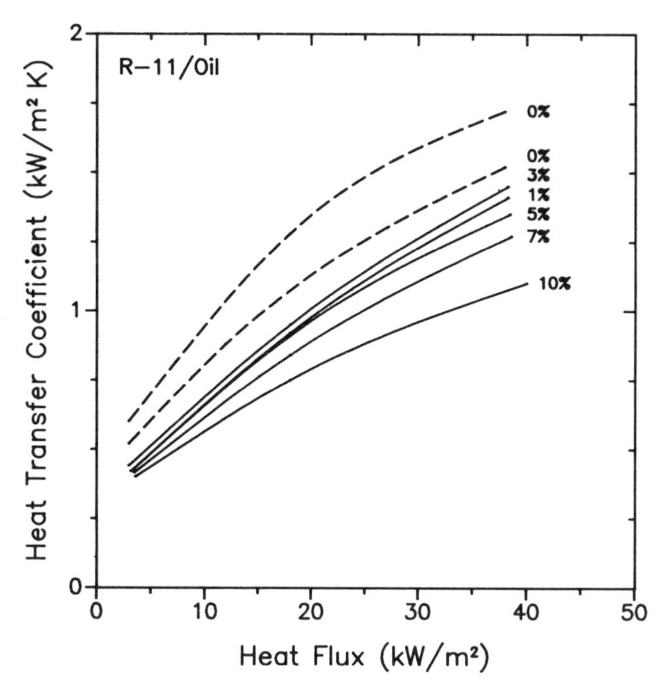

Figure 9-16 Effect of oil on finned tube boiling performance.

Table 9-3 Effect of oil on finned tube boiling nucleation

Oil concentration, weight percent	Nucleation superheat, K	
	33 cs	111 cs
0	10	17
1	14	18
3	18	13
5	20	19
7	32	26
10	41	32

on addition of oil to refrigerant-11 is expected from theory (e.g., Eq. [2-1]), because the surface tension increases with oil concentration.

Bell, Hewitt, and Morris (1987) studied nucleate pool boiling of refrigerant-oil mixtures on a smooth tube and a low-finned tube. R-113 was the test fluid. Shell Clavus-68 was the oil used, and several of its physical properties were measured. The plain tube was 25.4 mm in outer diameter, and the low-finned tube had 19 fpi (750 fins per meter) and an outside diameter over the fins of 25.4 mm. The fins were 1.6 mm high and 0.3 mm thick. Both tubes were made of brass. Using six thermocouples embedded around the circumference of the tubes, the investigators observed a much wider variation in the heat transfer coefficient around the finned tube than for the plain tube at an oil concentration of 5% (by weight). The heat transfer coefficient was higher at the top of the tube than at the bottom by as much as 50%. This apparently results from the additional effect of flow-induced convection between the fins. For oil concentrations up to 5%, the ratios of the heat duties per unit length of tubing of the finned tube to the plain tube were about the same as for pure R-113. At concentrations of 7% and 10%, the ratios were higher than the pure R-113 values, indicating less degradation in the finned tube's boiling performance.

Gewa-T tube. Stephan and Mitrovic (1981, 1982) studied the boiling of R-12–oil mixtures on the Gewa-T tube for oil concentrations ranging up to 9% by weight. The tube specifications were as follows: outside diameter, 18.1 mm; root diameter, 15.9 mm; slit opening, 0.25 mm; and fin density, 18.8 fpi (741 fins per meter). The oil used in the study was Shell Clavus-G68; its viscosity was not cited.

Figure 9-17 depicts the ratio of refrigerant-oil mixture heat transfer coefficients to the pure refrigerant value as a function of oil concentration at several different heat fluxes. These results are similar to those for the plain and finned tubes described above. The effect of oil on the boiling heat transfer coefficient is also shown to be strongly dependent on how vigorously the tube is boiling, as indicated by the improvement in performance with increasing heat flux. For the Gewa-T tube with this particular refrigerant-oil combination, the large fall-off in the boiling heat transfer coefficient is seen to occur at concentrations greater than about 5.5%. Thus Gewa-T tube evaporators may be able to function at close to normal capacity up to this oil concentration.

Wanniarachchi, Sawyer, and Marto (1987) found similar performance deterioration for the Gewa-T (and Thermoexcel-E) tube with oil–R-114 mixtures.

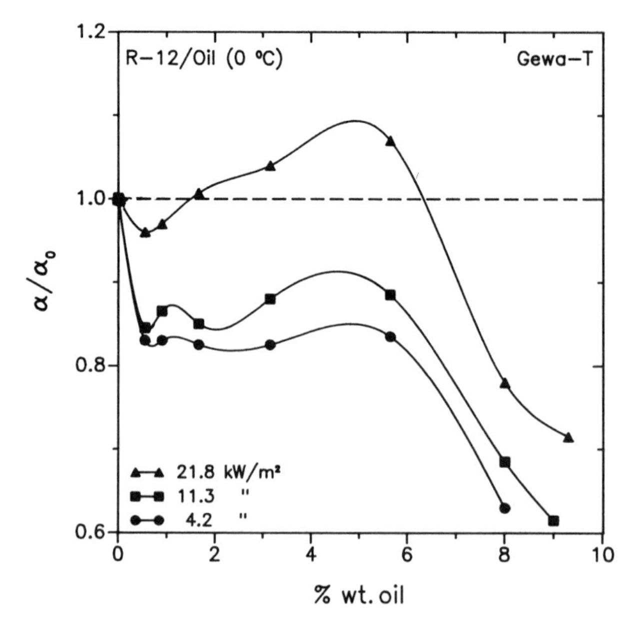

Figure 9-17 Boiling of refrigerant-oil mixtures on a Gewa-T tube.

Porous coated tubes. Gottzmann, O'Neill, and Minton (1973) obtained High Flux nucleate boiling curves for propylene-oil mixtures boiling at 1.0 bar; the oil was not specified. Figure 9-18 depicts the variation in the boiling heat transfer coefficient for different levels of oil concentration (in volume percent). One pure propylene boiling curve was obtained before the oil runs, and the other one was taken on the uncleaned test section after the oil runs. The difference between their curves is within normal experimental error and does not necessarily indicate that any fouling occurred. The results showed that even a small amount of oil will adversely affect the High Flux coating's boiling performance. In a heat exchanger evaporating propylene in a cascade refrigeration system, however, the overall thermal resistance is strongly controlled by the heating fluid's heat transfer coefficient, and thus overall performance is only marginally affected. The investigators noted that removing the oil in one of these types of evaporators quickly restored its performance to the original clean values.

In another study on the High Flux tube, Czikk et al. (1970) reported the effect of oil on the boiling performance of a 20-ton chiller evaporating refrigerant-11 on the shell side. Injecting refrigeration oil into the refrigeration loop at a concentration of 2% by weight of the refrigerant charge was found to have no effect on the chiller's performance.

Wanniarachchi, Marto, and Reilly (1986) tested an unspecified porous coated tube with oil concentrations up to 10% (by weight) in R-114 at −2.2°C. They measured nucleation superheats of 4, 8, and 24 K for oil concentrations of 0%, 3%, and 10%, respectively, which are similar to the low-finned tube data in Table 9-3. The presence of the oil markedly decreased the heat transfer coefficient. For instance, 3% oil resulted in a 25% reduction on average, and 10% oil decreased performance by 25% to 85% as the heat flux increased from 1 to 98 kW/m^2.

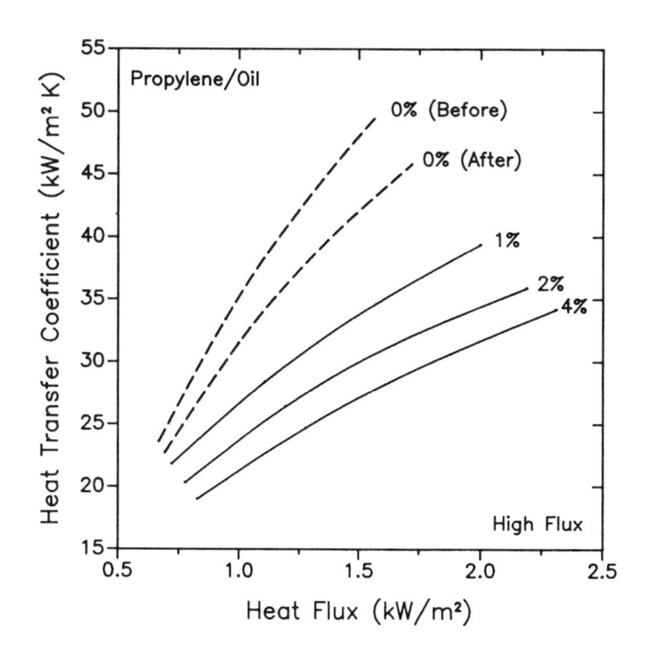

Figure 9-18 Effect of oil on the performance of a High Flux tube.

Thermoexcel-E tube. Arai et al. (1977) investigated the effect of oil on the boiling performance of R-12 on an electrically heated, Thermoexcel-E tube bundle with a thermal capacity of 70.3 kW. The evaporator was made with 225 tubes of 16.4 mm outside diameter with 10 horizontal rows of tubes on a triangular pitch. Boiling was on the shell side. The investigators obtained individual tube performances for 89 tubes within this arrangement for oil concentrations up to 3.4% by weight.

Figure 9-19 shows their experimental results. The test conditions were varied as follows: (1) for the oil concentrations of 0.5%, 1.8%, and 3.4% equal heat loads were applied to all the tubes in the bundle, and all the tubes were immersed below the foaming region in the bundle; and (2) for the 2.8% oil level a nonuniform heat load was applied, and the level of refrigerant was controlled to place the top three tube rows in the foaming region. As can be seen from the data, adding oil reduced the bundle's thermal performance when it was immersed totally in liquid (condition 1), whereas adding oil augmented performance when the top of the bundle was in the foaming region (condition 2). The investigators concluded that, with proper control of the level of the refrigerant in the shell to exploit this effect, high performance of the Thermoexcel-E evaporator could be maintained even in the presence of low oil concentrations. In addition, tube bundle performance was better for the bundle with 0.5% oil contamination than for the single tube boiling in pure refrigerant.

Summary. Experimental work for nucleate pool boiling and bundle boiling conditions has shown that oil can have a substantial effect on the boiling heat transfer coefficient.

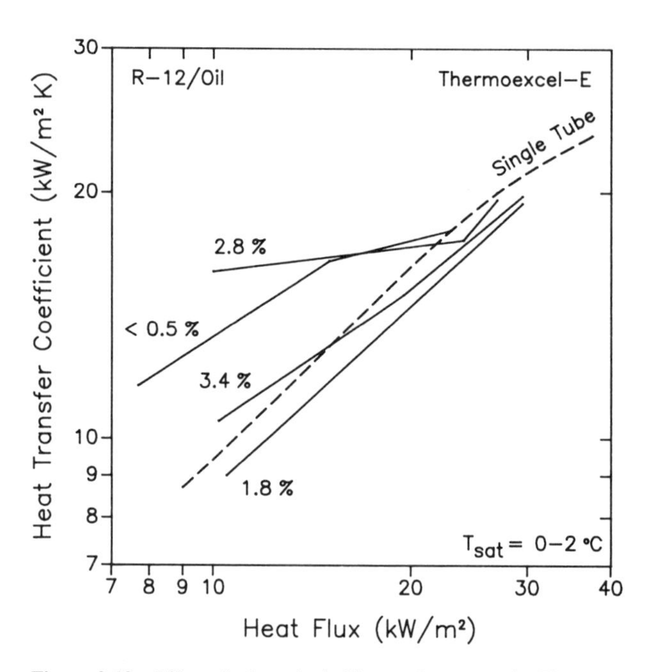

Figure 9-19 Effect of oil on the boiling performance of a Thermoexcel-E tube bundle.

For very small concentrations, however, these effects can probably be ignored. At high concentrations, the adverse effect of mass diffusion (as described above under nucleate pool boiling of mixtures on smooth surfaces) tends to lower substantially the boiling heat transfer coefficient because the refrigerant-oil mixture behaves as a mixture with a wide boiling range.

9-6 NOMENCLATURE

B_o	empirical scaling factor (= 1.0)
C	mixture convection factor
C'	mixture boiling factor
F'	modified two-phase multiplier
Δh_v	latent heat of vaporization (J/kg)
i	component i
n	number of components
q	heat flux (W/m^2)
S	boiling suppression factor
T_{sat}	saturation temperature
ΔT	wall superheat (K)
ΔT_{bp}	boiling range (K)
ΔT_i	wall superheat of component i (K)
ΔT_I	ideal wall superheat (K)
ΔT_{sat}	boiling nucleation superheat (K)
x	liquid mole fraction
x_i	liquid mole fraction of component i
y	vapor mole fraction
α	heat transfer coefficient [$W/(m^2 \cdot K)$]
α_I	ideal heat transfer coefficient [$W/(m^2 \cdot K)$]
α_L	liquid-only heat transfer coefficient [$W/(m^2 \cdot K)$]
α_p	pool boiling heat transfer coefficient [$W/(m^2 \cdot K)$]
α_o	pure refrigerant heat transfer coefficient [$W/(m^2 \cdot K)$]
β	contact angle (degrees)
β_L	liquid mass diffusion coefficient (= 0.0003 m/s)
$\Delta\theta$	mixture effect superheat (K)
λ_L	liquid thermal conductivity [$W/(m \cdot K)$]
μ_L	liquid dynamic viscosity ($N \cdot s/m^2$)
ρ_L	liquid density (kg/m^3)
ρ_v	vapor density (kg/m^3)
σ	surface tension (N/m)

REFERENCES

Ali, S. M., and J. R. Thome. 1984. Boiling of ethanol-water and ethanol-benzene mixtures on an enhanced boiling surface. *Heat Transfer Eng.* 5(3–4):70–81.

Antonelli, R., and P. S. O'Neill. 1981. Design and application considerations for heat exchanger with enhanced boiling surfaces. Paper read at International Conference on Advances in Heat Exchangers, September, Dubrovnik, Yugoslavia.

Arai, N., T. Fukushima, A. Arai, T. Nakajima, K. Fujie, and Y. Nakayama. 1977. Heat transfer tubes enhancing boiling and condensation in heat exchangers of a refrigerating machine. *ASHRAE Trans.* 83(part 2): 58–70.

Bajorek, S. M. 1988. An experimental and theoretical investigation of multicomponent pool boiling on smooth and finned surfaces. Ph.D. diss., Michigan State University, East Lansing.

Bell, K. I., G. F. Hewitt, and S. D. Morris. 1987. Nucleate pool boiling of refrigerant/oil mixtures. *Exp. Heat Transfer* 1:71–86.

Bennett, D. L., and J. C. Chen. 1980. Forced convective boiling in vertical tubes for saturated pure components and binary mixtures. *AIChE J.* 26:454–61.

Chaddock, J. B. 1976. Influence of oil on refrigerant evaporator performance. *ASHRAE Trans.* 82(part 1):474–86.

Chen, J. C. 1963. *A correlation for boiling heat transfer to saturated fluids in convective flow* (ASME paper 63-HT-34).

Chongrungreong, S., and H. J. Sauer, Jr. 1979. Nucleate boiling performance of refrigerants and refrigerant-oil mixtures. *J. Heat Transfer* 102:701–705.

Chu, J. C. 1950. *Distillation equilibrium data.* New York: Reinhold.

Czikk, A. M., C. F. Gottzmann, E. G. Ragi, J. G. Withers, and E. P. Habdas. 1970. Performance of advanced heat transfer tubes in refrigerant-flooded liquid coolers. *ASHRAE Trans.* 76(part 1):96–109.

Czikk, A. M., P. S. O'Neill, and C. F. Gottzmann. 1981. Nucleate pool boiling from porous metallic films: Effect of primary variables. In *Advances in Enhanced Heat Transfer*, HTD Vol. 18, 109–22. New York: American Society of Mechanical Engineers.

Dougherty, R. L., and H. J. Sauer, Jr. 1975. Nucleate pool boiling of refrigerant-oil mixtures from tubes. *ASHRAE Trans.* 80(part 2):175–93.

Fredenslund, A., J. Gmehling, and P. Rasmussen. 1979. *Vapor-liquid equilibria using UNIFAC*. Amsterdam: Elsevier Scientific.

Gmehling, J., U. Onken, W. Arlt, P. Grenzheuser, U. Weidlich, and B. Kolbe. 1984. VLE data collection. In *Chemistry data series*. 2d ed., ed. D. Behrens and R. Eckermann, vol. 1. Amsterdam: Elsevier.

Gottzmann, C. F., P. S. O'Neill, and P. E. Minton. 1973. High efficiency heat exchangers. *Chem. Eng. Progr.* 69(7):69–75.

Grigor'ev, L. N. 1962. Studies of heat transfer to two component mixtures. *Teplo Masso Perenos* 2:120–27.

Hala, E., R.Pick, B. Fried, and R. J. Vilim. 1968. *Vapour-liquid equilibrium data at normal pressures.* Oxford: Pergamon.

Hui, T. O., and J. R. Thome. 1985. A study of binary mixture boiling: Boiling site density and subcooled heat transfer. *Int. J. Heat Mass Transfer* 28:919–28.

Kirschbaum, E. 1948. *Distillation and rectification. New York: Chemical Publishing.*

Mostinski, I. L. 1963. Application of the rule of corresponding states for the calculation of heat transfer and critical heat flux. Teploenergetika 4:66.

Ponter, A. B., and W. Peier. 1978. *Int. J. Heat Mass Transfer* 21:1025–28.

Prausnitz, J. M. 1969. *Molecular thermodynamics of fluid-phase equilibria*. Englewood Cliffs, N.J.: Prentice-Hall.

Reid, R. C., J. M. Prausnitz, and B. E. Poling. 1987. *The properties of gases and liquids,* 4th ed. New York: McGraw-Hill.

Sardesai, R. G., J. W. Palen, and J. R. Thome. 1986. Nucleate pool boiling of hydrocarbon mixtures. Paper read at AIChE National Winter Meeting, November 2–7, Miami Beach (paper 127a).

Sauer, Jr., H. J., G. W. Davidson, and S. Chongrungreong. 1980. *Nucleate boiling of refrigerant-oil mixtures from finned tubing* (ASME paper 80-HT-111).

Sauer, Jr., H. J., R. K. Gibson, and S. Chongrungreong. 1978. Influence of oil on the nucleate boiling of refrigerants. *Proc. 6th Int. Heat Transfer Conf.* 1.

Schlager, L. M., M. B. Pate, and A. E. Bergles. 1987. A survey of refrigerant heat transfer and pressure drop emphasizing oil effects and in-tube augmentation. *ASHRAE Trans.* 93(part 1):392–416.

Schlünder, E. U. 1982. Uber den wärmeubergang bei der blasenverdampfung von gemischen. *Int. Chem. Eng.* 23:589–99.

Shakir, S. 1987. Boiling incipience and heat transfer on smooth and enhanced surfaces. Ph.D. diss., Michigan State University, East Lansing.

———. and J. R. Thome. 1986. Boiling nucleation of mixtures on smooth and enhanced boiling surfaces. *Proc. 8th Int. Heat Tranfer Conf.* 4:2081–86.

———. and J. R. Lloyd. 1985. Boiling of methanol-water mixtures on smooth and enhanced surfaces. In *Multiphase flow and heat transfer,* ed. V. K. Dhir, J. C. Chen, and O. C. Jones. HTD vol. 47, 1–6. New York: American Society of Mechanical Engineers.

Shock, R. A. W. 1977. Nucleate boiling in binary mixtures. *Int. J. Heat Mass Transfer* 20:701–709.

Stephan, K. 1963. Influence of oil on heat transfer of boiling Freon 12 and Freon 22. *Proc. 11th Int. Congr. Refrig.* 1:Bulletin 3.

———. and M. Korner. 1969. Calculation of heat transfer in evaporating binary liquid mixtures. *Chem. Ing. Tech.* 41(7):407–17.

Stephan, K., and J. Mitrovic. 1981. Heat transfer in natural convective boiling of refrigerants and refrigerant-oil mixtures in bundles of T-shaped finned tubes. In *Advances in Enhanced Heat Transfer*, HTD Vol. 18, 131–46. New York: American Society of Mechanical Engineers.

———. 1982. Heat transfer in natural convective boiling of refrigerant-oil mixtures. *Proc. 7th Int. Heat Transfer Conf.* 4:73–87.

Stephan, K., and P. Preusser. 1979. Heat transfer and critical heat flux in pool boiling of binary and ternary mixtures. *Germ. Chem. Eng.* 2:161–69.

Sternling, C. V., and L. J. Tichacek. 1961. Heat transfer coefficients for boiling mixtures. *Chem. Eng. Sci.* 16:297–337.

Thome, J. R. 1981. Nucleate pool boiling of binary mixtures—An analytical equation. *AIChE Symp. Ser.* 77(208):238–50.

———. 1982. Latent and sensible heat transport rates in the boiling of binary mixtures. *J. Heat Transfer* 104:474–78.

———. 1983. Prediction of binary mixture boiling heat transfer coefficients using only phase equilibrium data. *Int. J. Heat Mass Transfer* 26:965–74.

———. 1986*a*. Prediction of in-tube boiling of mixtures in thermosyphon reboilers. Paper read at AIChE National Winter Meeting, November 2–7, Miami Beach (paper 127c).

———. 1986*b*. Boiling of hydrocarbon mixtures on a Gewa-TX tube. Paper read at AIChE National Winter Meeting, November 2–7, Miami Beach (paper 127e).

———. 1987. Enhanced boiling of mixtures. *Chem. Eng. Sci.* 42:909–17.

———. 1989. Nucleate pool boiling of hydrocarbon mixtures on a Gewa-TX tube. *Heat Transfer Eng.* 10(1):37–44.

———. and S. Shakir. 1987. A new correlation for nucleate pool boiling of aqueous mixtures. *AIChE Symp. Ser.* 83(257):46–51.

———. and C. Mercier. 1982. Effect of composition on boiling incipient superheats in binary mixtures. *Proc. 7th Int. Heat Transfer Conf.* 4:95–100.

Thome, J. R., and R. A. W. Shock. 1984. Boiling of multicomponent mixtures. *Adv. Heat Transfer* 16:59–156.

Toral, H., D. B. R. Kenning, and R. A. W. Shock. 1982. Flow boiling of ethanol/cyclohexane mixtures. *Proc. 7th Int. Heat Transfer Conf.* 4:255–60.

Uhlig, E., and J. R. Thome. 1985. Boiling of acetone-water mixtures on smooth and enhanced surfaces. In *Advances in Enhanced Heat Transfer*, HTD Vol. 43, 49–56. New York: American Society of Mechanical Engineers.

Van Wijk, W. R., A. S. Vos, and S. J. D. Van Stralen. 1956. Heat transfer to boiling binary liquid mixtures. *Chem. Eng. Sci.* 5:68–80.

Wanniarachchi, A. S., P. J. Marto, and J. T. Reilly. 1986. The effect of oil contamination on the performance of R-114 from a porous-coated surface. *ASHRAE Trans.* 92(part 2B):525–38.

Wanniarachchi, A. S., L. M. Sawyer, and P. J. Marto. 1987. Effect of oil on pool boiling performance of R-114 from enhanced surfaces. *Proc. 1987 ASME-JSME Therm. Eng. Joint Conf.* 1:531–37.

CHAPTER

TEN

ENHANCED FLOW BOILING

Enhancement under nucleate pool boiling conditions has been discussed almost exclusively in the preceding chapters. In practice, however, the performance characteristics of enhanced boiling tubes under flow boiling conditions are required, whether for forced circulation boiling inside horizontal or vertical tubes or for natural circulation boiling on the shell side of horizontal tube bundles. The present survey of enhanced boiling under flow conditions is divided into two major sections, Sec. 10-1 dealing with boiling inside tubes and Sec. 10-2 with boiling on the outside of tube bundles.

10-1 INTUBE ENHANCED BOILING

Introduction

Flow boiling inside plain tubes has been studied extensively because of the practical importance of this process in the power, refrigeration and air conditioning, petroleum, and chemical processing industries. A brief review of the subject is presented in Chapter Two. In the present survey of enhanced boiling inside tubes, it is assumed that the reader has a working knowledge of the subject.

The state of the art of enhanced intube boiling is much less advanced than that of conventional boiling, where essentially all the variables affecting the heat transfer and two-phase flow processes have been extensively investigated and some relatively accurate and reliable correlations developed. Attaining the same level of understanding of enhanced boiling in tubes is a formidable task because of the effect of each change in the enhancement geometry on the boiling process and pressure drop. At present, the optimum geometries for most types of enhancements are still being sought, usually by systematic testing of the many variations that can be produced with existing tooling. Thus in-depth studies are still not time effective or cost effective for enhancements that may soon become outdated by new and better ones. Nevertheless, improved understanding of the augmentation process should reduce research and development costs to obtain a suitably optimum geometry in less time and at lower cost.

Producing complicated geometries on the inner walls of tubes is much more difficult than producing them on the exterior. The size and expense of the machinery and the expertise required to make the cutting tools have effectively limited the development of new intube boiling enhancements to well-equipped industrial laboratories. Therefore, university and government laboratories primarily have been active only in testing commercially available tubes or prototype tubes obtained through industrial contacts. These various commercial enhancements for intube boiling are described in Chapter Four.

Intube boiling enhancement geometries can be categorized as follows: internally finned tubes, corrugated and spirally fluted tubes, twisted tape inserts, star-shaped inserts, mesh or brush inserts, and internally porous-coated tubes. Only the first two categories are integral type enhancements. Boiling augmentation resulting from coiling plain tubes or by virtue of complicated channel geometries, such as in reactor fuel rod assemblies, is not considered here.

Presenting a unified review of enhanced intube boiling research is a formidable task because of the large number of geometric parameters involved and the difficulty of relating the conclusions of one study to another. Therefore, the experimental work for each category of enhancement noted above is reviewed separately for an in-depth look at the factors affecting boiling augmentation and pressure drop. Boiling performances are compared where possible to determine the relative effectiveness of the various types of enhancements. This is followed by a survey of the existing heat transfer and pressure drop correlations available for some of the enhancements. Finally, critical heat flux augmentations are reviewed.

A word of caution is needed before interpreting the results of intube enhanced boiling studies. No consistent convention for defining the surface area to use in calculating the heat flux or boiling heat transfer coefficient for intube boiling enhancements exists. For instance, some investigators define this area as the total wetted area, others as the nominal area at the base of the enhancement, still others as the nominal area at the tip of the enhancement, and yet others have not defined the area that they used at all. Thus one must recalculate data from different studies using one common definition to make a valid comparison.

Intube Experimental Studies

Intube boiling studies are of two types: those that measure local heat transfer coefficients by use of electrical heating and local thermocouples, and those that measure average heat transfer coefficients for the entire test section, which is usually heated with a hot single-phase fluid. Because of the growing use of these enhancements in the refrigeration and air-conditioning industries, most experimental studies to date have tested refrigerants, such as R-12, R-22, and R-113. For a comprehensive bibliographic listing of these studies, the reader is referred to Bergles et al. (1983).

Internally finned tubes. Internally finned tubes are classified into three groups: microfinned tubes, high-finned tubes, and ribbed tubes. Microfinned tubes have helical fins characterized by their small fin heights (or groove depths), which typically range from 0.1 to 0.3 mm. They are classified separately from the higher-finned tubes because they are purported to be able to augment heat transfer without increasing the pressure drop under some operating conditions. High-finned tubes have larger fin heights than microfins (it may be appropriate to define another category between microfinned and high-finned tubes as low-finned tubes). Ribbed tubes have spiral or longitudinal fins with a large width-to-height ratio and are used primarily to augment the critical heat flux in fossil-fuel boilers rather than to increase the boiling heat transfer coefficient.

Microfinned tubes. Figures 4-13 and 4-14 depict several different microfinned tube geometries. The primary geometric parameters that can be varied to affect performance are the cross-sectional profile of the fins, their height and width, the spiral or helix angle of the finning, the pitch between adjacent fins, the additional wetted surface area, and the tube diameter. These factors affect not only the boiling augmentation but also the two-phase pressure drop through the tube. Thus the objective is to obtain the maximum heat transfer augmentation with as little increase in pressure drop as feasible while minimizing the increase in manufacturing cost and tube weight per unit length.

Several comprehensive studies have been performed on microfinned tubes. At Hitachi, parametric studies were run by Ito and Kimura (1979), Kimura and Ito (1981), and Shinohara and Tobe (1985). At Iowa State University, comparative studies were performed by Khanpara, Bergles, and Pate (1986), Reid, Pate, and Bergles (1987), and Khanpara, Pate, and Bergles (1987).

Ito and Kimura (1979) tested R-22 with a wide variety of microfin sizes for fins whose basic geometric form was that shallow grooves as depicted in Fig. 10-1. The tube specifications are given in Table 10-1. They have calculated their heat transfer

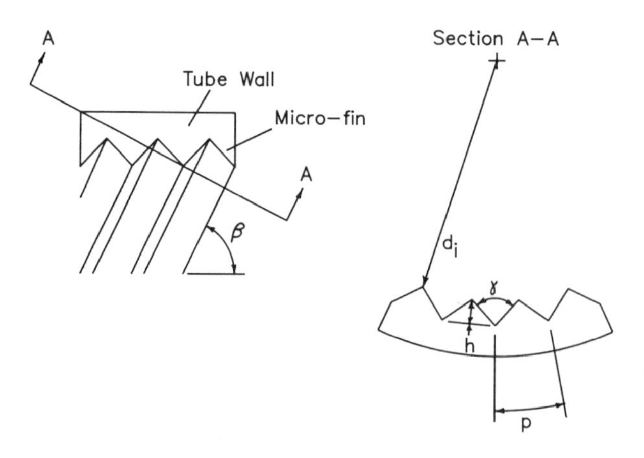

Figure 10-1 Geometry of microfinned tubes studied by Ito and Kimura (1979) and Kimura and Ito (1981).

Table 10-1 Ito and Kimura (1979) microfinned tube specifications

Tube material	Spiral angle β, degrees	Fin height h, mm	Fin pitch p, mm	Minimum inside diameter d_i, mm	Surface area ratio n
Copper	Smooth tube			11.10	1.00
Copper	7	0.06	0.52	11.60	1.07
	7	0.10	0.54	11.58	1.10
	7	0.16	0.52	11.46	1.26
	7	0.12	0.43	11.54	1.24
	7	0.10	0.32	11.64	1.17
Copper	90	0.16	0.50	11.45	1.42
	90	0.20	0.50	11.45	1.58
	90	0.30	1.00	11.80	1.38
	90	0.40	1.00	11.45	1.65
	90	0.24	1.75	11.80	1.14
Aluminum	Smooth tube			11.20	1.00
Aluminum	0	0.20	0.50	11.20	1.21
	3	0.20	0.50	11.20	1.21
	7	0.20	0.50	11.20	1.21
	15	0.20	0.50	11.20	1.21
	30	0.20	0.50	11.20	1.21
	75	0.20	0.50	11.20	1.21

data on the nominal surface area at the minimum inside diameter, (i.e., that at the tip of the fins). The area ratio cited in Table 10-1 is the ratio of the wetted surface area of the spiral grooved tube to that of a plain tube with the same diameter as the grooved tube's minimum inside diameter. This ratio is calculated as:

$$n = 1 + \frac{2h}{p}\left[\frac{1 - \sin(\gamma/2)}{\cos(\gamma/2)}\right] \tag{10-1}$$

The geometric factors are defined in Figure 10.1.

Kimura and Ito (1981) continued this work with R-12 for the additional fin sizes listed in Table 10-2. All tubes were copper. They again calculated their heat transfer data with the nominal area at the minimum inside diameter.

Figures 10-2 and 10-3 compare R-12 boiling data for otherwise identical plain and microfinned tubes, respectively, obtained by Kimura and Ito (1981). The data were taken for 1.0-m long, horizontal tube test sections that were electrically heated and instrumented over a central 0.3-m length. The mean vapor quality in the test sections was held constant at 55%. Figure 10-2 for the smooth tube depicts the variation in the heat transfer coefficient with heat flux and mass velocity that is also typical for larger-diameter smooth tubes. Figure 10-3 demonstrates that the situation is much different for the microfinned tube with a 7° spiral angle. Here the heat transfer coefficient goes through a minimum at low mass fluxes, whereas at higher mass fluxes the behavior is similar to that of the smooth tube. The heat transfer performance of the microfinned

Table 10-2 Kimura and Ito (1981) microfinned tube specifications

Tube	Spiral angle β, degrees	Fin height h, mm	Fin pitch p, mm	Minimum tube inside diameter d_i, mm	Tube outside diameter d_o, mm	Area ratio
Smooth				4.75	6.35	1.00
Microfinned	4	0.10	0.50	4.75	6.35	1.15
Microfinned	7	0.10	0.50	4.75	6.35	1.15
Microfinned	15	0.10	0.50	4.75	6.35	1.15
Microfinned	30	0.10	0.50	4.75	6.35	1.15

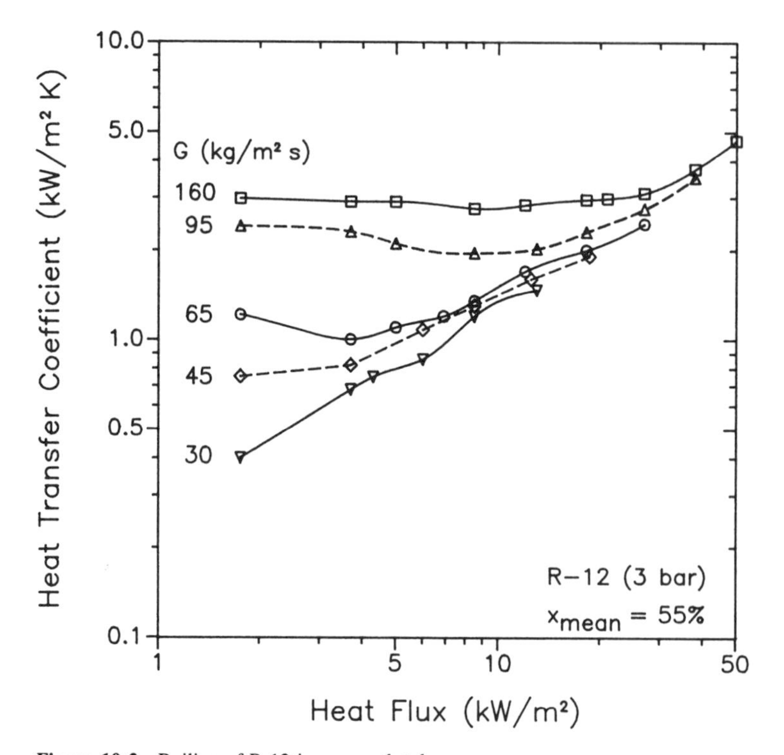

Figure 10-2 Boiling of R-12 in a smooth tube.

tube was notably higher than that for the smooth tube and was greater than its 15% increase in surface area.

Kimura and Ito (1981) explained the unexpected minima as being the result of capillary forces acting on the liquid film in the shallow grooves formed by the microfins. They reasoned that at low flow rates the capillary forces were sufficiently strong to pull liquid up from the stratified liquid to wet the entire circumference of the tube, thus producing annular flow that increased the heat transfer coefficient. As the heat flux increased at low flow rates, they hypothesized, the capillary films dried out before

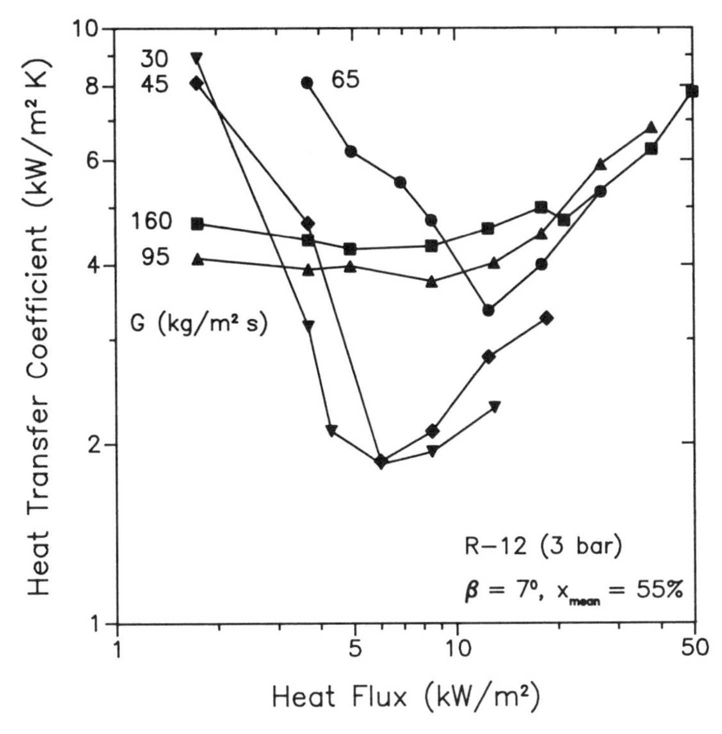

Figure 10-3 Boiling of R-12 in a microfinned tube.

reaching the top of the tube, thus reducing the heat transfer coefficient to that of stratified flow. Then, as the heat flux increased further, the nucleate boiling dominated regime was entered, and the heat transfer coefficient increased. This explanation is similar in some respects to the Grigorig (1954) phenomenon for film condensation on the outside of low-finned tubes.

The minima in Fig. 10-3 were not isolated to this particular spiral angle. They also occurred with spiral angles of 4°, 15°, and 30° (specified in Table 10-2) at mass fluxes less than 65 kg/(m^2·s). As shown in Fig. 10-4, the heat transfer coefficient was further augmented by this apparent capillary effect as the spiral angle increased. The heat transfer coefficient at the lowest heat fluxes was increased by a factor of 10 or more relative to the smooth tube. It is not clear whether these minima occur only for small microfinned tubes (4.75 mm in diameter), however. Additional research is required to investigate this phenomenon.

The effect of spiral angle on the heat transfer coefficient for R-12 boiling in microfinned tubes at a fixed heat flux of 8.5 kW/m^2 is shown in Fig. 10-5 (from Kimura and Ito [1981]); see Table 10-2 for the tube specifications. A maximum in the heat transfer coefficient occurred at high mass fluxes at a spiral angle of about 15°. A maximum did not occur at the lower mass fluxes. The optimal spiral angle for R-12 could not be determined, however, because the data only represent one vapor quality and heat flux; in practice the quality varies from 10% to 30% at the tube inlet and up

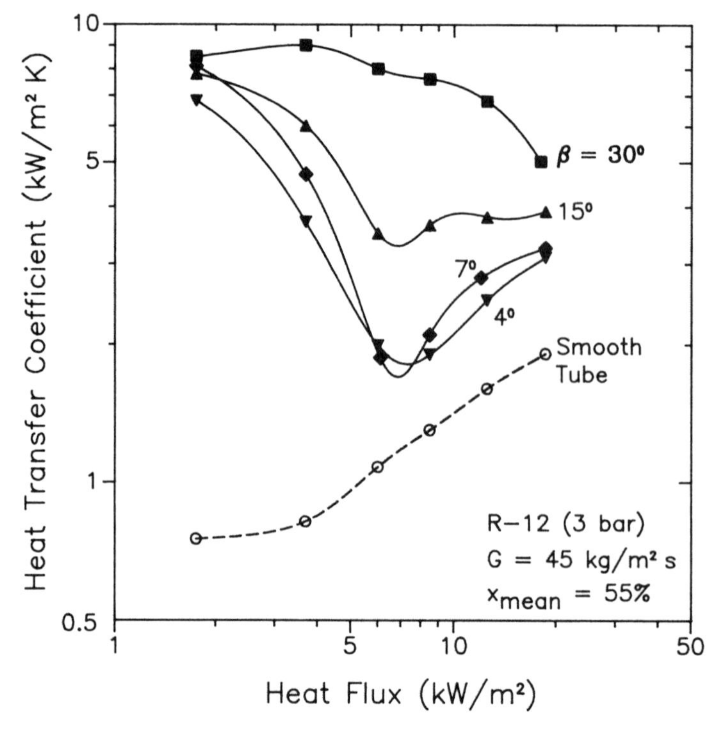

Figure 10-4 Effect of heat flux and spiral angle on R-12 boiling in microfinned tubes.

to 100% at the outlet. Heat transfer was augmented significantly relative to the smooth tube data points shown near the left axis.

The effect of spiral angle was also investigated by Ito and Kimura (1979). Figure 10-6 shows their heat transfer and pressure drop data for R-22 (see Table 10-1 for the tube specifications). Here the maximum in the heat transfer coefficient occurred at a spiral angle of about 7° for grooved aluminum test sections 11.2 mm in diameter. Heat transfer augmentation increased with increasing mass flux, although the level of augmentation was not all that substantial, as can be noted from the smooth tube data points. The pressure drops for horizontal 1.0-m long test sections were nearly identical to those for the smooth tube for spiral angles varying from 0° to 90°.

The effect of fin height on intube boiling in microfin tubes has been studied by Ito and Kimura (1979). Figure 10-7 depicts their boiling data as a function of heat flux for two different fin heights (see Table 10-1 for tube specifications) compared to a smooth tube. The mass flux ranged from 113 to 123 kg/($m^2 \cdot s$) and was calculated with the minimum inside diameter. At this mass flux level, no minimum in the heat transfer coefficient occurred as previously observed in Figs. 10-3 and 10-4. A fin height of only 0.1 mm provides substantial augmentation, considerably more than the 10% increase in surface area. The boiling heat transfer coefficient for the two microfin tubes behaved in a fashion quite similar to that displayed by the smooth tube. At low heat fluxes the heat transfer curves were flat, indicating that the process was convection controlled;

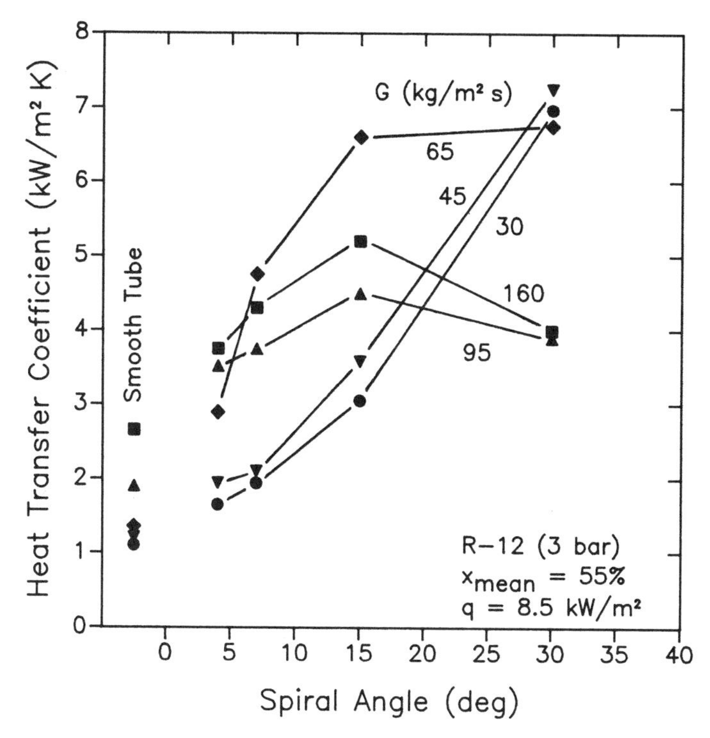

Figure 10-5 Effect of spiral angle on R-12 boiling in microfinned tubes.

at higher heat fluxes the curves rose, probably indicating that nucleate boiling was controlling.

Figure 10-8 shows the Ito and Kimura (1979) results plotted with the heat transfer coefficient and pressure drop per unit length of tube as a function of fin height at two different spiral angles and mass fluxes. The spiral pitch p and internal wetted surface area also varied as the fin height changed. The data at 199 kg/(m^2·s) have been approximately represented by the solid lines, and those at 80 kg/(m^2·s) are represented by the dashed lines. The boiling heat transfer coefficient increased with fin height for both spiral angles, more than the corresponding increase in wetted surface area. The pressure drop rose only slightly with fin heights up to 0.4 mm. For instance, the largest increase was only 19% for the spiral angle of 90° relative to the smooth tube values. The microfinned tubes with fin heights less than about 0.3 mm substantially augment the boiling heat transfer coefficient while behaving like smooth tubes in a fluid dynamic sense, at least under the present test conditions.

The effect of fin pitch on the heat transfer coefficient has been studied by Ito and Kimura (1979). Figure 10-9 for R-22 depicts their data for two different pitches (see Table 10-1 for the tube specifications). The fin height is the same for each of these 7° spiral angle tubes, but the tube with the smaller pitch has a 6% larger wetted surface area. The tube with 0.54-mm pitch gave better performance than the 0.32-mm pitch tube at low heat fluxes (which are convection dominated) and at high fluxes (which are

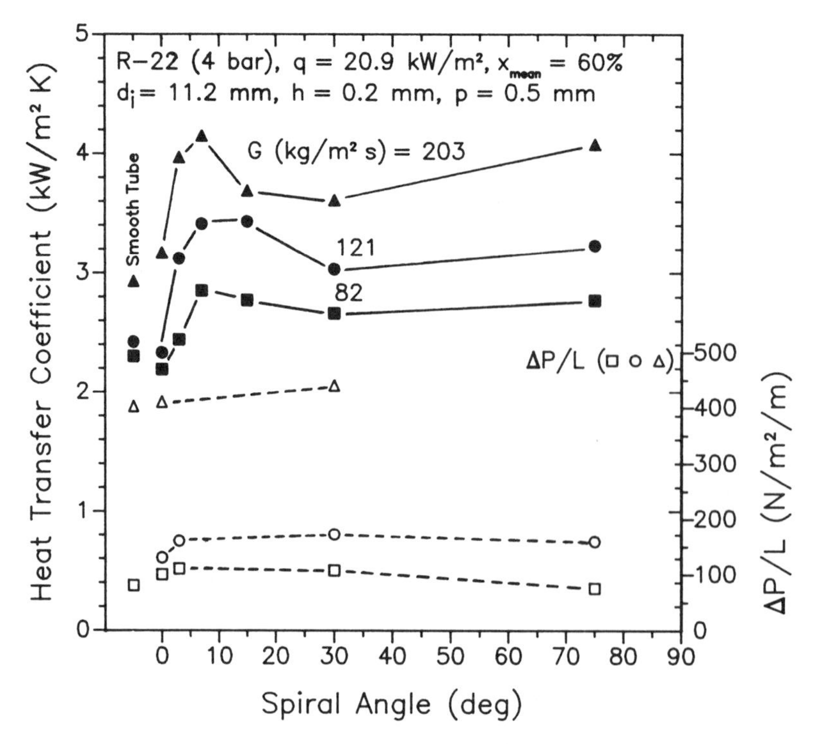

Figure 10-6 Effect of spiral angle on heat transfer and pressure drop for R-22 boiling in microfinned tubes.

nucleate boiling dominated), even though this tube has a smaller wetted surface area. In contrast, R-12 data obtained by Ito and Kimura (1979) (not shown) for a spiral angle of 90° displayed a decrease in the heat transfer coefficient as the fin pitch increased from 0.5 to 1.75 mm.

Shinohara and Tobe (1985) ran tests with R-22 for the Hitachi Thermofin tubes whose fin shapes are shown in Fig. 4-13. The tube specifications are given in Table 10-3, where the more recent Thermofin-HEX specifications were provided by Nakayama (personal communication 1987). Here the ratios of the wetted surface area of the Thermofin tubes were calculated relative to a plain tube whose inside diameter was set equal to the maximum inside diameter of the microfinned tube. The heat transfer data are apparently evaluated from the nominal area calculated with the maximum inside diameter.

Shinohara and Tobe (1985) investigated several geometric effects. For a fixed fin height of 0.2 mm and triangular fins, they observed that the heat transfer performance relative to a smooth tube increased linearly with a decrease in the apex angle of the fins. The maximum performance obtainable was limited by the increased difficulty of manufacturing the fins as the apex angle decreased, however. They were able to make apex angles as small as about 35°. They also observed that the heat transfer augmentation increased as the ratio of the fin height to minimum inside tube diameter increased. For ratios greater than about 0.03, the pressure drop began increasing relative to that of a

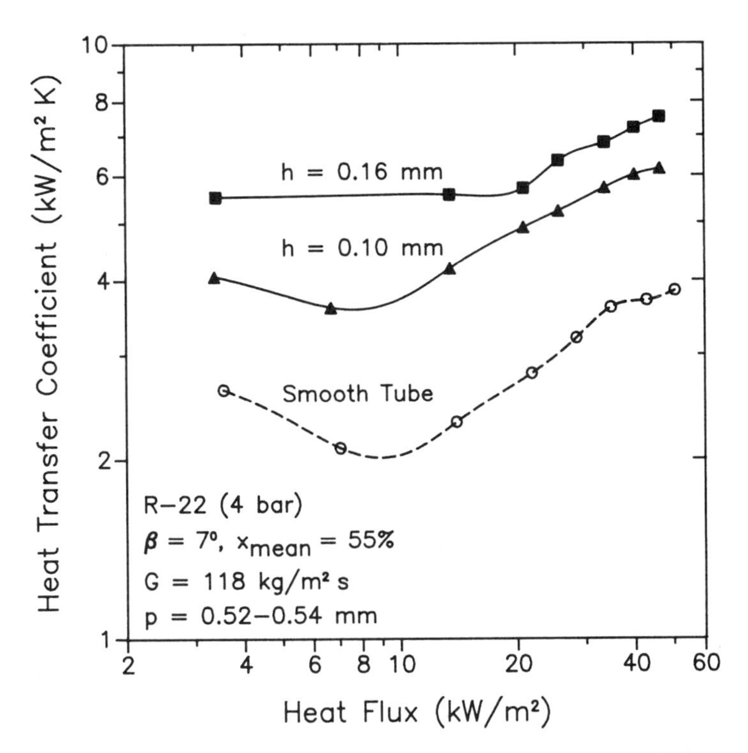

Figure 10-7 Effect of heat flux and fin height on boiling in microfinned tubes.

smooth tube. For triangular fins with a fixed fin height of 0.2 mm and a fin apex angle of 50°, a maximum in heat transfer augmentation was observed when plotted against the cross-sectional area of the groove between the fins.

Figure 10-10 from this study shows a comparison of the boiling performances for the various Thermofin tubes shown in Fig. 4-13 for evaporating R-22 in horizontal test sections (see Table 10-3 for the tube specifications). The more recent Thermofin-HEX data were provided by Nakayama (personal communication 1987). The maximum increase in the pressure drop was stated to be about 5% for the Thermofin-EX version compared to the smooth tube at a mass flux of 200 kg/(m^2·s), and the heat transfer coefficient was increased by 190%.

At Iowa State University, tests have been run with many different types of intube boiling enhancements. Reid, Pate, and Bergles (1987) conducted a performance comparison test for the tubes described in Table 10-4 (the results for the twisted tape enhancement and tubes with fin heights greater than 0.4 mm are described later). Triangular, rectangular, and trapezoidal fin cross-sections were represented. Two different tube diameters were tested, and smooth tube results were also obtained for reference purposes. Khanpara, Pate, and Bergles (1987) performed a comparison study for both R-22 and R-113 with one microfinned tube and a plain tube. The tube specifications are listed in Table 10-5, and a schematic drawing of the Thermofin tube tested is

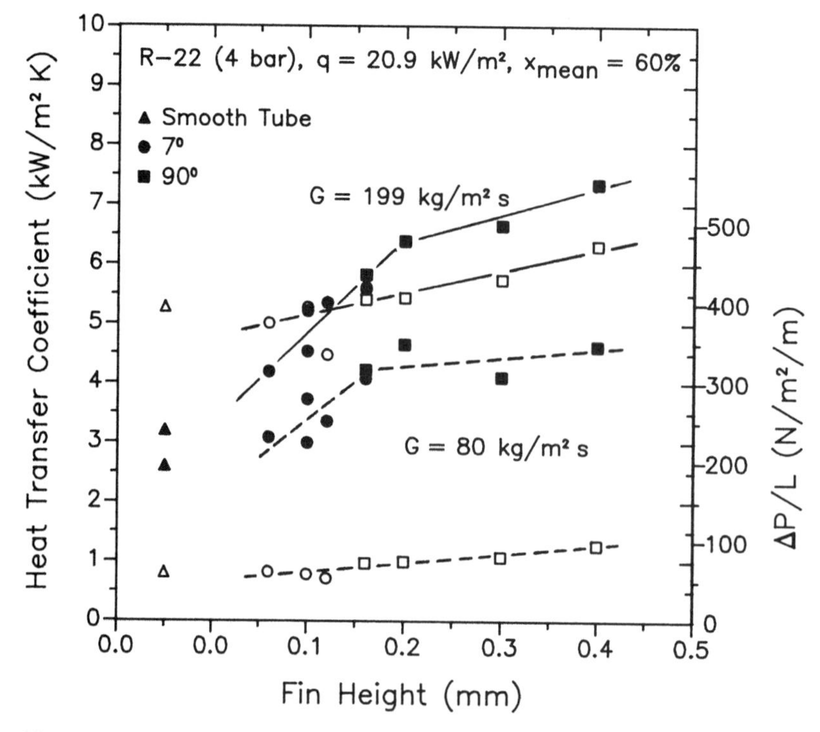

Figure 10-8 Effect of fin height on heat transfer (solid symbols) and pressure drop (open symbols) for boiling in microfinned tubes.

depicted in Fig. 10-11. The areas used for heat transfer calculations in these two studies were based on the maximum inside diameter of the microfinned tubes.

Figure 10-12 (from Reid, Pate, and Bergles [1987]) depicts local heat transfer coefficients as a function of local vapor quality at several heat fluxes for R-113 evaporating in a horizontal, electrically heated, 3.66-m long Hitachi Thermofin tube. At low vapor qualities the heat transfer coefficients were dependent on heat flux, which indicates a nucleate boiling regime; at high vapor qualities the heat transfer coefficient was less sensitive to heat flux (for the small range studied at least), indicating a convection controlled regime. At a fixed heat flux a minimum occurred in the heat transfer coefficient, the location of which moved toward higher vapor quality as the heat flux level increased. These minima, however, are shallow and may only be the result of experimental error. Actually, a maximum deviation of only about 20% occurs at each heat flux, demonstrating that the local heat transfer coefficient is fairly insensitive to quality in the range from 1% to 72%.

Figure 10-13 depicts the pressure drop of the Thermofin tube relative to the smooth tube obtained by Reid, Pate, and Bergles (1987) with five local pressure taps for conditions similar to those in Fig. 10-12. The pressure drops for the Hitachi Thermofin tube were as much as 180% greater than those for the smooth tube under the same operating conditions. The pressure drop ratio at nearly zero vapor quality was 1.6 times

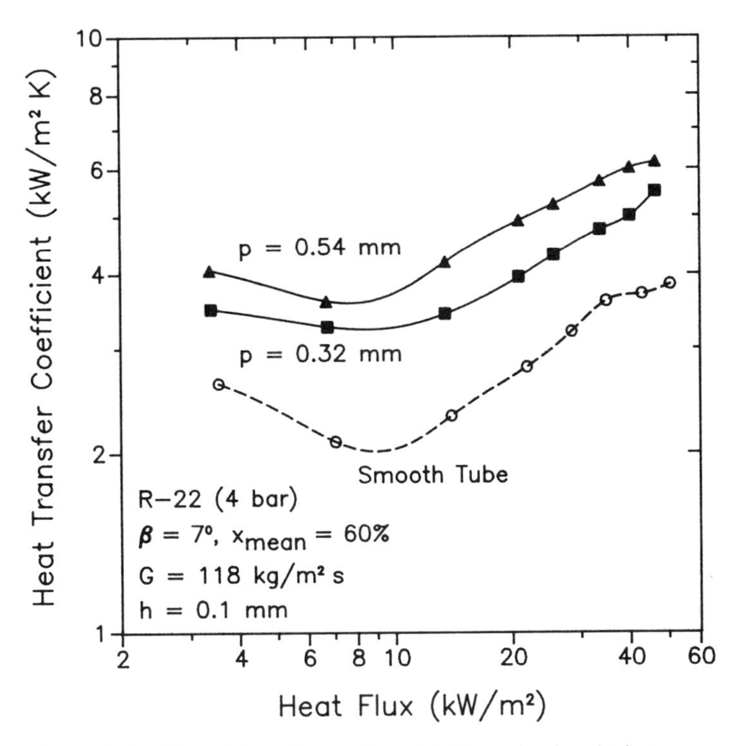

Figure 10-9 Effect of fin pitch on boiling of R-22 in microfinned tubes.

Table 10-3 Shinohara and Tobe (1985) Thermofin tube specifications

Tube type	Fin shape	Outside diameter, mm	Average wall thickness, mm	Minimum wall thickness, mm	Fin height, mm	Number of fins	Spiral angle, degrees	Weight per unit length, g/m	Area ratio
Plain		9.52	0.30					78	1.00
Thermofin	A	9.52	0.35	0.30	0.12	60	7	90	1.34
Thermofin	B	9.52	0.37	0.30	0.15	65	25	95	1.28
Thermofin-EX	C	7.94	0.36	0.30	0.20	50	18	77	1.52
Thermofin-EX	C	9.52	0.36	0.30	0.20	60	18	93	1.51
Thermofin-EX	C	12.70	0.48	0.40	0.25	75	18	166	1.61
Thermofin-EX	C	15.88	0.59	0.50	0.30	75	18	255	1.58
Thermofin-HEX	C′	9.52	0.36	0.30	0.20	60	18	93	1.60

that of the smooth tube and reached a maximum at a vapor quality of 20%. It then decreased rapidly as quality increased, leveling off at a ratio of 1.3. These ratios are much larger than those reported by Shinohara and Tobe (1985) and were for a ratio of fin height to minimum inside diameter of only 0.025. Shinohara and Tobe had observed no significant increase for ratios less than 0.03 for R-22.

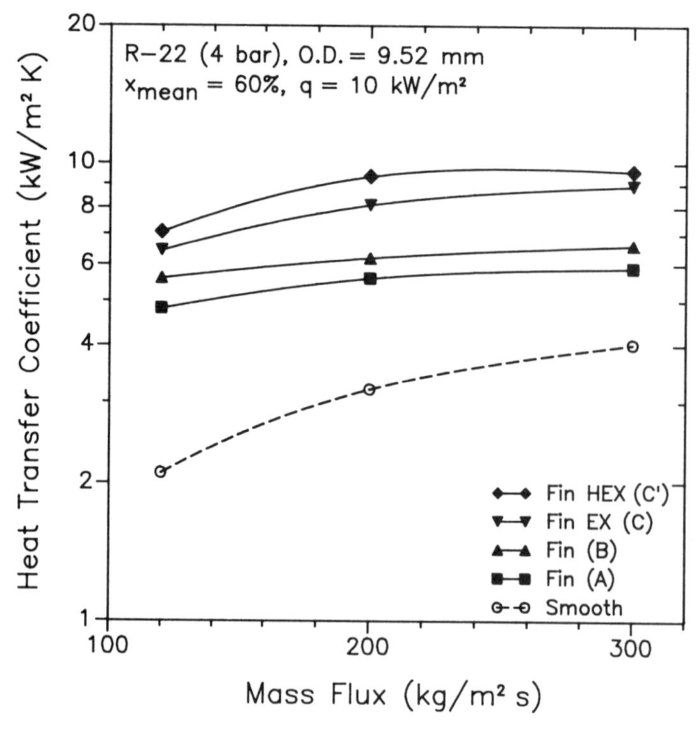

Figure 10-10 Comparison of Thermofin tube boiling performances for R-22 at 4 bar.

The results in Fig. 10-13 are similar to those obtained by Ito and Kimura (1979) for microfin tubes with fin heights of 0.22 mm and a ratio of fin height to minimum inside diameter of 0.017. They observed small increases in the friction factor compared to a smooth tube for a 7° spiral angle but much larger increases as the spiral angle increased further. The spiral angle of the Thermofin tube tested by Reid and co-workers was 17.5°. Ito and Kimura (1979) also observed that the friction factor for circulating water through their test apparatus varied as a function of Reynolds number, similar to the situation for rough tubes in turbulent single-phase flow.

Khanpara, Pate, and Bergles (1987) obtained data for R-22 in a horizontal copper microfinned tube with a 3.83-m electrically heated length. Some of their results are shown in Fig. 10-14 (see Table 10-5 for the tube specifications). No appreciable variation in the local heat transfer coefficient with quality was evident over the range 5% to 77% at the mass fluxes shown; this was similar to the smooth tube's behavior. The heat transfer coefficient did increase with increasing mass flux at a fixed vapor quality but to a lesser degree than was observed for the smooth tube. Their results for R-113 evaporating at 3.3 bar were similar to those for R-22.

The above tests were limited to maximum local vapor qualities of about 75%. Because the exit condition in refrigeration equipment is typically either 100% vapor quality or several degrees of superheating, the performance of microfinned tubes in the high–vapor quality range is of practical interest. Figure 10-15 (from Ito and Kimura [1979]) depicts results for vapor qualities up to 95% for two microfinned tubes with

Table 10-4 Reid, Pate, and Bergles (1987) tube specifications

Tube (material)	Fin shape	Outside diameter, mm	Inside diameter, mm	Number of fins	Spiral angle, degrees	Fin height, mm	Base width, mm	Tip width, mm	Wall thickness, mm
Smooth (copper)		9.53	8.71						0.41
Hitachi Thermofin* (copper)	Triangular	9.53	8.71	65	17.5	0.21	0.29	0.0	0.41
Noranda T0038† (copper)	Rectangular	9.53	8.51	21	30	0.38	0.43	0.43	0.51
Smooth (stainless steel)		12.70	10.92						0.89
Wieland† (copper)	Trapezoidal	12.58	10.92	32	0	0.55	0.40	0.24	0.76
Wieland† (copper)	Trapezoidal	12.55	10.92	32	16	0.55	0.40	0.24	0.76
Noranda T0050† (copper)	Rectangular	12.70	11.43	26	31	0.51	0.43	0.41	0.64
Twisted tape‡ (stainless steel)		12.70	10.92						0.89

*Microfin
†High fin
‡Helical tape.

Table 10-5 Khanpara, Pate, and Bergles (1987) tube specifications

Tube (material)	Fin shape	Outside diameter, mm	Inside diameter, mm	Number of fins	Spiral angle, degrees	Fin height, mm	Fin pitch, mm	Valley width, mm	Area ratio
Smooth (copper)		9.53	8.83						1.00
Microfin (copper)	Trapezoidal	9.53	8.83	60	16 to 17.5	0.22	0.46	0.20	1.54

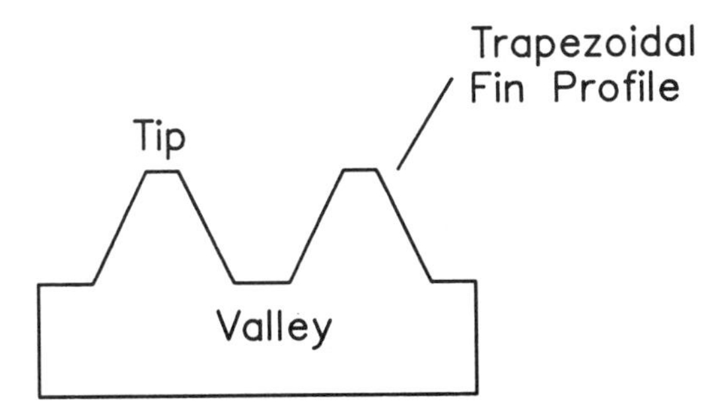

Figure 10-11 Geometry of microfinned tube tested by Khanpara, Pate, and Bergles (1987).

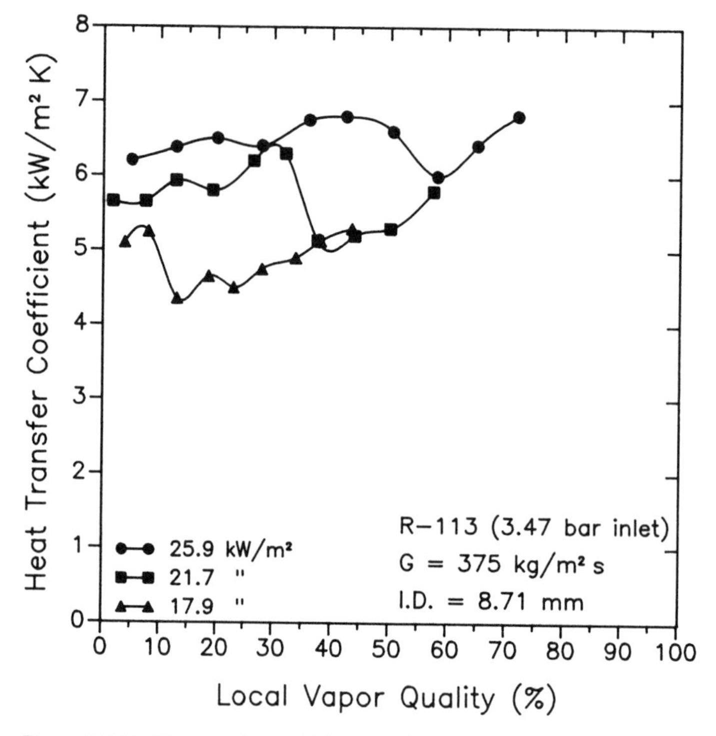

Figure 10-12 Heat transfer coefficient as a function of heat flux and local vapor quality for a Thermofin tube.

spiral angles of 7° and for a smooth tube (see Table 10-1 for the tube specifications). The vapor quality shown is the average value for a 0.2-m long heated test section. Hence the 95% vapor quality probably represents complete vaporization at the exit. The investigators found that the heat transfer coefficient increased with vapor quality, even in the high quality region, at the test conditions noted. The greater fin height produced more augmentation.

Two phenomena could explain the high heat transfer performance observed at large vapor qualities. First, the centrifugal force imparted to the vapor phase by the spiral fins may be sufficient to force the entrained liquid droplets to the wall, where they are readily evaporated. Second, the surface tension acting on the liquid film in the shallow grooves between the fins may be strong enough to prevent the film from being broken up into droplets. Obtaining high–vapor quality data for a spiral angle of 0° (i.e., for longitudinal fins) would be valuable for studying this mechanism because no centrifugal force would be imposed on the flow.

In summary, the following conclusions can be made for microfins:

1. Local heat transfer coefficients can be augmented by as much as a factor of 10 under some special conditions but more typically by about 50% to 250% (including the additional surface area) compared to a smooth tube.

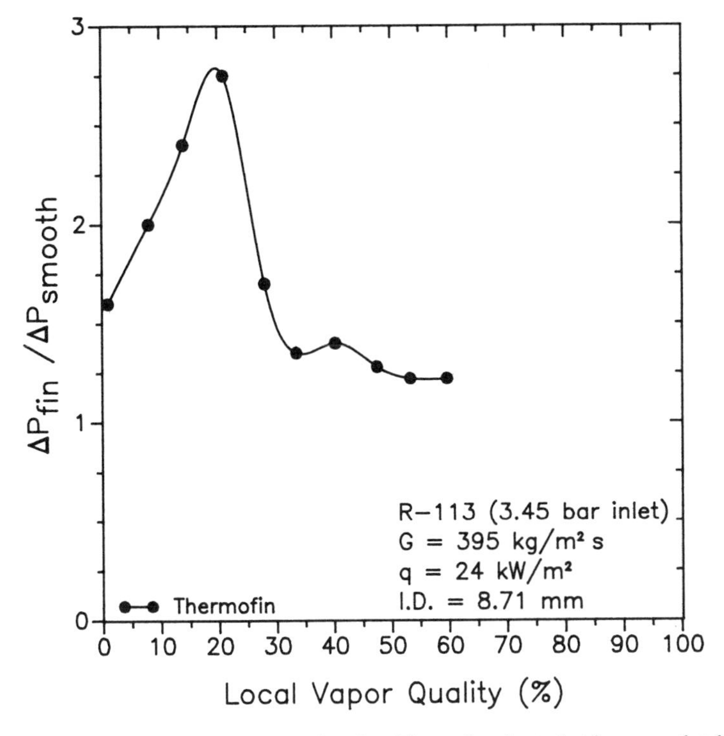

Figure 10-13 Ratio of pressure drop in a Thermofin tube to that in a smooth tube.

2. Heat transfer augmentation tends to increase with spiral angle up to about 10° to 15°, after which it tends to deteriorate, especially at high mass fluxes.
3. Microfin heat transfer coefficients increase with fin height for spiral angles from 7° to 90°.
4. Performance increases with fin pitch for small spiral angles; at large spiral angles, an increase in fin pitch is detrimental.
5. The heat transfer coefficient is a strong function of mass flux and heat flux but less so of vapor quality.
6. Microfinned tubes augment the heat transfer coefficient relative to a smooth tube by three means: (i) more surface area per unit length of tubing, (ii) increased circumferential wetting of the tube wall by capillary forces at low flow rates, and (iii) increased convection from the turbulence promoted by the fins.

The reader is also referred to a general survey of Japanese research on boiling in microfinned tubes by Tatsumi et al. (1982). They report some specific performance data for R-22 in a commercial microfinned tube with a 7° spiral angle. This tube was reported to provide up to twice the heat transfer coefficient of a plain tube with no appreciable increase in pressure drop. Tojo et al. (1984) also found similar results for R-22 in a microfin tube.

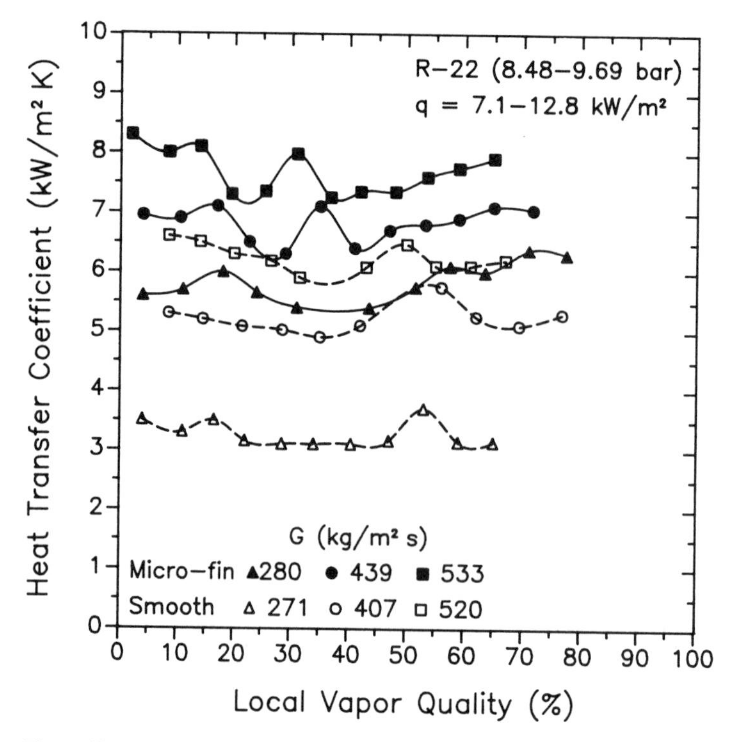

Figure 10-14 Heat transfer coefficient as a function of mass flux and local vapor quality for a microfinned tube.

High-fin effects. The effects of fin geometry of high-finned tubes on boiling heat transfer augmentation have not been investigated as thoroughly as those of microfinned tubes. Fin heights studied range from about 0.4 mm up to as high as 8 mm. Many different fin profiles have been tested, but it is not usually possible to compare the experimental results from these different studies because tube diameters, fluids, test conditions, and so on are not the same. Several comparative studies have been made: Lavin and Young (1965), Kubanek and Miletti (1979), and Reid, Pate, and Bergles (1987). The tube specifications for the last study are given in Table 10-4.

Lavin and Young (1965) conducted an extensive investigation with R-12 and R-22 for boiling in a plain tube, an internal screw tube (a tube with internal screw threads), longitudinal and helical splined tubes, and a cruciform tube, all in both horizontal and vertical orientations. Schematic diagrams of the tubes are shown in Fig. 10-16, and the tube specifications are given in Table 10-6. The tubes were made by Wolverine Tube and Hecla, Inc. The longitudinal and helical spline tubes differed only in spiral angle. The microfinned tubes tested by Ito and Kimura (1979) were similar to the screw tube tested here, except for the fin sizes. The test sections were 0.30 m long and heated electrically. Wall temperatures were measured at the midpoint of each test section. The local heat flux and heat transfer coefficient were determined at that point with the total wetted internal surface area rather than the nominal inside area. The

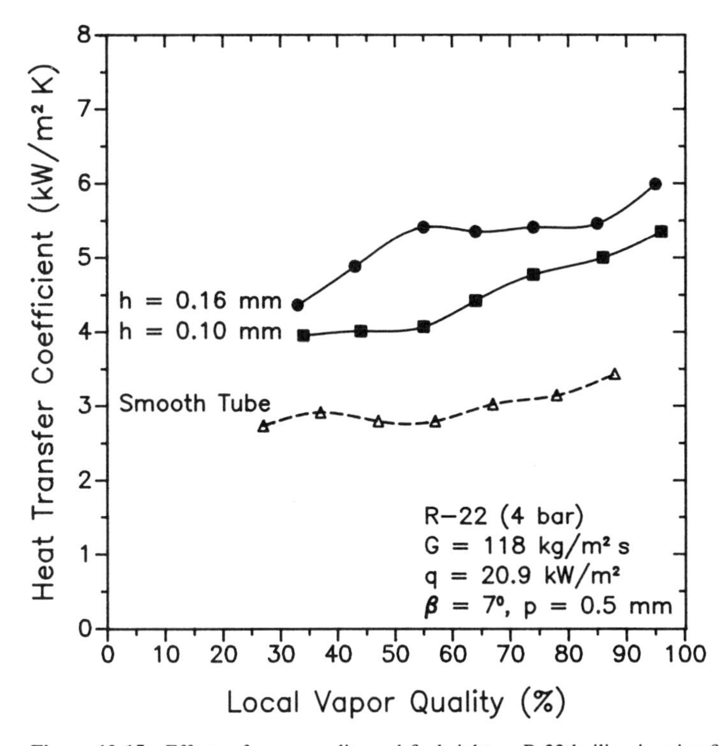

Figure 10-15 Effects of vapor quality and fin height on R-22 boiling in microfinned tubes.

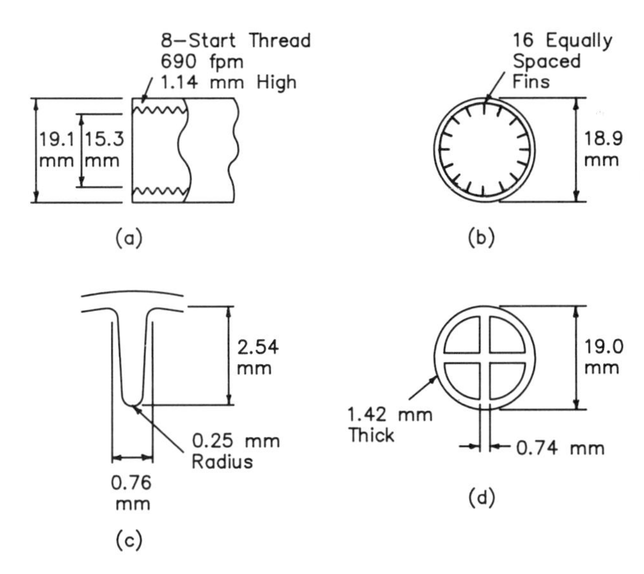

Figure 10-16 Finned tubes tested by Lavin and Young (1965). (*a*) Screw tube, (*b*) cross-section of splined tube, (*c*) splined tube fin profile, and (*d*) cruciform tube.

Table 10-6 Lavin and Young (1965) tube specifications

Tube type (material)	Outside diameter, mm	Maximum inside diameter, mm	Fin height, mm	Fin thickness at base, mm	Fin tip radius, mm	Number of fins	Spiral angle, degrees	Wetted area per unit length, m^2/m	Area ratio*	Flow area, $m^2 \times 10^3$	Equivalent hydraulic diameter, mm
Smooth tube (copper)	20.4	18.6						0.0582	1.00	0.272	18.6
Screw tube, triangular fins (copper)	19.1	17.6	1.14	1.34		19†	79	0.103	1.86	0.183	8.26
Longitudinal splined tube, tapered fins (copper)	18.9	16.2	2.54	0.76	0.25	16		0.125	2.46	0.182	5.82
Helical splined tube, tapered fins (copper)	19.0	16.5	2.54	0.76	0.25	16	10	0.126	2.43	0.188	5.97
Cruciform tube, inner cross fins (aluminum)	19.0	16.2	7.72	0.74		4		0.104	2.04	0.162	6.28

*Relative to nominal area at tube's maximum internal diameter.
†19 fins per inch (fpi) (750 fins per meter).

investigators noted that the fin efficiencies were close to 100% for all the tubes. The cruciform tube was apparently made by an extrusion process because the drawing shows the inner cross as an integral part of the tube not as an insert.

Lavin and Young (1965) presented most of their data in correlated form, so that only a few observations can be made from their results. For instance, they found no difference in the performances of the longitudinal and helical splined tubes in the nucleate boiling regime of flow boiling, both giving about an 80% improvement in the heat transfer coefficient beyond that of the increased surface area (this represents a 340% improvement compared to the smooth tube when calculated from the nominal inside surface area of the finned tubes). Thus the boiling process was augmented by the presence of the fins. No effect of tube orientation, vertical or horizontal, was observed for the two splined tubes, but a horizontal position gave better smooth tube performance. For local vapor qualities ranging from 25% to 75% in the annular flow regime, again no substantial difference occurred in the performances of the two splined tubes in vertical orientations. Heat transfer augmentation in the transition between the annular flow and mist flow regimes with the tubes in a vertical orientation was also found to be the same for the two splined tubes, the improvement being about 200% compared to the smooth tube (based on the total wetted surface area of the splined tubes) at the same Reynolds number.

The boiling performance for the cruciform tube in R-12 was less than the performances for the screw and splined tubes in all three flow regimes studied. For the nucleate and annular flow regimes the screw tube performed the same as the two splined tubes, but in the transition between the annular flow and mist flow regimes its performance was about 20% worse.

A performance comparison study with R-22 was conducted by Kubanek and Miletti (1979). They tested several Noranda Metals internally finned tubes, two smooth tubes of different diameters, and a tube with a star insert. Table 10-7 lists the tube specifications. The star insert had five splines. Average heat transfer data were obtained with warm water as the heating source. The boiling-side heat transfer coefficients were calculated from a modified Wilson plot. Two vapor quality changes, 20% and 70%, between the inlet and outlet were studied. The different inlet qualities were achieved by mixing saturated vapor and liquid streams before the inlet to the test section. Two test section lengths were tested, 0.80 and 2.44 m. The heat transfer coefficients were calculated with the nominal inside surface area at the maximum inside diameter. Pressure drops were measured with pressure taps spaced 1.07 m apart for the 0.80-m test sections and 2.74 m apart for the 2.44-m sections.

Kubanek and Miletti (1979), contrary to Lavin and Young (1965), observed a notable difference with spiral angle for their two high-finned tubes (22 and 25). Figures 10-17 and 3-22 show better heat transfer performance for tube 25 than for tube 22 with spiral angles of 16.9° and 8.6°, respectively. The effect of larger spiral angle was thus to increase performance, at least in this range of angles. Boiling data for tube 25 at different vapor qualities are also depicted in Fig. 10-17. No consistent effect of vapor quality on heat transfer was found, but there was a large standard deviation in the data, apparently because a wide range of heat fluxes are represented.

For fin heights similar to those of Kubanek and Miletti (1979), Reid, Pate, and

Table 10-7 Kubanek and Miletti (1979) tube specifications

Tube type (material)	Outside diameter, mm	Maximum inside diameter, mm	Fin height, mm	Fin pitch, mm	Number of fins	Spiral angle, degrees	Wetted surface area, m^2/m	Area ratio*	Flow area $m^2 \times 10^3$	Equivalent hydraulic diameter, mm	Heated length, m	
											Test length 1	Test length 2
Smooth tube 24A (copper)	19.1	16.9					0.0531	1.00	0.224	16.9	2.44	
Smooth tube 24B (copper)	15.9	14.4					0.0453	1.00	0.163	14.4	0.80	
Star insert 24C (aluminum)	15.9	14.4			5	4.2†	0.0908	2.01	0.093	4.09	0.80	2.44
High-finned tube 22 (copper)	15.9	14.7	0.635	1.14	32	8.6	0.0870	1.88	0.163	7.57	0.80	2.44
High-finned tube 25 (copper)	15.9	14.7	0.635	1.14	32	16.9	0.0872	1.89	0.163	7.57	0.80	2.44
High-finned tube 30 (copper)	12.7	11.9	0.508	0.76	30	20.1	0.0680	1.82	0.107	6.30	0.80	

*Relative to nominal area at tube's maximum internal diameter.
†Twist angle.

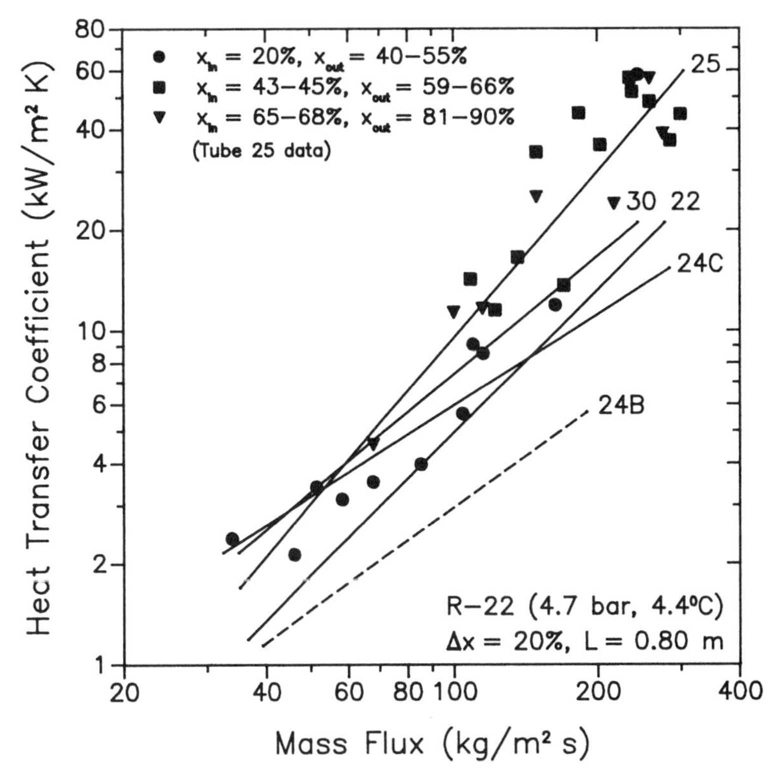

Figure 10-17 Boiling R-22 comparative study with 0.80-m test sections for an inlet-to-outlet vapor quality change of 20%.

Bergles (1987) found better heat transfer performance for a Wieland tube with a 16° spiral angle than for its identical longitudinal version. Figure 10-18 depicts their heat transfer and heat transfer–pressure drop performances relative to a smooth tube (see Table 10-4 for the tube specifications; the data for the other enhancements shown are discussed later). The improvement in heat transfer with spiral angle was more notable in the 20% to 50% local vapor quality range than at higher or lower ones. Pressure drops for the 16° tube were greater than those of the longitudinal tube, however, and thus the enhancement performance ratios shown in the upper half of Fig. 10-18 were higher for the longitudinal finned tube than for the spiral one.

The effect of fin shape on heat transfer performance has not received systematic attention. One of the few comparisons that can be made is shown in Fig. 10-18. Here, the Noranda rectangular fins are about the same size as the two Wieland trapezoidal finned tubes but have different spiral angles. The former gave similar heat transfer augmentation but slightly better enhancement performance ratios. Apparently the effect of fin shape is therefore less important for high fins than for microfins.

The variation in heat transfer augmentation with local vapor quality appears to be substantial in Fig. 10-18, more than was evident for microfinned tubes. This variation,

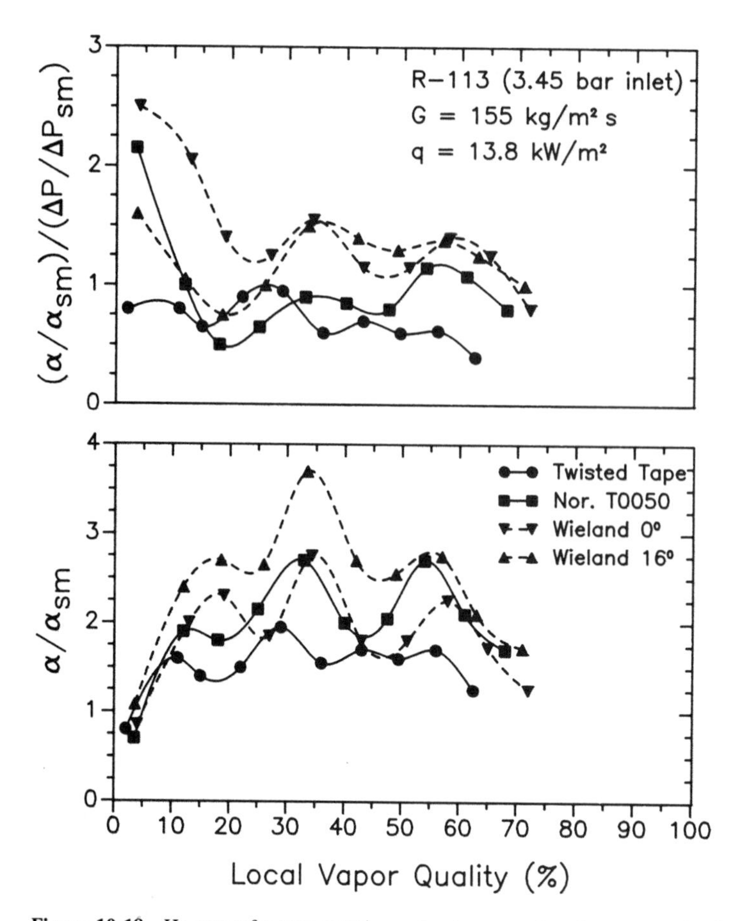

Figure 10-18 Heat transfer augmentation and enhancement performance ratios for R-113.

however, is actually a result of fluctuations in the smooth tube heat transfer values, not in the high-finned tube data. The effects of mass velocity and heat flux on the high-fin heat transfer coefficient were similar to those for smooth tubes for both the small- and large-diameter Noranda tubes (T0038 and T0050) tested by Reid, Pate, and Bergles (1987).

In summary, the spiral angle for high-finned tubes appears only to affect performance at fin heights of about 0.5 mm, augmenting the heat transfer coefficient with increasing spiral angle. The pressure drop also increases as the spiral angle increases, however. The optimum fin type is triangular or rectangular rather than screw or cruciform. All these conclusions, however, are based on a small sample of test results. Parametric studies for high fins need to be made to determine the interrelationships among fin height, thickness, pitch, spiral angle, shape, and so forth. Even so, the optimum fin for one refrigerant may give poor augmentation when applied to another.

Corrugated and spirally fluted tubes. Corrugated and spirally fluted tubes, the latter of which can be considered deeply corrugated tubes, are reasonably simple to fabricate and provide heat transfer augmentation on both the inside for boiling and on the outside for single-phase convection or condensation. A conventional corrugated tube is shown in Fig. 4-19, and one with internal fins for further augmentation of intube boiling is depicted in Fig. 4-20. A spiral fluted tube is shown in Fig. 4-18.

The geometric characteristics of these types of tubes are shown in Fig. 10-19. The helical pitch (width of corrugation), ridge height, spiral angle, inside tube diameter, and ridge contour can be varied to optimize the boiling performance. To quantify the geometrical relationship for a given ridge contour and spiral angle, Withers and Habdas (1974) defined a dimensionless parameter called the severity factor, which is given as

$$\phi = \frac{e^2}{pd_i} \tag{10-2}$$

Their rational for this definition was that increasing the ridge height or decreasing the pitch or inside diameter while holding the other two parameters fixed results in a more severe geometry relative to a plain tube. When the ridge height is zero as for a plain tube, the severity factor goes to zero. Instead, for a fixed ridge height e, reducing the tube inside diameter or the pitch increases the severity factor.

Withers and Habdas (1974) undertook an extensive experimental program to determine the optimum geometry for boiling R-12 inside horizontal corrugated tubes. The fluid entered as a saturated liquid ($x = 0\%$) and exited as a saturated vapor ($x = 100\%$). The test sections were heated by countercurrent flow of warm water. Overall heat transfer coefficients were determined, and the average inside heat transfer coefficients were then calculated from a Wilson plot analysis derived from separate water-to-water tests. A pump was used in the flow circuit rather than a compressor to prevent contamination of the refrigerant with lubricating oil. The 3.96-m long heated test sections had essentially the same outside diameter of 19 mm (¾ inch), the pitch was varied from 3.2 to 15.9 mm (⅛ to ⅝ inch), and the ridge height was varied from 0.4 to 1.6 mm (1/64 to 1/16 inch). Tubes with severity factors ranging from 0 (i.e., a smooth tube) up to 0.0092 were made and tested. The overall pressure drops between inlet and outlet were measured over a test section length of 4.27 m. The tubes were made in copper by Wolverine Tube.

Figure 10-20 depicts the test data as a function of the severity factor. The average boiling-side heat transfer coefficient and the boiling capacity w_{FV} were found to pass through a maximum value at a severity factor of about 0.004, where the boiling capacity was defined as the flow rate (in kilograms per hour) of R-12 fully vaporized. For this

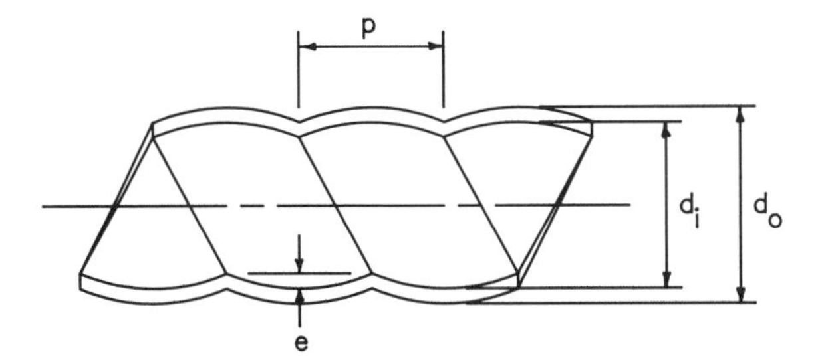

Figure 10-19 Diagram of a corrugated tube.

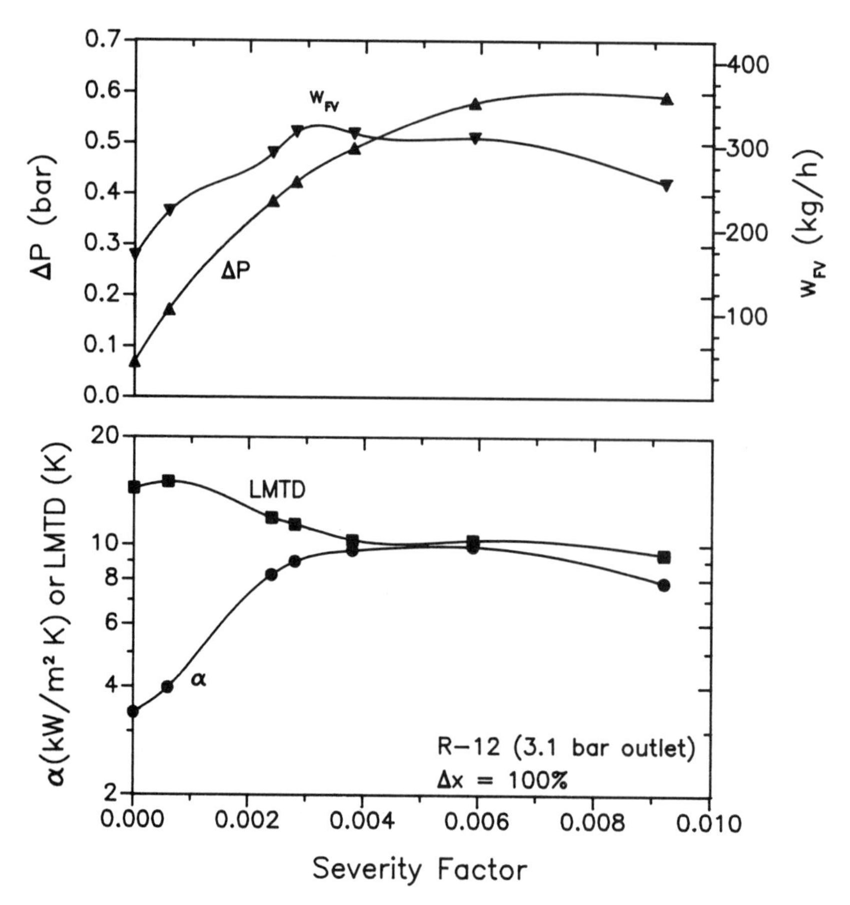

Figure 10-20 Corrugated tube optimization results for R-12. LMTD, log mean temperature difference (in Kelvins).

corrugated tube geometry the heat transfer coefficient was augmented by 190%, and the boiling capacity increased by 85%, relative to the plain tube. The boiling-side pressure drop increased with increasing severity until leveling off at a value of 0.6 bar at a severity factor of 0.006. The maximum boiling capacity of 314 kg/h corresponds to a mass flux of 383 kg/(m^2·s) and an average heat flux of 63.8 kW/m^2 based on a nominal inside diameter of 16.3 mm (0.67 inch).

Figure 10-21 depicts similar R-12 data for two corrugated tubes of different diameters given in a Wolverine Tube Koro-Chil product sheet (see Bell and Mueller [1984]). The mass flux was based on the inside diameter of the uncorrugated portion of the tubing. The ¾- and ⅝-inch (19.0- and 15.9-mm) Koro-Chil tubes both had groove pitches of ⅜ inch (9.5 mm) and groove depths of 0.041 inch (1.04 mm) with internal diameters of 0.714 and 0.589 inch (18.1 and 15.0 mm), respectively, for severity factors of 0.0063 and 0.0076. The ¾-inch Koro-Chil tube out-performs both

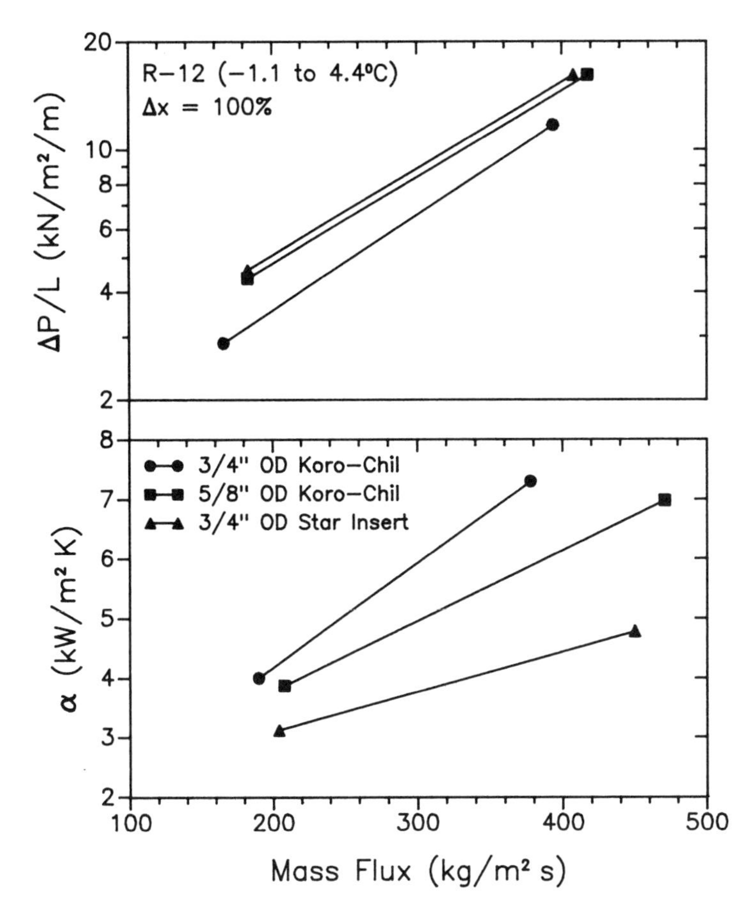

Figure 10-21 Heat transfer and pressure drop comparison for corrugated and star-insert tubes boiling R-12. OD, outer diameter.

the ⅝-inch version and a five-spline, aluminum star-insert tube while having a substantially smaller pressure drop.

Two other studies have been performed. Kawai and Machiyama (1975) and Kawai and Yamada (1977) have tested R-12 and R-22, respectively, in several corrugated tubes. They found that their overall heat transfer performances were better by about a factor of 2 relative to a high-finned tube also tested.

The spiral fluted tube of GA Technologies has been tested in vertical upward flow with R-11 and R-114. This work has been described by Panchal et al. (1986) and in more detail by Panchal and France (1986). The test unit was a vertical shell-and-tube heat exchanger. The tube bundle was 356 mm in diameter and contained 84 tubes arranged in a 30° triangular pattern with a ratio of tube pitch to tube diameter of 1.35. The total outside surface area of the tube bundle was 51 m^2. The tube specifications are given in Table 10-8. From Eq. (10-2) the severity ratio is calculated to be 0.034,

Table 10-8 Spiral fluted tube specifications

Tube material	Aluminum (Al 6063)
Wall thickness	1.65 mm
Tube flow area	563.2 mm^2
Mean inside diameter*	26.77 mm
Mean outside diameter	30.10 mm
Inside perimeter	137.5 mm
Outside perimeter	136.4 mm
Hydraulic diameter	16.38 mm
Flute angle to tube axis	30°
Pitch of flutes	2.63 mm
Depth of flutes	1.54 mm
Effective tube length	4.45 m
Inside diameter area ratio†	1.635
Outside diameter area ratio‡	1.443

*Based on cross-sectional flow area.
†Relative to smooth tube with 26.77 mm inside diameter.
‡Relative to Smooth Tube with 30.10 mm outside diameter.

which is much larger than the value for the corrugated tubes described above. The spiral angle was 30°. A photograph of the tube bundle is shown in Fig. 10-22.

The tube bundle was heated on the shell-side by condensing steam. The intube boiling heat transfer coefficients were determined from the measured overall heat transfer coefficients by using condensation data obtained in another part of the project. Because the refrigerants R-11 and R-114 entered the tube bundle subcooled, the tube-side heat transfer coefficients were further adjusted by subtracting the single-phase heat transfer contribution by using previously measured single-phase test data. The tube-side heat transfer coefficients thus obtained were mean values for the boiling zone and were calculated from the total inside wetted surface area.

Figure 10-23 from (Panchal et al. [1986]) depicts a comparison between the boiling performance of the spiral fluted tube bundle and that of a smooth tube, the latter of which was predicted by a heat transfer correlation of Herd, Goss, and Connell (1985) developed with R-22, R-113, and R-114 data. Heat fluxes for R-11 ranged from 1.35 to 14.8 kW/m^2, and those for R-114 varied from 9.8 to 17.5 kW/m^2. Mass fluxes for R-11 ranged from 60 to 182 $kg/(m^2 \cdot s)$, and those for R-114 varied from 132 to 231 $kg/(m^2 \cdot s)$. The tests showed that for similar operating conditions the spiral fluted tube augmented the boiling heat transfer coefficient by an average of 80% and 170% for R-11 and R-114, respectively, in addition to the extra surface area of 63.5%, which represents an overall increase of 194% and 340%. The spiral fluted tube was observed to perform well even at exit vapor qualities up to 90%. Pressure drop data are available in Panchal and France (1986) for the entire height of the tube, including the sensible heating region. They were on the order of 60 kPa.

In summary, corrugated tubes and spiral fluted tubes have been shown to provide substantial heat transfer augmentation relative to a plain tube and a star-insert tube. The heat transfer augmentation is primarily affected by the pitch and depth of the

Figure 10-22 Spiral fluted tube bundle (courtesy GA Technologies).

corrugation for a given tube size. The pressure drop was found to increase with increasing severity of the corrugation, which probably limits practical applications to low severities with less heat transfer augmentation than the maximum obtained. No information about the effect of spiral angle is apparently available for corrugated tubes.

Twisted-tape inserts. Twisted-tape inserts can be used to increase the thermal capacity of existing evaporators without replacing the tube bundle. Because the tapes are typically only pulled into the tube rather than being tightly fitted by stretch-reducing the tube onto the insert, the integrity of the thermal contact between a twisted-tape and the tube wall is not ensured. Therefore, heat transfer may only be augmented on the tube wall by the swirl flow created by the tape rather than by heat transfer from the tape itself. This may be sufficient for such applications as eliminating a hot spot in a steam generator tube, however.

Many studies on heat transfer and pressure drop have been done with twisted-tape inserts. To name just a few, Allen (1961) ran tests on R-114 in a natural circulation loop; Blatt and Alt (1963) studied heat transfer and pressure drop for R-11 and water; Gambill (1965) studied subcooled boiling of water; Herbert and Sterns (1968) investigated swirl flow boiling of water at low pressures; Lopina and Bergles (1973) studied

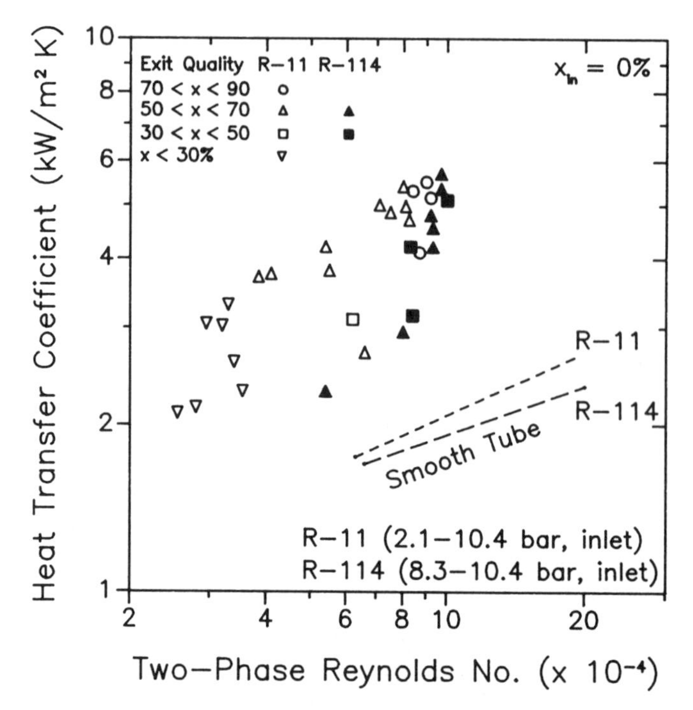

Figure 10-23 Boiling of R-11 and R-114 inside a vertical spiral fluted tube bundle.

subcooled boiling of water; and Cumo et al. (1974) ran tests with R-12. Several recent investigations have provided an in-depth look at the effect of twist ratios of twisted tapes on heat transfer and pressure drop; two independent studies are described here.

Agrawal, Varma, and Lal (1982, 1986) studied twist ratios varying from 3.76 to 10.15 for the evaporation of R-12 in horizontal test sections. The twist ratio is defined as the ratio of half the pitch of the helix to the inside tube diameter, that is the length for a 180° twist of the tape divided by the tube inside diameter. Thus the lower the twist ratio, the higher the spiral angle of the tape. The test sections were made with 10-mm internal diameter stainless steel tubing. The electrically heated length was 2.1 m. The twisted tapes were made from 0.5-mm thick stainless steel flats and were reported to fit snugly inside the test section tube. The heat transfer coefficients were based on the total heat dissipated divided by the surface area of the inside of the tube, not including the twisted tape's surface area. About 12% of the total resistance heating was generated in the twisted tape.

Figure 3-24 shows some of the results for a heat flux of 10.7 kW/m^2 and a mass flux of 236 $kg/(m^2 \cdot s)$. The heat transfer coefficient was a function of vapor quality and increased as the twist ratio decreased. Heat transfer augmentation thus increased with increasing swirl, as expected. The mean augmentation at a twist ratio of 3.76 was about 80%. At the same conditions but a heat flux of 8.0 kW/m^2, the augmentation was only about 30% for the two best twist ratios, 3.76 and 5.58. At a heat flux of 13.8 kW/m^2 for identical flow conditions, the twist ratio of 3.76 was superior to the others by a wide margin, providing an augmentation on the average of about 100%. For the largest twist ratio studied, $y = 10.15$, which represents about half a twist of the tape for every meter of tubing, the placement of the tape inside the tube was actually detrimental to heat transfer. The pressure drop data in Agrawal, Varma, and Lal (1982) showed that the pressure drop ratio for twisted tapes relative to the smooth tube increased with decreasing twist ratio, as expected. For the twist ratio of 3.76, the pressure drop increased from 70% to 270% with respect to the smooth tube, depending on the test conditions.

Jensen and Bensler (1986) investigated the effect of twist ratio on boiling R-113 in vertical upward flow. Their test sections were constructed with stainless steel tubes with inside diameters of 8.10 mm and heated lengths of 1.2 m. Three twisted tapes were made of 304 stainless steel strips 0.254 mm thick and were nearly as long as the total length of the tubing (i.e., longer than the heated test section), so that swirl flow was produced in the unheated entrance region. Heat fluxes and heat transfer coefficients were calculated on the basis of the inside tube surface area after subtracting the heat generated in the tapes from the total heat generated.

In their experiments, Jensen and Bensler (1986) observed an "entrance effect" similar to that which occurs in single-phase flow. The reduction in the heat transfer coefficient in the first 10 to 20 cm of the heated section was less for the twisted tapes than for axial flow in their smooth tube, and the flow became thermally developed over a shorter length. The magnitude of the entrance effect on heat transfer decreased with increasing heat flux, vapor quality, and mass flux but increased with pressure. For example, at a mass flux of 686 kg/(m^2·s), a pressure of 5.64 bar, a heat flux of 32.3 kW/m^2, and an inlet vapor quality of 32.3%, the heat transfer coefficient dropped by 8% before rising along the rest of the test section for the tape with a twist ratio of 8.94. This reduction in the heat transfer coefficient near the inlet could, however, also be explained by the effect of swirl flow on the nucleate boiling process. The increased heat transfer would tend to suppress bubble formation in the liquid film.

Figure 10-24 depicts a heat transfer performance comparison between tubes with twisted tapes and a plain tube plotted as a function of position along the tube. For the particular test conditions shown, there was little difference among the twisted-tape performances at the first test location. The performances then improved along the tube as an inverse function of the twist ratio. Jensen and Bensler (1986) observed from their many experiments that:

1. the heat transfer coefficient increased with twist ratio, and augmentation was largest at the high vapor qualities and mass fluxes;
2. at low vapor qualities and mass fluxes, there was almost no difference between the swirl flow and axial flow heat transfer values, apparently because the nucleate boiling controlled regime was unaffected by the low magnitude of the swirl;
3. as heat flux increased, the difference between the swirl flow and axial flow heat transfer coefficients decreased because the flow was transformed from convection controlled to nucleate boiling controlled;
4. for all other conditions held the same, increasing the pressure reduced the differences in performance of the tapes as shown in Fig. 10-25, an effect that was

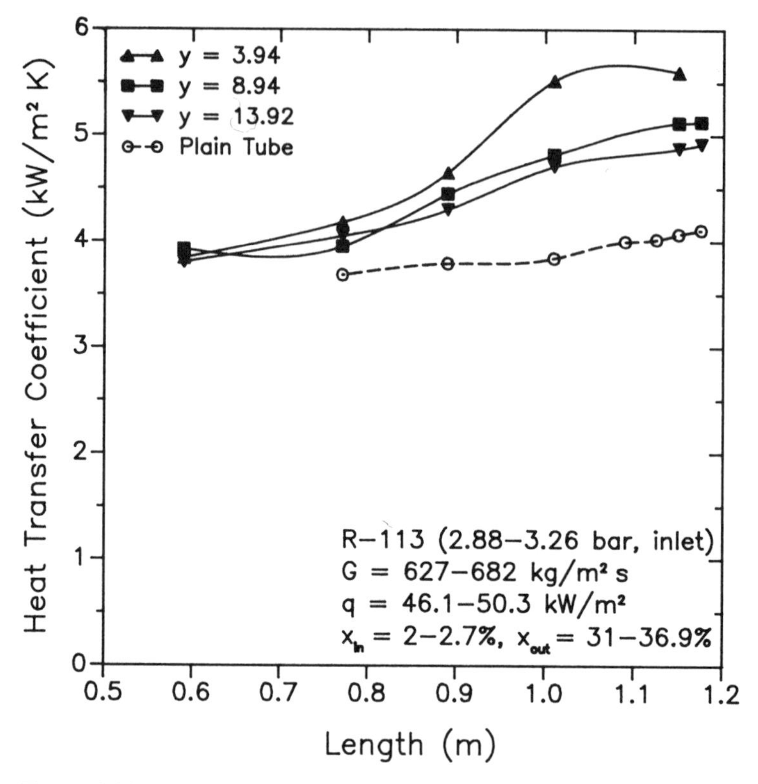

Figure 10-24 Effect of twist ratio on the flow boiling of R-113.

attributed to the reduction in the two-phase velocity with increasing pressure and thus to its diminishing influence on convection; and

5. overall, heat transfer augmentation ranged from about −5% under favorable conditions to as high as 100% under the most favorable conditions.

In conclusion, it is seen that the application of twisted tapes in vertical flows is most appropriate at relatively low pressures, low heat fluxes, high mass fluxes, and vapor qualities greater than about 50%. In a parametric study by Jensen (1985), however, these conditions produced high pressure drops at the twist ratios most effective for heat transfer augmentation.

Star inserts. Star-shaped inserts, so called because early versions typically had five splines forming a shape like that of a star, have been and still are used extensively in the refrigeration industry for augmentation of intube evaporation in expansion water chillers. Figures 4-21 and 4-22 depict several of the versions available commercially. The main advantages of star inserts are (1) their large surface areas, similar in some respects to the plate-fin geometries, and (2) the relative ease of manufacturing extruded

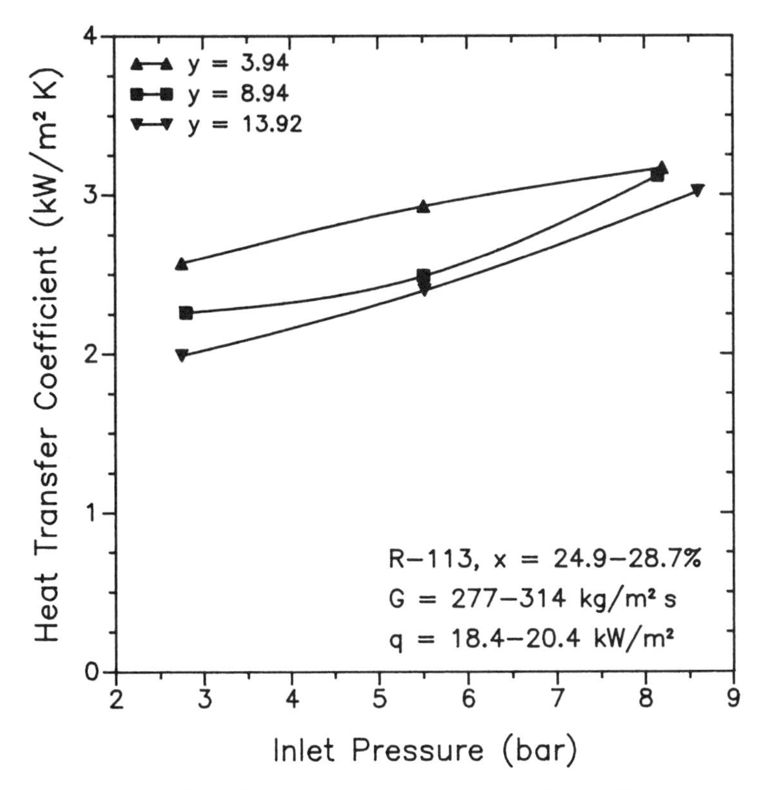

Figure 10-25 Effect of pressure on twisted-tape heat transfer in R-113.

aluminum inserts and positioning them inside copper tubes before stretching the tube to shrink it firmly onto the insert.

Schlünder and Chawla (1967) performed an extensive study on R-11 boiling inside horizontal copper tubes with four different star inserts: one with two splines, two with four splines of different thicknesses, and one with eight splines. Their specifications are given in Table 10-9, and schematic diagrams of the four inserts are shown in Fig. 10-26. The aluminum inserts were tightly fitted inside copper tubes of 14 mm inside diameter. A smooth tube was also tested for comparison purposes. The heat fluxes and heat transfer coefficients were calculated with the inside surface area of the tube without including the surface area of the inserts. All the experiments were run at 10°C, which corresponds to a saturation pressure of 0.6 bar.

Figure 10-27 depicts the heat transfer coefficient and pressure drop as a function of heat flux at a fixed vapor quality of 10% at the midpoint of the test section with insert B. The heat transfer curves are similar to those for boiling inside smooth tubes, being flat in the convection dominated regime and increasing with heat flux in the nucleate boiling dominated regime. The pressure drop was found to increase with mass flux and heat flux, varying linearly with heat flux for a given mass flux.

Schlünder and Chawla (1967) also obtained data for other test conditions. From these, they concluded that:

Table 10-9 Schlünder and Chawla (1967) star insert specifications

Tube type (material)	Insert material	Number of splines	Tube inside diameter, mm	Tube outside diameter, mm	Hub diameter, mm	Fin thickness, mm	Hydraulic diameter, mm	Area ratio
Plain (copper)			14.0	20.0			14.0	1.00
Insert A (copper)	Aluminum	2	14.0	20.0	5.0	1.0	6.8	1.77
Insert B (copper)	Aluminum	4	14.0	20.0	5.0	1.0	5.3	1.99
Insert C (copper)	Aluminum	4	14.0	20.0	5.0	0.5	5.4	2.08
Insert D (copper)	Aluminum	8	14.0	20.0	5.0	1.0	3.4	2.38

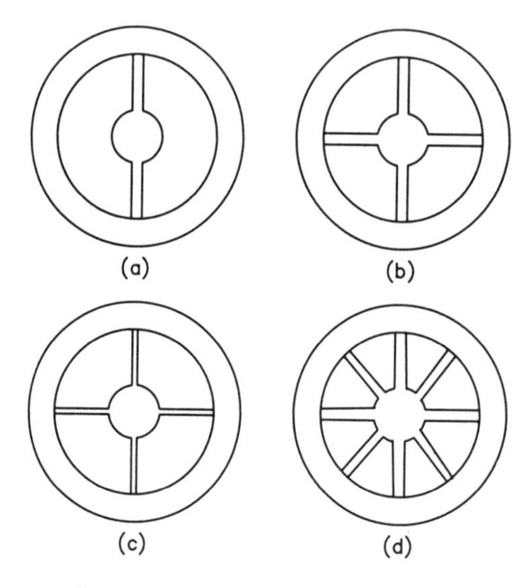

Figure 10-26 Star insert geometries.

1. increasing the number of splines increased the heat transfer coefficient more than percentage increase in the wetted surface area;
2. decreasing the thickness of the spline from 1.0 to 0.5 mm reduced heat transfer performance slightly, suggesting a small fin efficiency effect;
3. the transition from convective controlled boiling to nucleate boiling occurred at higher heat fluxes as the number of splines increased, the mass velocity increased, or the hydraulic diameter decreased; and
4. heat transfer augmentation in the convective controlled regime relative to the plain tube was greatest at low mass fluxes and small heat fluxes.

D'Yachkov (1978) investigated heat transfer and pressure drop for R-22 boiling in horizontal copper tubes with five- and ten-spline aluminum star inserts. The tube inside

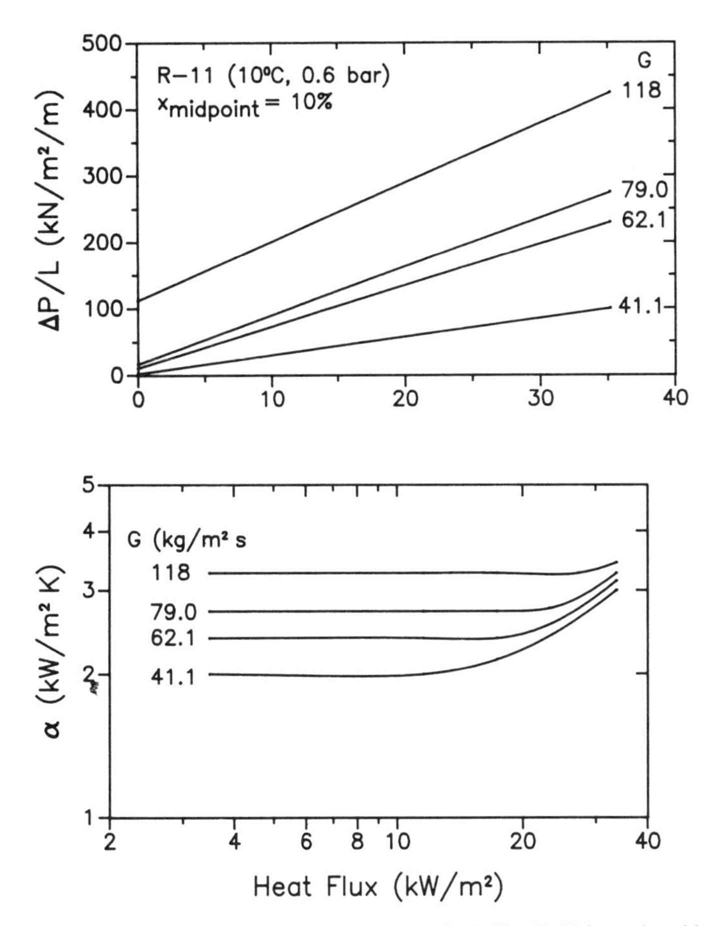

Figure 10-27 Heat transfer and pressure drop for boiling R-11 in a tube with a four-spline insert.

diameter was 15.80 mm. The inserts had fins 5.50 mm high and 1.25 mm thick on a 4.80-mm diameter hub. The internal wetted surface areas per unit length were 0.1075 and 0.1496 m^2/m for the five- and ten-spline test sections, respectively. These tube inserts were reported to be standard configurations used in Soviet refrigeration equipment. Heating was supplied with Nichrome tape heaters wrapped around the tubes. Thermocouples were embedded in the insert at the tips of its splines and at the center of its hub. Glass tube sections at the inlet and outlet were utilized to observe the flow.

The following flow patterns were observed for the five-spline test section: slug flow at the lower range of flow rates [$G < 120$ to 130 kg/($m^2 \cdot s$)] with transition to wavy flow with bridging, annular flow, dispersed annular flow, and dispersed flow. The transitions occurred over a range of vapor qualities, and no clear boundary between the various flow patterns was established. With ten splines, these transitions were shifted to lower vapor qualities. At heat fluxes from 10 to 15 kW/m^2, entrained droplets were observed up to exit vapor superheats of 10 to 12 K.

The local heat transfer coefficient varied appreciably with local vapor quality at heat fluxes less than 7 kW/m^2 for a mass flow rate of 95 $kg/(m^2 \cdot s)$ for the five-spline insert at saturation temperatures of $-5°$ and $-15°C$. Under these conditions, there was no effect of heat flux. Two maxima and one minimum occurred in the heat transfer coefficient along the tube; these were attributed to a change in flow patterns with increasing vapor quality. At higher heat fluxes, the heat transfer coefficient was observed to vary little along the length of the tube with rising vapor quality, similar to the microfin results with R-22 shown in Fig. 10-14. The local heat transfer coefficients dropped off sharply at local vapor qualities greater than 80%. The heat transfer coefficients were observed to decrease with a decrease in saturation temperature, a phenomenon that was attributed to the higher thermal contact resistance between the spline and the tube with decreasing temperature.

With the ten-spline star insert the heat transfer coefficient varied in approximately the same manner as for the five-spline insert, except that no stratified flow regime was observed. At a mass flux of 200 $kg/(m^2 \cdot s)$, the heat transfer coefficient was observed to increase substantially with local vapor quality from 0% to 65% before dropping off rapidly at higher qualities.

Mesh, brush, and spring inserts. Wire meshes, brushes, and springs can be used as inserts to augment the performance of existing evaporators and reboilers or to eliminate hot spots in the radiation section of steam generators. One example of a coiled wire insert is the Heatex insert of Cal Gavin, Ltd. Their basic application in steam generators is to direct the liquid droplets in annular mist flow to the wall to improve wetting and thus to reduce the wall temperature. In other applications, such as in microwave power tubes and high-field electromagnets, their use is to augment boiling heat transfer by inducing swirl flow and to extend wetted wall boiling to higher wall superheats without reaching the critical heat flux.

Wire springs can be snugly fitted inside tubes by stretching the tubes during insertion. The principal parameters that can be varied to augment their intube boiling performance are the wire diameter and material, the pitch, and the helix angle. Larson, Quaint, and Bryan (1949) were apparently the first to run systematic tests. For R-12, they found about a factor of 2 increase in heat transfer and also a substantial increase in pressure drop. Later, however, Bryan and Quaint (1951) observed no heat transfer augmentation for R-11 with a helical wire insert. In subsequent experiments with R-11, Bryan and Siegel (1955) systematically varied the wire diameter and pitch and obtained an optimum configuration that augmented performance in R-11 by about a factor of 2. In their flow visualization study, they observed that the springs caused the top portion of the tube to remain wetted for conditions that produced stratified flow in a similar plain tube.

Megerlin, Murphy, and Bergles (1974) studied water boiling inside short, horizontal tubes with mesh and brush inserts. The test sections were made from 347 stainless steel tubing of 5.3 mm internal diameter. The mesh inserts were constructed from commercial felt metal pads made of 0.1-mm diameter 430 stainless steel fibers. These pads were copper plated and then inserted into the tube to produce a porosity of about 80%. The assembly was then heated in a furnace to braze the mesh to the inside wall of the tube. These investigators also tested commercial metallic spiral brushes made by Spiral Brush, Inc. The central stem of the brush inserts was fabricated from four twisted 302 stainless steel wires (20 gauge), which were wrapped with 0.076-mm diameter 302 stainless steel wire to form the bristles. The nominal outside diameter of the brushes was 6.35 mm, so that they were spring fitted inside the tube. The test sections were electrically heated but only 50.8 mm long. The experiments were for subcooled flow boiling, but the amount of subcooling was not mentioned. Test pressures ranged from 2.4 to 7.9 bar, and mass velocities varied from 2,100 to 4,000 kg/(m^2·s).

The boiling results demonstrated substantial heat transfer augmentation compared to the plain tube tested under similar conditions. For instance, at a wall superheat of 3 K, the tube with the mesh insert had a heat transfer coefficient four times that of the plain tube. The brush insert demonstrated even better heat transfer performance than the mesh insert, but it was only studied at wall superheats and heat fluxes larger than the highest values tested with the plain tube, preventing a direct comparison. The pressure drops for the two inserts were much greater than those for the plain tube.

Danilova et al. (1976) ran R-113 tests with a vertical annulus in which wire and fiber glass fabrics were wrapped around an inner electrically heated tube. Heat transfer was reported to increase from 20% to 100%. Dembi, Dhar, and Arora (1979) also tried various wire meshes fitted tightly against the inner wall of a tube. For R-22, thermal performance increased up to 80% for some test conditions.

Mentes et al. (1982, 1983*a*, 1983*b*) investigated flow boiling instabilities inside six tubes with various types of enhancement: a plain tube, three tubes with springs of different diameter and pitch, a threaded tube, and a porous coated (High Flux) tube. Their experimental results indicated that the porous coated tube was the most stable.

Internal porous coated tubes. The last type of intube boiling enhancement to be discussed is thin porous metallic coatings. These tubes are made either by spraying on the coating, as for the High Flux tube, or by plating copper particles, as described briefly by Ikeuchi et al. (1984).

Czikk, O'Neill, and Gottzmann (1981) reported some of their results for intube boiling in the High Flux tube for liquid oxygen, ammonia, and R-22. Figure 3-23 depicts a comparison among their measured results for flow boiling of liquid oxygen, nucleate pool boiling on the same High Flux coating, and plain tube boiling performance predicted with the Chen (1963) correlation. The experiments were performed on a vertical tube with 18.7 mm internal diameter that was heated electrically. Apparently, the tests were conducted at atmospheric pressure. The tube had thermocouples installed at nine locations along the heated section. No other geometrical information was cited. From their data they concluded that:

1. the local vapor quality had no effect on the boiling heat transfer coefficient, as indicated by the uniformity of the wall temperatures measured along the tube and by the narrow variation in the wall superheats for the wide range of exit vapor qualities tested;
2. the mass flux had no significant effect on heat transfer for a vertical High Flux tube;

3. flow boiling performance for the High Flux tube was essentially identical to that for nucleate pool boiling; and
4. boiling heat transfer augmentation compared to a plain tube was typically an order of magnitude.

These porous coating boiling results differ from all the other results described so far in that the intube boiling performance was increased by augmenting the nucleate boiling heat transfer process rather than the convection process. Therefore, large heat transfer augmentations were observed for essentially the entire range of mass fluxes, heat fluxes, and vapor qualities.

For flow boiling inside horizontal High Flux tubes, the situation is more complex. Boiling ammonia inside an aluminum tube 25 mm in outside diameter at a saturation temperature of 51.7°C (21.6 bar), Czikk, O'Neill, and Gottzmann (1981) observed that the mass flux had a sizeable influence. The boiling performance improved with increasing mass flux over the entire range studied, 30 to 450 kg/(m^2·s). The High Flux performance at 51.7°C was observed to be more than an order of magnitude better than that of a plain tube boiling ammonia at 0°C (4.4 bar) at a mass flux of 230 kg/(m^2·s).

Ikeuchi et al. (1984) studied boiling inside two horizontal porous coated tubes. One tube had the coating on the inside surface of the tube, and the other tube had the coating applied to a flat metal insert installed inside the tube, like a twisted-tape insert with no twist. The performances of these two tubes were compared to those of a plain tube and of a plain tube with an uncoated insert identical to the coated one. The copper tubes all had an internal diameter of 17.05 mm and were heated by concurrent flow of water in an annulus. The heated length of the plain tube and the porous coated tube was 3.38 m, and the tubes with the inserts were heated over 3.01 m. The insert was 1.6 mm thick and divided the flow passage into two identical halves. The porous layers were made by fixing three layers of 0.115-mm diameter copper particles to the surface and then plating with copper. The tests were run with a fixed exit pressure of 5.9 bar. An oil separator was used to minimize the oil content in the refrigerant that was introduced into the system by a reciprocating compressor.

Figure 10-28 depicts a performance comparison between the smooth and porous coated tubes without inserts for two different test conditions, one with incomplete evaporation (exit vapor quality ranging from 70% to 95%) and one with complete evaporation (5 K of exit superheat). The inlet vapor quality was held fixed at 14% for all runs. The heat transfer coefficients shown are average values for the entire heated length of tubing. The porous coated tube augmented heat transfer by a factor of about 3 for complete evaporation, which is representative of actual performance in a R-22 evaporator. The performance was much higher for incomplete evaporation, about 5 times better or more. Use of a porous coating on the insert was observed to improve performance by a factor of 2 for incomplete evaporation relative to the tube with the plain insert but was only marginally better when there was complete evaporation. Adding the porous coating to only the insert rather than to the inner tube wall (with no insert) gave inferior results.

The effect of exit vapor quality on heat transfer for R-22 boiling in horizontal tubes has been investigated by both Ikeuchi et al. (1984) for their plain and porous coated

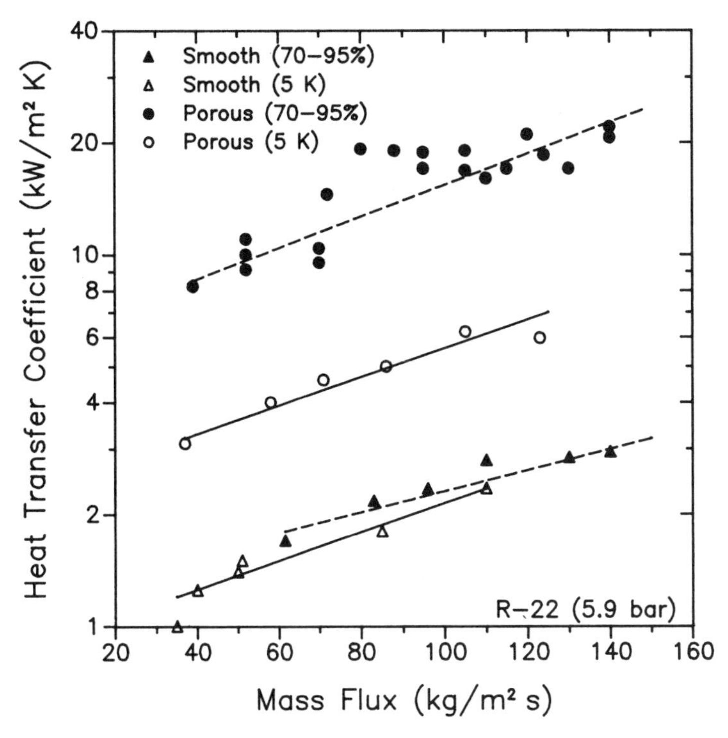

Figure 10-28 R-22 boiling performance comparison between a porous coated tube and a plain tube.

tubes without inserts and by Czikk, O'Neill, and Gottzmann (1981) for 14.5-mm internal diameter High Flux and plain tubes. Their respective results are compared in Fig. 10-29, where the data of Ikeuchi et al. (1984) are average values for the entire tube with a fixed inlet vapor quality of 15% and the High Flux data are apparently local values. No test pressure was cited for the High Flux data. The data agree quite well quantitatively considering the two different types of porous coatings involved. The average heat transfer coefficients obtained by Ikeuchi and colleagues did not drop off in the high exit vapor quality range, and thus the reduction for the 5-K superheating exit conditions in Fig. 10-28 must have resulted from superheating the vapor. The High Flux tube's local heat transfer coefficients decreased at vapor qualities greater than 80%. Because the transition from annular to stratified flow on a Baker type flow regime map for a smooth tube would be expected to occur at about 95% for these mass velocities, Czikk and co-workers hypothesized that this drop was caused by a partial dryout of the upper half of the tube, rendering the area ineffective for nucleate boiling.

Effect of oil on refrigerants boiling inside tubes. The effect of oil contamination in refrigeration systems is an important design and operational problem with these units' evaporators. The oil affects the evaporation process in several ways: (1) the physical properties of the fluid, most notably the viscosity, are altered; (2) the local saturation temperature tends to rise along the tube as the oil concentration increases, which reduces

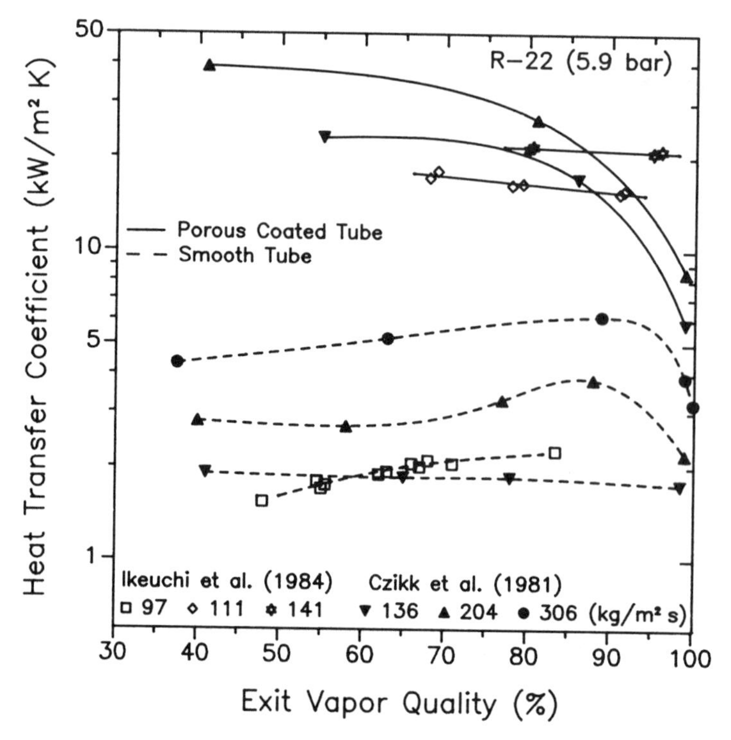

Figure 10-29 Effect of exit vapor quality on R-22 boiling performance in porous coated and plain tubes.

the log mean temperature difference of the evaporator; (3) the two-phase pressure drop tends to increase, which decreases the circulation rate of the refrigerant and also reduces the local saturation temperature; (4) the flow patterns are altered by the increased wettability of the fluid; and (5) the average heat transfer coefficient along the tube may either increase or decrease. In some instances, depending on the particular oil, its concentration, and temperature, the oil may be immiscible and foul the tubes.

The rise in the bubble point of the refrigerant with increasing oil concentration is often neglected in the thermal design of evaporators. In addition, the concentration of oil in an operating unit is typically unknown. As an approximate rule of thumb for R-22, the bubble point temperature increases by 0.1 to 0.2 K per 1% of oil by weight (phase equilibrium diagrams for binary mixture systems are shown in Chapter Nine). Thus for a 5% oil concentration at the inlet the saturation temperature will only rise by about 0.5 to 1.0 K; at an exit vapor quality of 88%, however, about 40% of the remaining liquid will be oil, representing a rise from 4 to 8 K in the exit temperature. The larger pressure drop of the refrigerant-oil mixture will partially compensate for this. The temperature approach in this section of an evaporator will narrow accordingly and may approach the heating fluid's temperature. One further complicating problem is that the oil may not be miscible at these high local concentrations near the exit for some refrigerants.

One other point concerning refrigeration practice should also be made. It is common to design refrigeration systems to provide from 2 to 5 K of superheat at the exit of the evaporator, based on evaporation of a pure refrigerant. In thermodynamic terms, the exit vapor of a refrigerant-oil mixture cannot actually become superheated because the fluid leaves below the pure oil's boiling point temperature and is thus a two-phase fluid with a fraction of the refrigerant still in the liquid. Therefore, it is the exit bubble point temperature that would be from 2 to 5 K higher than the pure refrigerant's saturation temperature. Consequently, a part of the heat duty of the evaporator is consumed as sensible heating of the liquid. The higher the oil concentration and rise in bubble point temperature, the more energy that will be utilized as sensible heat. Thus the refrigeration capacity is reduced.

A review of boiling inside plain and enhanced boiling tubes is presented below. For pool boiling of refrigerant-oil mixtures, the reader is referred to Chapter Nine, Sec. 9-5.

Plain tubes. For boiling refrigerant-oil mixtures in plain tubes, many of the above characteristics have been investigated experimentally. Nevertheless, correlations to predict the effect of oil on heat transfer and pressure drop with various refrigerant-oil combinations are generally not available.

Worsoe-Schmidt (1960) made visual observations of boiling on the inside of a plain, horizontal glass tube. He observed that the addition of oil to R-12 changed the wetting characteristics. The predominant flow pattern that he observed at mass fluxes from 240 to 330 kg/($m^2 \cdot s$) without oil was stratified flow. Adding even small amounts of oil changed the flow pattern to annular flow, in which the increased wetted circumference of the tube enlarged the measured heat transfer coefficients. Because oil is nonvolatile, however, it built up in the liquid film as the evaporation process progressed along the tube and the local heat transfer coefficients decreased to less than the pure R-12 values in the latter portion of the tube. For example, for a 1.9% (by weight) oil concentration the local heat transfer coefficient increased by about 50% over most of the tube; at higher concentrations its average value for the entire tube was less than that for pure R-12.

Chaddock and Mathur (1980) ran tests with R-22 and confirmed the results of Worsoe-Schmidt (1960). A 1% oil concentration increased the local heat transfer coefficient by about 20% to 30% over most of the tube. They found the optimum oil concentration based on the average heat transfer coefficient for the entire tube to be about 1% to 2.9% oil. Chaddock and Buzzard (1986) also conducted experiments with the azeotropic refrigerant R-502 and ammonia. For 1% oil in R-502, which were completely miscible in each other, the heat transfer performance was identical to that obtained in tests with no oil. On increasing the oil concentration past the maximum miscibility concentration of 2%, the heat transfer coefficients dropped by up to 50%. For ammonia with mineral oil, the immiscible mixture demonstrated heat transfer coefficients 50% to 90% less than those for pure ammonia.

Readers interested in more details about the effects of oil on refrigerants evaporating inside plain tubes are referred to Chaddock (1976) and Schlager, Pate, and Bergles (1987).

Enhanced boiling tubes. The effect of oil on boiling augmentation inside tubes is of considerable industrial importance because the expected boiling enhancement may not be attainable. Yet apparently only two studies have investigated boiling of refrigerant-oil mixtures inside enhanced boiling tubes.

Schlager, Bergles, and Pate (1988) performed boiling and condensation tests with a microfinned tube, and Schlager, Pate, and Bergles (1988) tested a low-finned tube. They also tested a plain tube for comparison purposes. Both studies were for R-22 with a naphthenic-based mineral oil with a viscosity of 150 Seybolt Universal Seconds (SUS). The refrigerant and oil were reported to be completely miscible over the concentration range studied (up to 5% by weight). The tube specifications are listed in Table 10-10. All test sections were 3.67 m long and were mounted horizontally. They were heated by a countercurrent water flow through an outer annulus.

The experimental heat transfer coefficients were average values for the entire tube, which were determined from a Wilson plot. They were based on the nominal internal surface area at the maximum internal diameter of the tube. They were calculated with the assumption that the evaporation temperature was constant throughout the test section and equal to the saturation temperature of pure R-22 at the average test section pressure. Thus no allowance was made for the effect of oil on the local saturation temperature. Consequently, the exit vapor quality was maintained at less than 88%.

The test conditions for the two comparison studies were as follows: mass flux, 125 to 400 kg/(m^2·s); oil concentration, 0% to 5.1% (by weight); pressure, 5 to 6 bar; saturation temperature, 0° to 6°C; inlet vapor quality, 0.1 to 0.2; and outlet vapor quality, 0.8 to 0.88.

Figure 10-30 depicts the low-finned tube heat transfer data. The heat transfer coefficient with 1.3% oil is almost identical to that of pure R-22. Increasing the oil concentration to 2.4% and 4.9% resulted in some degradation in performance.

Figure 10-31 shows two different methods for comparing the results. In the upper graph, the ratio of the refrigerant-oil coefficient to that for pure R-22 is shown for each tube at two mass fluxes. The smooth tube's ratio increased with oil concentrations up to about 2.5%, similar to the trend observed in other smooth tube studies. The microfinned tube's ratio ranged only from 1.1 to a low of 0.96, and the low-finned tube's ratio varied from 1.01 to a low of 0.86. In the lower graph, the ratio of the enhanced tube's coefficient to that of the smooth tube at the same oil concentration is depicted. Most

Table 10-10 Tube specifications for refrigerant-oil tests

Tube type	Tube outside diameter, mm	Maximum inside diameter, mm	Flow area, mm^2	Fin height, mm	Fin pitch, mm	Number of fins	Spiral angle, degrees	Area ratio*
Smooth	9.52	8.0	50.3					
Microfinned	9.52	8.72	57.0	0.20	0.44	60	18	1.5
Low-finned	9.52	8.51	53.3	0.38	1.21	21	30	1.8

*Ratio of wetted surface area to nominal surface area at maximum internal diameter.

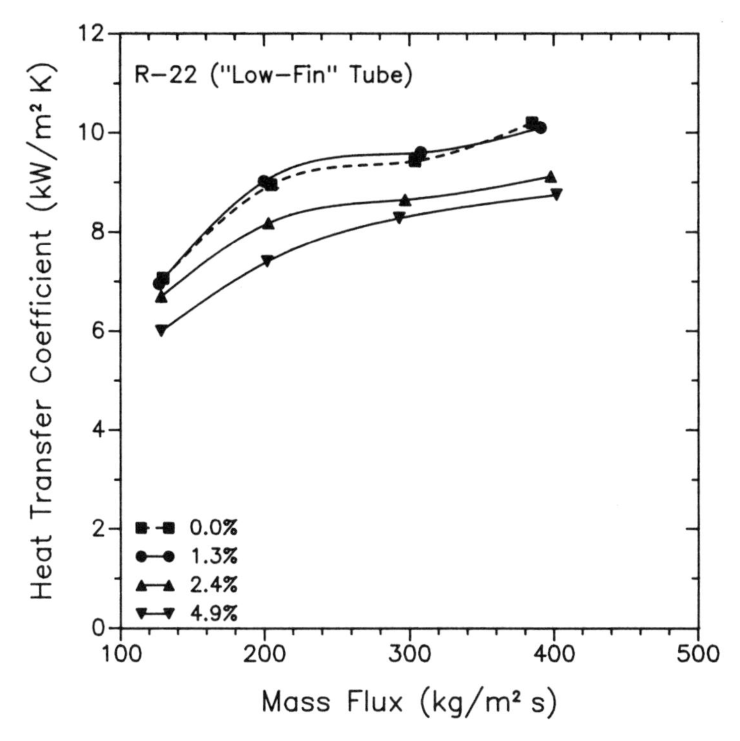

Figure 10-30 Effect of oil on enhanced boiling inside a tube.

of the variation in the performance ratio with oil concentration is due to the smooth tube values. The augmentation for both enhanced boiling tubes decreased with increasing oil concentration before leveling off at about 2.4%.

A few comments should be made here to interpret these (and other) results. The actual average saturation temperatures of the refrigerant-oil mixtures are higher than the pure R-22 values used to reduce the data, and consequently their actual average boiling wall superheats are smaller than otherwise estimated. As a result the mixture heat transfer coefficients calculated with the actual saturation temperatures would be higher than those shown, and the degradation in the heat transfer coefficients would be less than indicated in Fig. 10-30 and in the upper graph of Fig. 10-31.

The data as presented in Figs. 10-30 and 10-31 are an indication of the thermal performance that would be achieved in practice because most refrigeration engineers would simplify the thermal design procedure by including the oil effect in the heat transfer coefficient rather than use vapor-liquid equilibrium theory to determine the bubble point temperature along the tube. In a refrigeration system the exit conditions would be more severe than those in the present tests because the exit vapor quality would be higher, resulting in a higher elevation in the local bubble point temperature at the exit. Thus the actual degradation in practice may be worse than indicated here.

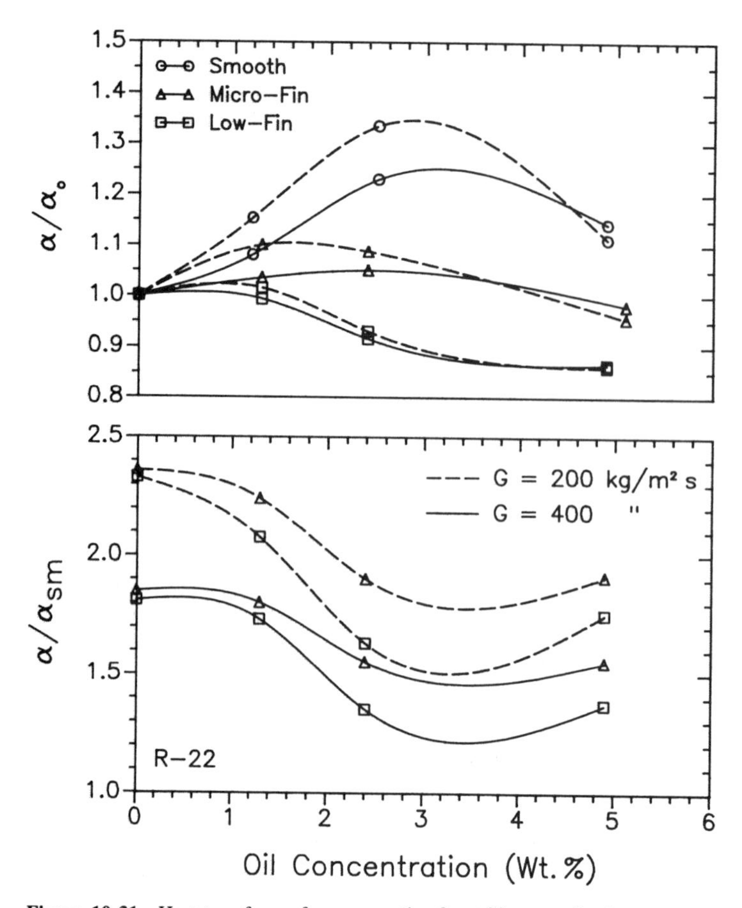

Figure 10-31 Heat transfer performance ratios for refrigerant-oil mixtures.

Intube Enhanced Boiling Correlations

A number of correlations have been developed for enhanced boiling inside tubes. It should be pointed out beforehand, however, that each of these correlations was developed with data for only one fluid. Thus, even if the correlation has dimensionless groups (such as a Reynolds number or boiling number) giving an appearance of universal applicability, it is not recommended that the expression be used for other fluids or perhaps even for tube diameters other than those of the original one unless it has been verified with some additional test data.

Internally finned tubes. Boiling in internally finned tubes is discussed in detail above, where the types of internally finned tubes are subdivided into microfins and high fins. At present, apparently no correlation is available for predicting the heat transfer performance of microfinned tubes. For high-finned tubes, Lavin and Young (1965) presented several correlations for the tubes described in Table 10-6 for nucleate boiling, annular flow, and upper transition regimes of boiling in horizontal and vertical orientations. The test conditions, such as mass velocities, heat fluxes, and vapor qualities, were not well documented, however, and therefore the application range of these correlations is not known.

Azer and Sivakumar (1984) developed heat transfer and pressure drop correlations for R-113 boiling in horizontal copper high-finned tubes. The conditions covered in their test of three different internally finned tubes are given in Table 10-11.

The Azer and Sivakumar (1984) correlation for the average heat transfer coefficient for a tube was developed with the Pierre (1957, 1964) plain tube correlation used as a starting point. The Azer and Sivakumar correlation for R-113 is

$$\frac{\overline{\alpha} D_i}{\lambda_L} = 12.24\,(\mathrm{Re}_{Di}^2 K_f)^{0.146}\,(1 + 0.0024 F_1^{3.72} F_2^{-8.88}) \tag{10-3}$$

where the Reynolds number is defined with the maximum internal diameter D_i as

$$\mathrm{Re}_{Di} \equiv \frac{G D_i}{\mu_L} \tag{10-4}$$

and Pierre's boiling parameter K_f is defined as

$$K_f = J(x_{out} - x_{in})\,\Delta h_v g_c/(gL) \tag{10-5}$$

The finned tube modifying factors developed by Carnavos (1980) for liquids were used to model the effect of the fins on flow boiling. The first of these factors is defined as follows:

$$F_1 = \frac{A_{fa}}{A_{fc}} = \frac{1 - [(4Nht)/(\pi D_i^2 \cos\beta)]}{[1 - (2h/D_i)]^2} \tag{10-6}$$

which represents the ratio between the actual flow area and the open core flow area. The second factor is given as

Table 10-11 Azer and Sivakumar (1984) test conditions

Fin height	0.7 to 2.0 mm
Maximum inside diameter	14.2 to 20.4 mm
Spiral angle	0° to 16°
Heated length	1.35 m
Number of fins	10 to 32
Mass velocity	44.5 to 184 kg/(m^2·s)
Heat flux	5.1 to 13.2 kW/m^2
Saturation temperature	47.2° to 63.2°C
Inlet vapor quality	0%
Outlet vapor quality	11% to 81.5%

$$F_2 = \frac{A_n}{A_a} = \frac{\pi D_i}{\pi D_i + (2Nh/\cos \beta)} \tag{10-7}$$

and is the ratio of the nominal internal surface area (calculated with the maximum internal diameter) to the total wetted surface area. The boiling data were predicted to within about 30% by this method. For predicting the total pressure drop for a tube of length L (including both frictional and accelerational contributions), Azer and Sivakumar (1984) developed the following expression:

$$\Delta P = F\left[f_{TP} + \frac{D_i(x_{out} - x_{in})}{x_m L}\right]\left(\frac{G^2 v_m L}{g_c D_i}\right) \tag{10-8}$$

where the two-phase friction factor is given by

$$f_{TP} = 0.0011\left(\frac{K_f}{Re_{D_i}}\right)^{0.112} \quad \text{for} \quad \frac{Re_{D_i}}{K_f} > 1 \tag{10-9}$$

The mean vapor quality is obtained from the inlet and outlet values:

$$x_m = \frac{x_{in} + x_{out}}{2} \tag{10-10}$$

The mean specific volume is calculated from the liquid and vapor densities:

$$v_m = \frac{x_m}{\rho_v} + \frac{1 - x_m}{\rho_L} \tag{10-11}$$

The finned tube multiplying factor F is given as

$$F = 0.628 F_3^{4.89} F_4^{-12.29} \tag{10-12}$$

where the modifying parameters F_3 and F_4 are

$$F_3 = \cos \beta \tag{10-13}$$

$$F_4 = \frac{A_{fa}}{A_{fn}} = 1 - [4Nht/(\pi D_i^2 \cos \beta)] \tag{10-14}$$

These two parameters account for the effects of the spiral angle of the fins on the flow and the ratio of the actual flow area to the nominal flow area (calculated with the maximum inside diameter), respectively. The experimental pressure drops were predicted to within about 40% by this method.

Corrugated tubes. Withers and Habdas (1974) developed heat transfer and pressure drop correlations for R-12 from data for two ¾-inch (19-mm) outside diameter corrugated tubes, one with a severity factor of 0.00234 and the second with a severity factor of 0.00382. The severity factor is defined in Eq. (10-2). The heat transfer and pressure drop correlations were developed by modifying the Pierre (1957, 1964) correlations for smooth tubes. The mean heat transfer coefficient for the tube was correlated as

$$\frac{\overline{\alpha} D_i}{\lambda_L} = c_1 (\mathrm{Re}_{D_i}^2 K_f)^{n_1} \tag{10-15}$$

where values for c_1 were 0.000585 and 0.0118 and the exponents n_1 were 0.526 and 0.424, respectively, for tubes 1 and 2. The two-phase friction factor for Eq. (10-8) without the leading factor F was correlated as

$$f_{TP} = c_2 \left(\frac{K_f}{\mathrm{Re}_{D_i}} \right)^{n_2} \tag{10-16}$$

The leading empirical constants were found to be 0.0434 and 0.084 and the exponents 0.152 and 0.305, respectively, for tubes 1 and 2. The test conditions were as described earlier (under corrugated and spirally fluted tubes) for a complete quality change of 100% from inlet to outlet and for saturation temperatures varying from $-1°$ to 4.4°C. No information was given about the range of heat fluxes or mass fluxes covered in the experiments.

Twisted-tape inserts. For twisted-tape inserts, the effect of swirl flow on heat transfer and pressure drop must be modeled. Agrawal, Varma, and Lal (1986) developed the following correlation for R-12 boiling in horizontal tubes, which gives the average heat transfer coefficient for the tube wall as a function of the twist ratio relative to the Pierre smooth tube heat transfer coefficient:

$$\frac{\overline{\alpha}}{\overline{\alpha}_{BP}} = 0.001877\, \mathrm{Re}_s^{2.228}\, \mathrm{Bo}^{1.615} y^{-0.3565} \tag{10-17}$$

The Pierre 1964 heat transfer coefficient is calculated as

$$\overline{\alpha}_{BP} = c_3 \frac{\lambda_L}{D_i} (\mathrm{Re}_{D_i}^2 K_f)^{n_3} \tag{10-18}$$

where the empirical constant c_3 and exponent n_3 are 0.0009 and 0.5, respectively, for exit vapor qualities less than 90%; their values are 0.0082 and 0.4 for exit vapor qualities greater than 90% and up to 6 K of superheating. The swirl Reynolds number Re_s in Eq. (10-17) is defined as

$$\mathrm{Re}_s \equiv \mathrm{Re}_a \left[\frac{(\pi^2 + 4y^2)^{1/2}}{2y} \right] \tag{10-19}$$

and the axial flow Reynolds number Re_a is defined as

$$\mathrm{Re}_a \equiv \frac{u_a D_i}{\nu_L} \tag{10-20}$$

The axial flow velocity u_a is calculated from the actual cross-sectional flow area:

$$u_a = \frac{\dot{m}}{\rho_L[(\pi/4)(D_i^2) - tD_i]} \tag{10-21}$$

The boiling number Bo is defined as

$$\mathrm{Bo} = \frac{q}{G\Delta h_v} \tag{10-22}$$

This correlation is applicable to the conditions cited earlier (for twisted-tape inserts) and was able to correlate the swirl flow data to within a standard deviation of 28%.

Agrawal, Varma, and Lal (1982) developed a pressure drop correlation similar to that of Blatt and Alt (1963) for R-12 evaporating in horizontal tubes for essentially the same flow conditions as in their boiling correlation. The swirl flow pressure drop, normalized by the plain tube pressure drop calculated by the Martinelli and Nelson (1948) method, is given as

$$\frac{\Delta P_s}{\Delta P_{MN}} = \frac{5.12}{y^{0.509}} \tag{10-23}$$

This expression predicted their experimental data to within about 20%.

Jensen and Bensler (1986) approached the correlation of their R-113 data for vertical upward flow in a different manner. They modified the Chen (1966) plain tube correlation to include the effects of swirl flow. The basic premise of the Chen model is the summation of the nucleate boiling and convection contributions to heat transfer, which gives the total local heat transfer coefficient as

$$\alpha = \alpha_L F + \alpha_{nb} S \tag{10-24}$$

The liquid-only heat transfer coefficient was calculated with the forced convection expression in the Lopina and Bergles (1969) correlation for single-phase swirl flow:

$$\alpha_L = 0.20 \frac{\lambda_L}{D_h} \mathrm{Re}_s^{0.8} \mathrm{Pr}_L^{0.4} \tag{10-25}$$

and their swirl flow Reynolds number Re_s was defined as:

$$\mathrm{Re}_s \equiv \mathrm{Re}_h \left[\frac{(\pi^2 + 4y^2)^{1/2}}{2y} \right] \tag{10-26}$$

where the hydraulic diameter D_h was used for calculating the axial flow Reynolds number Re_h. The other parameters in the Chen (1966) correlation were left unmodified and are given by Eqs. (2-29), (2-30), (2-32), and (2-33). The method predicted the

2,682 data points for the conditions cited earlier (for twisted-tape inserts) with an average error of 24%, although the scatter band was about 50%. Jensen (1985) has shown that this method adequately predicted the vertical intube nucleate boiling data of Cumo et al. (1974) for R-12 and a twist ratio of 4.4.

Star inserts. Schlünder and Chawla (1967) modified the Chawla (1967) heat transfer and pressure drop correlations for plain horizontal tubes to predict their R-11 data described above (for star inserts) for the four different inserts shown in Fig. 10-26. The total pressure drop per unit length of horizontal tube is the summation of the frictional and accelerational pressure drops:

$$\frac{\Delta P}{L} = \frac{\Delta P_{f}}{L} + \frac{\Delta P_{a}}{L} \tag{10-27}$$

The frictional pressure drop is predicted with the following expression:

$$\frac{\Delta P_{f}}{L} = \frac{0.3164}{(\mathrm{Re}_{h})_{v}^{1/4}}\left(\frac{G^{2}x^{7/4}}{2D_{h}\rho_{v}}\right)\left[1 + \frac{\rho_{v}(1-x)}{\rho_{L}x\varepsilon^{*}}\right]^{19/8} \tag{10-28}$$

The original empirical constant and exponent for the two-phase flow parameter were modified to obtain

$$\varepsilon^{*} = 0.54\left(\frac{1-x}{x}\right)\left(\frac{\mu_{v}}{\mu_{L}}\right)[(\mathrm{Re}_{h})_{L}\mathrm{Fr}_{L}]^{-1/4} \tag{10-29}$$

where the liquid-phase Reynolds number was defined with the hydraulic diameter of one of the flow passages:

$$(\mathrm{Re}_{h})_{L} \equiv \frac{G(1-x)D_{h}}{\mu_{L}} \tag{10-30}$$

and the liquid Froude number Fr_{L} was defined as

$$\mathrm{Fr}_{L} \equiv \frac{G^{2}(1-x)^{2}}{\rho_{L}^{2}gD_{h}} \tag{10-31}$$

The accelerational pressure drop was correlated with the modified two-phase flow parameter as

$$\frac{\Delta P_{a}}{L} = \frac{Gxq\pi D_{i}}{\Delta h_{v}\rho_{v}A_{i}}\left\{\left(1 - \frac{\rho_{v}}{\rho_{L}\varepsilon^{*}}\right)\left[1 + \frac{(1-x)\varepsilon^{*}}{x}\right] + \left[1 + \frac{(1-x)\rho_{v}}{x\varepsilon^{*}\rho_{L}}\right](1-\varepsilon^{*})\right\} \tag{10-32}$$

The frictional pressure drop correlation was developed with adiabatic data obtained for vapor qualities ranging from 10% to 80% for each of the four inserts. The overall experimental pressure drops under boiling conditions were predicted to within about 25% by this method.

The local heat transfer coefficient at a fixed point in the channel for convection controlled boiling was correlated as

$$\frac{\alpha_h D_h}{\lambda_L} = 0.23 \left(\frac{x}{1-x} \right) \left(\frac{\mu_L}{\mu_v} \right) (\mathrm{Re}_h)_L^{0.575} \mathrm{Fr}_L^{0.775} \tag{10-33}$$

where the heat transfer coefficient has been averaged over the entire wetted wall of the channel at that point and includes the fin efficiency effect. For the nucleate boiling dominated regime, Schlünder and Chawla (1967) gave the following correlation:

$$\frac{\alpha_h}{\alpha_{nb}} = 29\, (\mathrm{Re}_h)_L^{-0.3} \mathrm{Fr}_L^{0.2} \tag{10-34}$$

where the nucleate boiling heat transfer coefficient is for pool boiling on the outside of a plain tube at the same conditions. Therefore, the correlation essentially represents a method for applying nucleate pool boiling data to boiling in tubes with an insert. These correlations were developed for straight, plain surface star inserts and are probably not applicable to ones with wavy surfaces or a helix (e.g., those shown in Figs. 4-21 and 4-22).

Chistyakov, Frolova, and Kuvshinov (1978) also correlated refrigerant pressure drop data obtained for a ten-spline aluminum star insert. Only pressure drop data in the convection dominated heat transfer regimes were included in the study, and the refrigerants tested were not specified.

Internal porous coated tubes. Ikeuchi et al. (1984) developed several simple correlations from their research described above for boiling of R-22 in horizontal tubes coated internally with a porous layer. Using the Pierre (1957) correlation as a starting point, they correlated their data for the tube without the insert as

$$\frac{\bar{\alpha} D_i}{\lambda_L} = 0.143\, (\mathrm{Re}_{D_i}^2 K_f)^{0.381} \tag{10-35}$$

for the incomplete evaporation test conditions and as

$$\frac{\bar{\alpha} D_i}{\lambda_L} = 0.125\, (\mathrm{Re}_{D_i}^2 K_f)^{0.343} \tag{10-36}$$

for complete evaporation. These correlations are compared to their porous coated tube data in Fig. 10-28 (represented by the solid and dashed lines).

For intube boiling in vertical porous coated tubes, O'Neill, King, and Ragi (1980) recommended using nucleate pool boiling data obtained with the same porous surface for modeling of the nucleate boiling zone inside the tube. As illustrated in Fig. 3-23, this gives a good prediction for boiling oxygen. This method, however, is only applicable up to locations in the tube where the nucleate boiling regime still dominates.

Intube Critical Heat Flux Augmentation

Various methods have been developed to impart a swirling motion to two-phase flows inside tubes to augment the critical heat flux relative to plain, smooth tubes operating under similar circumstances. One of the simplest devices is the twisted-tape insert, which can readily be installed in an existing smooth-bore tube experiencing critical heat flux problems. Gambill, Bundy, and Wansbrough (1961) completed an extensive study on subcooled boiling of water with inserts. They found that the critical heat flux could be augmented from 20% to 200% relative to a smooth tube, depending on the liquid velocity. The higher the liquid velocity, the higher the observed increase in the critical heat flux. Gambill (1963) explained the higher critical heat fluxes as being primarily the result of the swirl flow's beneficial effect on forced convection because high heat transfer coefficients (and hence heat fluxes) could be produced by the swirl flow before the initiation of boiling. Bergles, Fuller, and Hynek (1971) observed similar results with liquid nitrogen for several different twist ratios in a vertically oriented tube. Cumo et al. (1974) found that the augmentation in critical heat flux for R-12 in vertical upward flow passed through a slight maximum at a reduced pressure of 0.5 for a twist ratio of 4.4.

Jensen (1984) proposed the following correlation for predicting the critical heat flux for tubes with twisted tapes:

$$\frac{(q_{\mathrm{cr}})_{\mathrm{swirl}}}{(q_{\mathrm{cr}})_{\mathrm{smooth}}} = (4.597 + 0.09254y + 0.004154y^2)\left(\frac{\rho_{\mathrm{L}}}{\rho_{\mathrm{v}}}\right) + 0.09012 \ln\left(\frac{a}{g}\right) \quad (10\text{-}37)$$

where the smooth tube critical heat flux is calculated at the same inlet conditions and the radial acceleration a is obtained from the following expression:

$$a = \left(\frac{2}{D_{\mathrm{i}}}\right)\left(\frac{u_{\mathrm{a}}\pi}{2y}\right)^2 \quad (10\text{-}38)$$

Ribbed tubes, such as the one shown in schematic form in Fig. 4-15, are another effective type of enhancement and have the benefit of being an integral part of the tube itself. Swenson, Carver, and Szoeke (1962) observed at subcritical pressures with water that internal ribs inhibited film boiling and allowed steam-generating tubes to be used at lower mass velocities and higher heat fluxes than those at which film boiling would occur in smooth tubes. Ackerman (1970) studied supercritical boiling in a vertical ribbed tube with six helical ribs and found that the resulting swirl flow suppressed pseudofilm boiling and permitted operation at higher heat fluxes than were possible with the smooth tubes tested. Shiralkar and Griffith (1969) obtained similar results for supercritical boiling of carbon dioxide.

Watson, Lee, and Wiener (1974) noted that there was a minimum mass velocity that must be exceeded to induce water to swirl in multilead vertical ribbed tubes.

More recently, Kitto and Wiener (1982) reported on the effects of nonuniform circumferential heating and tube inclination on the critical heat flux of subcritical water in spirally ribbed tubes. For a vertical orientation with a ratio of peak nonuniform circumferential heat flux to average heat flux of 1.91, they found that the local critical heat flux at the hotter side of the tube was higher than that for uniform heating of the ribbed tube under the same conditions. For a 30° inclination of their ribbed tube from the vertical, they observed a decrease in the critical heat flux relative to the vertical ribbed tube of roughly 30%. The ribbed tubes increased the critical heat flux by as much as a factor of 3 relative to smooth tubes operating at the same quality, pressure, mass velocity, and orientation. The investigators explained this as being due to the swirling motion, which forces the liquid to the tube wall and the vapor to the center and also continuously feeds liquid to the hotter side of the tube, where the heat flux is higher for nonuniform heating. For the vertical orientation, circumferential liquid flow was thought to retard substantially the formation of a stable vapor film at the hotter side of the tube. The improved performance for the inclined ribbed tube compared to that of the smooth tube was thought to be due to the centrifugal force imparted to the liquid by the ribs, which tended to overcome the stratification effect caused by the gravitational and buoyancy forces. The ribs, however, had the negative effect of increasing the pressure drop through the boiler tube.

Several other geometries have also been investigated for augmenting the critical heat flux. Megerlin, Murphy, and Bergles (1974) studied the effects of steel meshes and brushes (see the discussion of such inserts, above) on subcritical water in horizontal tubes. The critical heat flux was increased by several times, but the pressure drop was increased by an order of magnitude, relative to a smooth tube. Withers and Habdas (1974) found that their horizontal corrugated tubes could evaporate R-12 without reaching the critical point at much higher heat fluxes than their plain tube. The corrugations apparently induced swirl flow, which kept the inner wall adequately wetted up to reported exit vapor qualities of 100%.

In summary, inducing a swirling motion to a two-phase flow or a subcooled liquid flow inside vertical or horizontal tubes can produce a sizeable augmentation of the critical heat flux relative to smooth tubes. The actual increase in performance obtained is dependent on the particular method and geometry utilized to impose swirl flow, the fluid involved, the pressure, the mass velocity, the local vapor quality or degree of subcooling, any nonuniformity in the imposed heat flux, and other parameters. The advantages of increasing the critical heat flux have to be weighed against the disadvantages of increasing the pressure drop.

10-2 ENHANCED BOILING ON TUBE BUNDLES

Bundle Boiling Studies

Enhanced boiling on tube bundles has been studied since the early 1940s, when Jones (1941) investigated the boiling of R-11 on a low-finned tube bundle. Compared to intube boiling, however, shell-side boiling is much more complicated because of the many possible variations in bundle geometry and the difficulty in making local measurements. Table 10-12 presents a representative list of enhanced bundle boiling investigations.

Shell-side bundle boiling studies typically provide much less specific information about the boiling process than intube boiling studies. For instance, local values of heat transfer coefficients, vapor qualities, mass flux, and so forth are not easily obtained, especially as the size of the bundle increases. Usually, not even the liquid circulation rate is cited. In the following sections, some of the experimental work on enhanced boiling on tube bundles is described. Several of the more in-depth studies are discussed in Chapter Twelve.

Low-finned tubes. Danilova and Dyundin (1972) tested two 19-tube bundles, 18 of which were heated by cartridge heaters. The bundles had equilateral triangular tube layouts on a ratio of tube pitch to tube diameter of 1.28. One tube bundle was made from low-finned tubing with 12.5 fins per inch (fpi) (493 fins per meter) and an outside diameter of 20.9 mm, and a second similar bundle was made with 27.7-fpi (1,095 fins per meter), 19.15-mm diameter tubing. The results showed considerable boiling augmentation for R-12 and R-22 compared to smooth tube bundles. The average bundle boiling heat transfer coefficients were larger than those for a single low-finned tube at low heat fluxes; thus the bundle boiling factor described in Eq. (2-18) was greater than 1.0. At higher heat fluxes, there was essentially no difference between the single tube and bundle values. In addition, the investigators reported that the second bundle only performed 5% to 10% better than the standard finned tube bundle, even though its single tube performance was about 20% higher. Therefore, the effort and expense to optimize the geometry or fin density of a low-finned tube for boiling a particular fluid with only single tube tests may not necessarily be worthwhile.

Palen, Yarden, and Taborek (1972) reported overall heat transfer data for two different low-finned tube bundles, one with 413 90-10 cupro-nickel low-finned tubes of 9.5 mm diameter and with a tube pitch of 14.3 mm and a second bundle with 1,008 of the same tubes. They observed, for an unspecified hydrocarbon, that the mean bundle boiling heat transfer coefficient was usually larger than that for a single tube.

Hahne and Muller (1983) boiled R-11 on various small tube bundles comprising from 2 to 18 tubes. The tubes had outside diameters of 18.9 mm and were placed in square arrays with tube pitches of 37.8 mm. The 18-tube bundle consisted of three vertical rows of 6 tubes each. The bundle boiling factor for this configuration was about 1.5 at low heat fluxes, diminishing to 1.0 at the highest heat fluxes tested.

Yilmaz and Palen (1984) reported overall heat transfer performance for a low-finned, U-tube reboiler installed inside a distillation column for evaporating pure *p*-xylene. The tubes were cupro-nickel of 15.8 mm diameter and had a tube pitch of 23.8 mm. The results at 1.03 bar (15 psia) are shown in Fig. 10-32. Heating was provided by condensing saturated steam on the tube side. The finned tube bundle performance was considerably better than that of the plain tube bundle when compared with reference to the nominal outside surface area; it was worse or nearly the same when based on the total wetted surface area of the finned tubes. The finned tube bundle

Table 10-12 Enhanced bundle boiling studies

Reference	Fluid(s)	Pressure or temperature	Tube	Number of tubes	Type of measurements	Bundle boiling factor
Low-finned tubes						
Jones (1941)	R-11	0.41 bar	16 fins per inch (fpi)		Overall	
Myers and Katz (1953)	R-12 and five others	Various	19.5 fpi	4	Each tube	> 1.0
Danilova and Dyundin (1972)	R-12	–20° to 20°C	12.5 and 27.7 fpi	19	Each tube	> 1.0
	R-22	–20° to 30°C	12.5 and 27.7 fpi		Mean	
Palen, Yarden, and Taborek (1972)	Hydrocarbon		19 fpi	413, 1,008	Overall	> 1.0
Nakajima and Shiozawa (1975)	R-11*	0.45 to 0.55 bar	19 fpi	185	Each tube and mean	> 1.0
Arai et al. (1977)	R-12*	0° to 2°C	19 fpi	225	Mean	> 1.0
Hahne and Muller (1983)	R-11	23°C	18.8 fpi	2 to 18	Each tube and mean	> 1.0
Yilmaz and Palen (1984)	*P*-Xylene	0.35 and 1.0 bar	18.7 fpi	168	Overall	
Windisch and Hahne (1985)	R-11	23°C	18.8 fpi	6, 18	Each tube and mean	> 1.0
Muller (1986)	R-11	23°C	18.8 fpi	18	Each tube	> 1.0
High-Performance tubes						
Czikk et al. (1970)	R-11*	0° to 6°C	High Flux	18, 20	Mean	
Palen, Yarden, and Taborek (1972)	Hydrocarbons	0.3 bar	High Flux	112, 569	Overall	
		6.9 bar		112, 569		
Arai et al. (1977)	R-12*	0° to 2°C	Thermoexcel-E	225	Mean	> 1.0
	R-12–oil	0° to 2°C				
Antonelli and O'Neill (1981)	R-11	0.4 bar	High Flux	104	Mean	> 1.0
Yilmaz, Palen, and Taborek (1981)	*p*-Xylene	0.35 bar	Themoexcel-E	168	Overall	< 1.0
		1.0 bar	Gewa-T	168	Overall and mean	> 0.8
Stephan and Mitrovic (1981)	R-12	–20°C	Gewa-T	8†	Each tube	1.0
	R-114	25°C	Gewa-T	8†	Each tube	1.0
	R-12/Oil	–10° and 0°C	Gewa-T	8†	Each tube	
Bukin, Danilova, and Dyundin (1982)	R-12	–40° to 0°C	13 porous coatings	19	Each tube	
	R-22					
Mochida, Takahata, and Miyoshi (1983)	R-22*	25°C	Flame sprayed coating	870	Overall	
Fujita et al. (1986)	R-113	1.0 bar	Porous, sintered coating	2, 10	Each tube	> 1.0

*The reader is referred to the case histories for these studies in Chapter Twelve.
†Only the central vertical row of three tubes was heated.

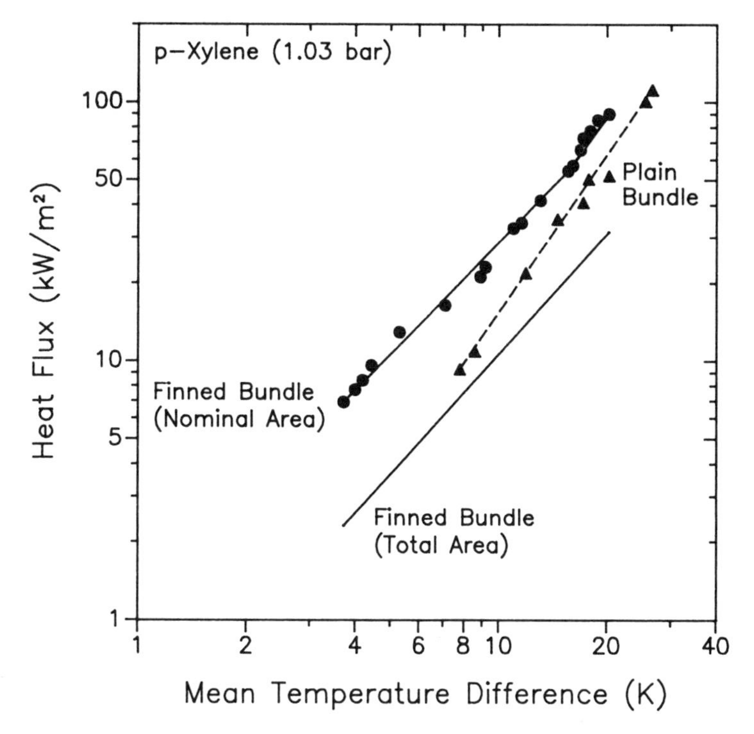

Figure 10-32 Comparison of overall performance of a low-finned tube bundle to that of a plain tube bundle for boiling *p*-xylene.

demonstrated another thermal advantage: it operated effectively down to lower mean temperature differences compared to the plain tube bundle.

High-performance boiling tubes. Palen, Yarden, and Taborek (1972) tested two different High Flux floating head kettle reboiler bundles for two unspecified hydrocarbons. The smaller bundle was made from 19-mm diameter tubes on a 25.4-mm pitch, and the second bundle was made from 9.5-mm tubes on a 12.6-mm pitch. Boiling performance at 0.3 bar (5 psia) with steam condensing on the tube side demonstrated considerable augmentation compared to a similar plain tube bundle. At a higher pressure no augmentation occurred, apparently because the steam side became the controlling thermal resistance.

Antonelli and O'Neill (1981) compared R-11 boiling performance of identical High Flux porous coatings on tube bundles to boiling on identical coatings on flat plates facing upward. One tube bundle consisted of 104 tubes of 25.4 mm outside diameter on a 29.4-mm triangular pitch and was rated as a 160-ton water chiller. The second bundle was identified only as a 20-ton water chiller. The two bundles displayed bundle boiling factors greater than 1.0 at wall superheats less than 2 K; at wall superheats greater than 2 K the bundle boiling performance coincided with that of the porous coated flat plate.

Yilmaz, Palen, and Taborek (1981) tested U-tube internal reboilers made from Thermoexcel-E and Gewa-T 90-10 cupro-nickel tubes. The Thermoexcel-E tubes were 17.6 mm in outside diameter with a tube pitch of 25.4 mm on a 90° tube layout. The Gewa-T bundle was made with 14.7-mm diameter tubes on a similar layout with a pitch of 23.8 mm. The investigators also constructed a plain tube bundle for comparison purposes. The performances for both enhanced boiling bundles were better than the performance of the plain tube bundle by about a factor of 1.5 to 2 when compared at the same heat flux and by about a factor of 1.5 to 2.5 when compared at the same overall temperature difference. (For the plain tube bundle at a heat flux of 10 kW/m^2 the thermal resistance of the boiling side was about twice that of the condensing steam, which limited the maximum possible increase in U to a factor of 3.) The bundle boiling factors were less than or equal to 1.0 for both types of enhanced tubes. This represents a deterioration in these tubes' performances when used in bundles. It was not stated, however, whether the dimensions of the enhancements of the cupro-nickel Thermoexcel-E and Gewa-T tubes used to make the tube bundles were the same as those of the larger-diameter copper versions tested as single tubes. Because the metalworking properties of the two metals are different, it would be rather difficult to manufacture identical enhancements. Therefore, a single tube study with the cupro-nickel tubes is needed to clarify the bundle boiling factors observed here.

Stephan and Mitrovic (1981) tested a small eight-tube bundle using 18.1-mm outside diameter Gewa-T tubes. Only the central vertical row of three tubes was heated. They observed a slight improvement in performance in the top two tubes compared to the bottom tube at low heat fluxes but none at heat fluxes greater than about 5 kW/m^2.

Bukin, Danilova, and Dyundin (1982) performed small tube bundle tests with 13 different kinds of metallic porous coated tubes. Several methods were employed to manufacture the coatings: metallic deposition, electric-arc spraying, sintering of powdered metal, and wrapping the tubes with either stainless steel or glass gauze. The investigators observed the best performance for the sintered porous layers, which also performed much better than low-finned tube bundles and smooth tube bundles in similar tests. No single tube data were given to determine the bundle boiling factors.

Fujita et al. (1986) studied several different multiple tube configurations with porous boiling enhancements. They made their porous coatings by sintering bronze particles on the outside of smooth tubes. For two enhanced tubes mounted one above the other, they observed a small augmentation for the upper tube at low heat fluxes that diminished as the heat flux increased. For a smooth tube placed above an enhanced tube, they found that the smooth tube's performance was increased by as much as a factor of 5 compared to its single tube performance at low heat fluxes for various levels of heat flux on the bottom tube. The upper tube's performance tended toward its single tube boiling curve as its heat flux increased, however. The investigators also reported studying a ten-tube bundle composed of nine enhanced tubes and one smooth tube located in the center of the bundle. The smooth tube's performance was stated to be similar to that of the two-tube configuration.

Bundle Geometry Effects

From the above experimental studies, some information is available about the effect of bundle geometry on the performance of enhanced boiling tubes (the reader is also referred to Chapters Eleven and Twelve for additional information about factors such as the maximum bundle heat fluxes, heat exchanger layouts, and circulation rates).

The effect of tube pitch has been studied in several investigations. Arai et al. (1977) found that there was no difference in R-12 boiling performance for two otherwise identical 225-tube bundles with triangular tube layouts made with 16.4-mm diameter Thermoexcel-E tubes, one with a horizontal pitch of 25 mm and a vertical pitch of 20 mm and the other with pitches of 21 and 17.5 mm, respectively. Muller (1986) observed that performance increased with smaller ratios of tube pitch to tube diameter for triangular layouts with low-finned tubes. For a ratio of 2.0, only the top tube row performed better than a single tube. Decreasing the tube pitch ratio to 1.6 and 1.3 increased boiling performance, however. These results appear to be reasonable because the contribution of convection can have a greater effect on a low-finned tube with its lower single tube performance.

The effect of tube layout on enhanced boiling bundle performance is not well understood. For a large tube pitch ratio of 2.0, a square tube layout tested by Hahne and Muller (1983) gave essentially the same performance as the triangular layout tested by Muller (1986). For High Flux tubes, Antonelli and O'Neill (1981) have stated that triangular tube layout is preferred, although the use of a square, diamond, or rotated square layout does not affect the boiling performance for typical tube pitch ratios of 1.25.

In all the above described studies in which individual low-finned tubes were instrumented, individual tube performance improved with increasing height up through the tube bundle but only at low to medium heat flux levels. For instance, Fig. 10-33 depicts the finned tube boiling data for the center row of 6 tubes in an 18-tube bundle tested by Windisch and Hahne (1985) for a triangular tube layout with a tube pitch ratio of 1.6. Tube 6 is seen to perform several times better than the bottom tube (1) up to heat fluxes of about 25 kW/m^2, after which the level of augmentation decreases rapidly. In large bundles of low-finned tubes, a lateral variation in the heat transfer performance can also be expected at low to medium heat fluxes; such variations are similar to those observed for a plain tube bundle by Leong and Cornwell (1979).

Summary

Enhanced boiling on the shell side of horizontal tube bundles still requires extensive research to investigate the influence of the many geometric factors involved. Single tube boiling correlations for the various enhanced boiling tubes are not generally available (refer to Chapter Eight for a review of existing methods), so that experimental boiling curves must be obtained for each particular application. A general method for predicting bundle boiling factors is also required (for low-finned tubes in particular), although it appears that simply using the single tube boiling curve will give a fairly

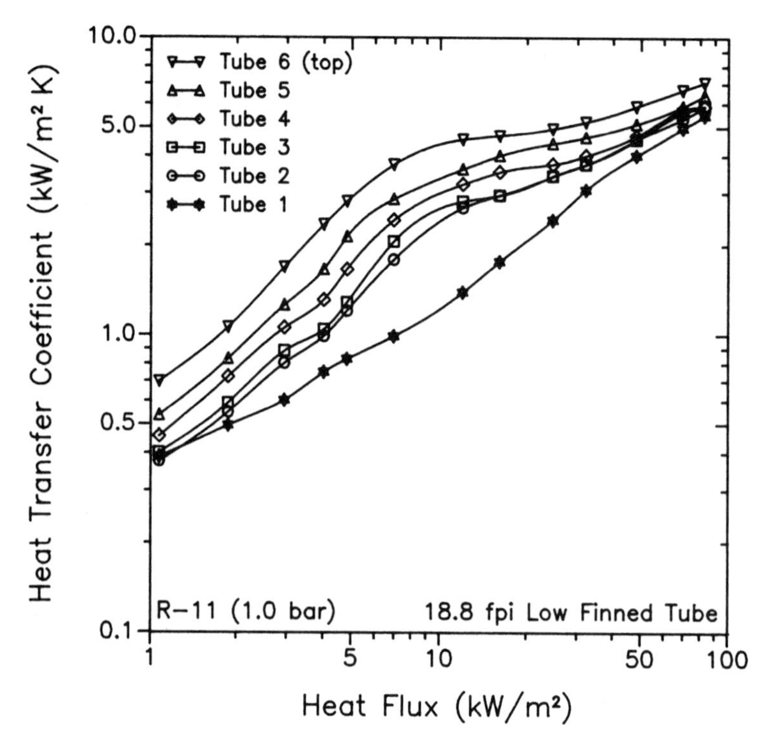

Figure 10-33 Local boiling heat transfer coefficients for the central row of a low-finned tube bundle.

accurate and conservative design. The maximum bundle heat fluxes must be studied experimentally to provide guidelines in this area.

10-3 NOMENCLATURE

a	radial acceleration (m/s^2)
A_a	heat transfer surface area per unit length (m^2/m)
A_{fa}	actual flow area (m^2)
A_{fc}	open core flow area (m^2)
A_{fn}	nominal flow area based on D_i (m^2)
A_i	internal flow area of tube with star insert (m^2)
A_n	nominal heat transfer surface area per unit length based on D_i (m^2/m)
Bo	Boiling number $[q/(G\Delta h_v)]$
c_1	empirical constant
c_2	empirical constant
c_3	empirical constant
d_i	tube minimum inside diameter (m)
d_i	inside diameter of corrugated tube (m)
d_o	tube outside diameter

D_h	hydraulic diameter (m)
D_i	tube maximum inside diameter (m)
e	ridge height of corrugated tube (m)
f_{TP}	two-phase friction factor
F	parameter defined in Eq. (10-12)
F_1	flow area ratio
F_2	heat transfer surface area ratio
F_3	parameter defined in Eq. (10-13)
F_4	nominal flow area ratio
Fr_L	liquid Froud number
g	acceleration due to gravity (m/s^2)
g_c	gravitational constant
G	mass flux [kg/(m^2·s)]
h	fin height (m)
Δh_v	latent heat of vaporization (J/kg)
J	mechanical equivalent of heat
K_f	Pierre parameter
L	tube length (m)
$\dot{m}$	mass flow rate (kg/s)
n	ratio of heat transfer surface area to nominal surface area based on d_i
n_1	exponent
n_2	exponent
n_3	exponent
N	number of internal fins
p	fin or ridge pitch (m)
Pr_L	liquid Prandtl number
ΔP	pressure drop (N/m^2)
ΔP_a	acceleration pressure drop (N/m^2)
ΔP_f	frictional pressure drop (N/m^2)
ΔP_{fin}	pressure drop for finned tube in Fig. 10-13
ΔP_{MN}	Martinelli-Nelson pressure drop (N/m^2)
ΔP_s	swirl flow pressure drop (N/m^2)
ΔP_{sm},	
ΔP_{smooth}	pressure drop for smooth tube (N/m^2)
q	heat flux (W/m^2)
$(q_{cr})_{swirl}$	critical heat flux of swirl flow (W/m^2)
$(q_{cr})_{smooth}$	critical heat flux of smooth tube (W/m^2)
Re_a	axial flow Reynolds number
Re_{D_i}	liquid only Reynolds number
Re_h	axial Reynolds number based on hydraulic diameter
$(\mathrm{Re}_h)_L$	liquid hydraulic Reynolds number
$(\mathrm{Re}_h)_v$	vapor hydraulic Reynolds number
Re_s	swirl flow Reynolds number
S	boiling suppression factor
t	thickness of fin (m)

u_a	axial flow velocity (m/s)
U	overall heat transfer coefficient [W/(m^2·K)]
v_m	mean two-phase specific volume (m^3/kg)
w_{FV}	boiling capacity (kg/h)
x	vapor quality
x_{in}	inlet vapor quality
x_m, x_{mean}	mean vapor quality
$x_{midpoint}$	vapor quality at midpoint of tube test section
x_{out}	outlet vapor quality
Δx	inlet-to-outlet vapor quality change
y	tape twist ratio
α	local heat transfer coefficient [W/m^2·K)]
$\overline{\alpha}$	average heat transfer coefficient [W/m^2·K)]
$\overline{\alpha}_{BP}$	average plain tube heat transfer coefficient from Pierre correlation [W/(m^2·K)]
α_h	average local heat transfer coefficient in channel [W/(m^2·K)]
α_L	liquid-only heat transfer coefficient [W/(m^2·K)]
α_{nb}	nucleate pool boiling heat transfer coefficient [W/(m^2·K)]
α_{sm}	smooth tube heat transfer coefficient [W/(m^2·K)]
α_o	heat transfer coefficient without oil [W/(m^2·K)]
β	spiral angle of helical fins (degrees)
γ	groove angle (radians)
ε^*	two-phase flow parameter
λ_L	liquid thermal conductivity [W(m·K)]
μ_L	liquid dynamic viscosity [kg/(m·s)]
μ_v	vapor dynamic viscosity [kg/(m·s)]
ν_L	liquid kinematic viscosity (m^2/s)
ρ_L	liquid density (kg/m^3)
ρ_v	vapor density (kg/m^3)
ϕ	severity factor

REFERENCES

Ackerman, J. W. 1970. Pseudoboiling heat transfer to supercritical pressure water in smooth and ribbed tubes. *J. Heat Transfer* 92:490–98.

Agrawal, K. N., H. K. Varma, and S. Lal. 1982. Pressure drop during forced convection boiling of R–12 under swirl flow. *J. Heat Transfer* 104:758–62.

———. 1986. Heat transfer during forced convection boiling of R-12 under swirl flow. *J. Heat Transfer* 108:567–73.

Allen, C. F. 1961. AEC R&D Report K-1487.

Antonelli, R., and P. S. O'Neill. 1981. Design and application considerations for heat exchangers with enhanced boiling surfaces. Paper read at International Conference on Advances in Heat Exchangers, September, Dubrovnik, Yugoslavia.

Arai, N., T. Fukushima, A. Arai, T. Nakajima, K. Fujie, and Y. Nakayama. 1977. Heat transfer tubes enhancing boiling and condensation in heat exchangers of a refrigerating machine. *ASHRAE Trans.* 83(part 2):58–70.

Azer, N. Z., and V. Sivakumar. 1984. Enhancement of saturated boiling heat transfer by internally finned tubes. *ASHRAE Trans.* 90(part 1A):58–73.

Bell, K. J. and A. C. Mueller. 1984. *Engineering data book II*. Decatur, Ala.: Wolverine Tube.

Bergles, A. E., W. D. Fuller, and S. J. Hynek. 1971. Dispersed flow boiling of nitrogen with swirl flow. *Int. J. Heat Mass Transfer* 14:1343–54.

Bergles, A. E., V. Nirmalan, G. H. Junkhan, and R. L. Webb. 1983. *Bibliography on augmentation of convective heat and mass transfer II* (report HTL-31). Ames, Iowa: Engineering Research Institute, Iowa State University.

Blatt, T. A., and R. R. Alt. 1963. *The effects of twisted-tape swirl generators on the heat transfer and pressure drop of boiling Freon-11 and water* (ASME paper 65-WA-42).

Bryan, W. L., and G. W. Quaint. 1951. Heat transfer coefficients in horizontal tube evaporators. *Refrig. Eng.* 59(1):67–72.

Bryan, W. L., and L. G. Seigel. 1955. Heat transfer coefficients in horizontal tube evaporators. *Refrig. Eng.* 63(5):36–45.

Bukin, V. G., G. N. Danilova, and V. A. Dyundin. 1982. Heat transfer from Freons in a film flowing over bundles of horizontal tubes that carry a porous coating. *Heat Transfer Sov. Res.* 14(2):98–103.

Carnavos, T. C. 1980. Heat transfer performance of internally finned tubes in turbulent flow. *Heat Transfer Eng.* 1(4):32–37.

Chaddock, J. B. 1976. Influence of oil on refrigerant evaporator performance. *ASHRAE Trans.* 82(part 1):474–86.

———. and G. H. Buzzard. 1986. Film coefficients for in-tube evaporation of ammonia and R-502 with and without small percentages of mineral oil. *ASHRAE Trans.* 92(part 1A):22–40.

Chaddock, J. B., and A. P. Mathur. 1980. Heat transfer to oil-refrigerant mixtures evaporating in tubes. *Proc. Multi-Phase Flow Heat Transfer Symp./Workshop* 2:861–84.

Chawla, J. M. 1967. Ph.D. diss., T-H Hannover, West Germany.

Chen, J. C. 1963. *A correlation for boiling heat transfer to saturated fluids in convective flow* (ASME paper 63-HT-34).

———. 1966. A correlation for boiling heat transfer to saturated fluids in convective flow. *I C Process Des. Dev.* 5(3):322–339.

Chistyakov, F. M., N. I. Frolova, and S. G. Kuvshinov. 1978. Determination of pressure drop in horizontal shell-in-tube evaporators operating with boiling of refrigerant inside the tubes. *Heat Transfer Sov. Res.* 10(2):1–10.

Cumo, M., G. E. Farello, G. Ferrari, and G. Palazzi. 1974. The influence of twisted tapes in subcritical, once-through vapor generators in counter flow. *J. Heat Transfer* 96:365–70.

Czikk, A. M., C. F. Gottzmann, E. G. Ragi, J. G. Withers, and E. P. Habdas. 1970. Performance of advanced heat transfer tubes in refrigerant-flooded coolers. *ASHRAE Trans.* 76(part 1):99–109.

Czikk, A. M., P. S. O'Neill, and C. F. Gottzmann. 1981. Nucleate pool boiling from porous metal films: Effect of primary variables. In *Advances in Enhanced Heat Transfer*, HTD Vol. 18, 109–22. New York: American Society of Mechanical Engineers.

Danilova, G. N., and V. A. Dyundin. 1972. Heat transfer with Freons 12 and 22 boiling at bundles of finned tubes. *Heat Transfer Sov. Res.* 4(4):48–54.

Danilova, G. N., E. I. Guygo, A. V. Borishanskaya, V. G. Bukin, A. V. Dyundin, A. A. Kozyrev, L. S. Malkov, and G. I. Malyugin. 1976. Enhancement of heat transfer during boiling of liquid refrigerants at low heat fluxes. *Heat Transfer Sov. Res.* 8(4):1–8.

Dembi, N. J., P. L. Dhar, and C. P. Arora. 1979. An investigation into the use of wire screens in DX-evaporators. *Proc. 15th Int. Congr. Refrig.* 2:485–88.

D'Yachkov, F. N. 1978. Investigation of heat transfer and hydraulics for boiling of Freon-22 in internally finned tubes. *Heat Transfer Sov. Res.* 10(2):10–19.

Fujita, Y., H. Ohta, S. Hidaka, and K. Nishikawa. 1986. Nucleate boiling heat transfer on horizontal tubes in bundles. *Proc. 8th Int. Heat Transfer Conf.* 5:2131–36.

Gambill, W. R. 1963. Generalized prediction of burnout heat flux for flowing, subcooled wetting liquids. *Chem. Eng. Progr. Symp. Ser.* 59(41):71–87.

———. 1965. Subcooled swirl-flow boiling and burnout with electrically heated twisted tapes and zero wall flux. *J. Heat Transfer* 97:342.

———. R. D. Bundy, and R. W. Wansbrough. 1961. Heat transfer, burnout, and pressure drop for water in swirl flow tubes with internal twisted tapes. *Chem. Eng. Progr. Symp. Ser.* 57(31):127–37.

Gregorig, R. 1954. Hautkondensation an fein gewellten oberflächen. *Z. Angew. Math. Phys.* 5:36–49.

Hahne, E., and J. Muller. 1983. Boiling on a finned tube and a finned tube bundle. *Int. J. Heat Mass Transfer* 26:849–59.

Herbert, L. S., and U. J. Sterns. 1968. An experimental investigation of heat transfer to water in film flow: Part II—Boiling runs with and without induced swirl. *Can. J. Chem. Eng.* 46:408–12.

Herd, K. G., W. P. Goss, and J. W. Connell. 1985. Correlation of forced-flow evaporation heat transfer coefficients in refrigerant systems. In *Heat exchangers for two-phase applications,* HTD vol. 27, 11–18. New York: American Society of Mechanical Engineers.

Ikeuchi, M., T. Yumikura, M. Fujii, and G. Yamanaka. 1984. Heat-transfer characteristics of an internal microporous tube with refrigerant-22 under evaporating conditions. *ASHRAE Trans.* 90(part 1A):196–211.

Ito, M., and H. Kimura. 1979. Boiling heat transfer and pressure drop in internal spiral-grooved tubes. *Bull. JSME* 22(171):1251–57.

Jensen, M. K. 1984. A correlation for predicting the critical heat flux condition with twisted-tape swirl generators. *Int. J. Heat Mass Transfer* 27:2171–73.

———. 1985. An evaluation of the effect of twisted-tape swirl generators in two-phase flow heat exchangers. *Heat Transfer Eng.* 6(4):19–30.

———. and H. P. Bensler. 1986. Saturated forced-convective boiling heat transfer with twisted-tape inserts. *J. Heat Transfer* 108:93–99.

Jones, W. 1941. Cooler and condenser heat transfer with low pressure refrigerant. *Refrig. Eng.* 41(6):413–18.

Kawai, S., and T. Machiyama. 1975. Experimental investigations on heat transfer characteristics of corrugated tubes on tube-side boiling and condensing heat transfer coefficients for refrigerant use (in Japanese). *Bull. Sci. Eng. Res. Lab. Waseda Univ.* 71:1–9.

Kawai, S., and H. Yamada. 1977. Experimental investigation of the heat transfer characteristics of corrugated tubes for tube-side boiling of refrigerants (in Japanese). *Reito* 52:449–56.

Khanpara, J. C., A. E. Bergles, and M. B. Pate. 1986. Augmentation of R-113 intube evaporator with micro-fin tubes. *ASHRAE Trans.* 92(part 2):506–24.

Khanpara, J. C., M. B. Pate, and A. E. Bergles. 1987. Local evaporation heat transfer in a smooth tube and a micro-fin tube using refrigerants 22 and 113. In *Boiling and condensation in heat transfer equipment,* ed. E. G. Ragi, HTD vol. 85, 31–39. New York: American Society of Mechanical Engineers.

Kimura, H., and M. Ito. 1981. Evaporating heat transfer in horizontal internal spiral-grooved tubes in the region of low flow rates. *Bull. JSME* 24:1602–1607.

Kitto, J. B., and M. Wiener. 1982. Effects of nonuniform circumferential heating and inclination on critical heat flux in smooth and ribbed bore tubes. *Proc. 7th Int. Heat Transfer Conf.* 4:303–308.

Kubanek, G. R., and D. L. Miletti. 1979. Evaporative heat transfer and pressure drop performance of internally finned tubes with refrigerant 22. *J. Heat Transfer* 101:447–52.

Larson, R. L., G. W. Quaint, and W. L. Bryan. 1949. Effects of turbulence promotors in refrigerant evaporator coils. *Refrig. Eng.* 57:1193–95.

Lavin, J. G., and E. H. Young. 1965. Heat transfer to evaporating refrigerants in two-phase flow. *AIChE J.* 11:1124–32.

Leong, L. S., and K. Cornwell. 1979. Heat transfer coefficients in a reboiler tube bundle. *Chem. Eng.* (343):219–21.

Lopina, R. F., and A. E. Bergles. 1969. Heat transfer and pressure drop in tape-generated swirl flow of single-phase water. *J. Heat Transfer* 91:434–42.

———. 1973. Subcooled boiling of water in tape-generated swirl flow. *J. Heat Transfer* 95:281–83.

Martinelli, R. C., and D. B. Nelson. 1948. Prediction of pressure drop during forced circulation boiling of water. *Trans. ASME* 70:695.

Megerlin, F. E., R. W. Murphy, and A. E. Bergles. 1974. Augmentation of heat transfer by use of mesh and brush inserts. *J. Heat Transfer* 96:145–51.

Mentes, A., H. Gurgenci, O. T. Yildirim, S. Kakac, and T. N. Veziroglu. 1982. Effect of heater surface configurations on two-phase flow instabilities in a vertical boiling channel. In *Proceedings of the 16th Southeastern Seminar on Thermal Sciences.* Washington, D.C.: Hemisphere.

Mentes, A., O. T. Yildirim, L. Q. Fu, S. Kakac, and T. N. Veziroglu. 1983*a*. Experimental investigation of boiling flow instabilities with heat transfer enhancement. In *Condensed Papers of the 3rd Multi-Phase Flow and Heat Transfer Symposium-Workshop,* 156.

Mentes, A., O. T. Yildirim, H. Gurgenci, S. Kakac, and T. N. Veziroglu. 1983*b*. Effect of heat transfer augmentation on two-phase flow instabilities in a vertical boiling channel. *Wärme Stoffuebertrag.* 17(3):161–69.

Mochida, Y., T. Takahata, and M. Miyoshi. 1983. Performance tests of an evaporator for a 100-kW (gross) OTEC plant. *Proc. ASME-JSME Therm. Eng. Joint Conf.* 2:241–45.

Muller, J. 1986. Boiling heat transfer on finned tube bundles—The effect of tube position and intertube spacing. *Proc. 8th Int. Heat Transfer Conf.* 4:2111–16.

Myers, J. E., and D. L. Katz. 1953. Boiling coefficients outside horizontal tubes. *Chem. Eng. Progr. Symp. Ser.* 49(5):107–14.

Nakajima, K., and A. Shiozawa. 1975. An experimental study on the performance of a flooded type evaporator. *Heat Transfer Jap. Res.* 4(3):49–66.

O'Neill, P. S., R. C. King, and E. G. Ragi. 1980. Application of high-performance evaporator tubing in refrigeration systems of large olefins plants. *AIChE Symp. Ser.* 76(199):289–300.

Palen, J. W., A. Yarden, and J. Taborek. 1972. Characteristics of boiling outside large-scale horizontal multitube bundles. *AIChE Symp. Ser.* 68(118):50–61.

Panchal, C. B., E. H. Buyco, J. Yampolsky, and K. J. Bell. 1986. A spirally fluted tube heat exchanger as a condenser and evaporator. *Proc. 8th Int. Heat Transfer Conf.* 6:2787–92.

Panchal, C. B., and D. M. France. 1986. *Performance tests on the spirally fluted tube heat exchanger for industrial cogeneration applications* (Argonne National Laboratory report ANL/CNSV).

Pierre, B. 1957. Varmeovergangen vid kokande koldmedier. *Kyltek. Tidschr.* 3:129.

———. 1964. Flow resistance with boiling refrigerants. *ASHRAE J.* 6(9):58–65; 6(10):73–77.

Reid, R. S., M. B. Pate, and A. E. Bergles. 1987. A comparison of augmentation techniques during in-tube evaporation of R-113. In *Boiling and condensation in heat transfer equipment,* ed. E. G. Ragi, HTD vol. 85, 21–30. New York: American Society of Mechanical Engineers.

Schlager, L. M., A. E. Bergles, and M. B. Pate. 1988. Evaporation and condensation of refrigerant-oil mixtures in a smooth tube and a micro-fin tube. *ASHRAE Trans.* 94(part 1).

Schlager, L. M., M. B. Pate, and A. E. Bergles. 1987. A survey of refrigerant heat transfer and pressure drop emphasizing oil effects and in-tube augmentation. *ASHRAE Trans.* 93(part 1):392–416.

———. 1988. Evaporation and condensation of refrigerant-oil mixtures in a low-fin tube. *ASHRAE Trans.* 94(part 1).

Schlünder, E. U., and J. Chawla. 1967. Local heat transfer and pressure drop for refrigerants evaporating in horizontal, internally finned tubes. *Proc. Int. Congr. Refrig.* (paper 2.47).

Shinohara, Y., and M. Tobe. 1985. Development of an improved "Thermofin tube." *Hitachi Cable Rev.* 4:47–50.

Shiralkar, B., and P. Griffith. 1969. *The effect of swirl, inlet conditions, flow direction and tube diameter on the heat transfer to fluids at supercritical pressure* (ASME paper 69-WA/HT-1).

Stephan, K., and J. Mitrovic. 1981. Heat transfer in natural convection boiling of refrigerants and refrigerant-oil mixtures in bundles of T-shaped finned tubes. In *Advances in Enhanced Heat Transfer*, HTD Vol. 18, 131–46. New York: American Society of Mechanical Engineers.

Swenson, H. S., J. R. Carver, and G. Szoeke. 1962. The effects of nucleate boiling versus film boiling on heat transfer in power boiler tubes. *J. Eng. Power Trans. ASME Ser. A* 84:365–71.

Tatsumi, A., K. Oizumi, M. Hayashi, and M. Ito. 1982. Application of inner groove tubes to air conditioners. *Hitachi Rev.* 32(1):55–60.

Tojo, S., K. Hosokawa, T. Arimoto, H. Yamada, and Y. Ohta. 1984. Performance characteristics of multigrooved tubes for air conditioners. *Aust. Refrig. Air Cond. Heat.* 38(8):45–61.

Watson, G. B., R. A. Lee, and M. Wiener. 1974. Critical heat flux in inclined and vertical smooth and ribbed tubes. *Proc. 5th Int. Heat Transfer Conf.* 4:275–79.

Windisch, R., and E. Hahne. 1985. Heat transfer for boiling on finned tube bundles. *Int. Commun. Heat Mass Transfer* 12:355–68.

Withers, J. G., and E. P. Habdas. 1974. Heat transfer characteristics of helical-corrugated tubes for intube boiling of refrigerant R-12. *AIChE Symp. Ser.* 70(138):98–106.

Worsoe-Schmidt, P. 1960. Some characteristics of flow pattern and heat transfer of Freon-12 evaporating in a horizontal tube. *J. Refrig.* 3(2):40–44.

Yilmaz, S., and J. W. Palen. 1984. *Performance of finned tube reboilers in hydrocarbon service* (ASME paper 84-HT-91).

———. and J. Taborek. 1981. Enhanced boiling surfaces as single tubes and tube bundles. In *Advances in Enhanced Heat Transfer*, HTD Vol. 18, 123–24. New York: American Society of Mechanical Engineers.

CHAPTER

ELEVEN

APPLICATION OF ENHANCED BOILING TUBES IN THE CHEMICAL PROCESSING INDUSTRIES

11-1 INTRODUCTION

Enhanced boiling tubes are used extensively in the refrigeration industry. In contrast, their consideration as a design option for reboilers and evaporators in chemical and petrochemical processing plants is still not standard practice. There are several reasons for this state of affairs: (1) engineering firms designing new facilities typically have little economic incentive to apply enhanced boiling tubes in place of plain tubes; (2) in existing facilities the application of enhanced boiling tubes is mostly limited to areas in which alternative, low-cost heat sources are available to take advantage of their increased heat transfer performance; (3) thermal design data are generally not available for the wide variety of fluids encountered in practice; and (4) fouling and plant reliability are concerns. This situation is expected to change in the future as thermal design methods for enhanced boiling tubes other than low-finned tubes become more widely available.

The increasing competition resulting from the internationalization of the hydrocarbon marketplace is forcing plant operators to utilize energy more efficiently, especially because energy can represent as much as 60% of the total operating expense of a plant. Therefore, a special review of the application of enhanced boiling tubes to reboilers and evaporators is presented here to bring together practical information for process heat transfer engineers, who will need to document the technical and the economic advantages to management to justify their use.

The subjects to be discussed are organized into the following categories: advantages of enhanced boiling tubes, enhanced boiling heat exchanger configurations, process applications, special thermal design considerations, mechanical design considerations, fouling, operating considerations, and economics and surface selection. Other pertinent information can be found elsewhere, including boiling data in Chapters Seven, Nine, and Ten and types of commercial enhanced boiling tubes in Chapter Four. Several design case histories for actual applications are presented in Chapter Twelve.

11-2 ADVANTAGES OF ENHANCED BOILING TUBES

Enhanced boiling tubes offer the process heat transfer designer several important technical advantages over and above conventional, plain tube design. They are:

1. a shell-side boiling heat transfer coefficient about 2 to 10 times that of a plain tube bundle or an intube boiling coefficient about 1.5 to 5 times that of a plain, vertical tube;
2. the ability to commence boiling and to operate at smaller wall superheats;
3. the option of augmenting the heating fluid heat transfer coefficient by a factor of up to 2.5 for single-phase fluids and about a factor of 2 to 3 for condensing vapors by use of doubly enhanced tubes; and
4. the improved fouling resistance offered by integral low-finned tubes.

These special characteristics can be utilized in a number of ways. For instance, when an existing plain tube bundle is being retubed with enhanced boiling tubes, either (1) the heat duty (and production capacity) can be increased, or (2) the log mean temperature difference (LMTD) can be decreased to allow the substitution of a cheaper, lower-temperature heat source, or (3) a combination of (1) and (2) can be used. For new installations, the size or number of reboilers in parallel can be reduced with associated savings in support structure and plot size. When energy usage is optimized, operation becomes possible at lower temperature approaches. For example, for an ethylene-ethane splitter reboiler, the temperature difference of about 15 K typical of older plants can be reduced to 5 K. With low-finned tubes in fouling applications, the time between cleanings can be lengthened and the manpower for cleaning the bundle reduced. Also, more heat can be extracted from a single-phase heating medium by use of a smaller temperature approach at the exit.

One area receiving much attention during the upgrading of existing facilities is heat integration programs, in which the waste heat from one unit process is used to drive another nearby unit operating at a slightly lower temperature (or even at a higher temperature with the introduction of a heat pump or vapor recompression system). The use of "pinch technology" to optimize energy usage results in smaller temperature approaches in these heat exchangers.

This trend toward closer temperature approaches in reboilers, vaporizers, reboiler-condensers, chillers, and so forth, however, is impeded by two physical limitations with conventional, plain tube designs: (1) extremely large surface areas in heat exchangers become necessary when combining large heat duties with small log mean temperature differences and small overall heat transfer coefficients, which also reduces the reliability of operation because of the increased possibility of maldistribution, and (2) the wall superheat driving the boiling process becomes smaller as the temperature approach is reduced and at some point becomes too small to support nucleate boiling on plain tubes, leading to flashing at the surface of the liquid pool and poor cyclic operation.

Enhanced boiling tubes are particularly well suited to these applications (as are the plate-fin heat exchangers described in Chapter Thirteen) by virtue of their larger overall heat transfer coefficients (typically on the order of two to five times those of conventional bundles) and their ability to activate and boil vigorously at low wall superheats. Thus enhanced boiling tubes, when applied correctly, can make previously impractical temperature approaches not only possible but thermally efficient.

When to utilize enhanced boiling tubes is another important consideration. One thermal rule of thumb which has been proposed is that the boiling-side heat transfer coefficient for the conventional, plain tube bundle design should be less than one-third that of the heating side before an enhanced boiling tube is considered. This rule is too conservative, however, because the heating medium heat transfer coefficient can be augmented by a factor of 2 to 3 by a doubly enhanced tube. Thus a more reasonable criterion is when the boiling-side heat transfer coefficient is less than or equal to the heating-side coefficient. Also, the economic benefits of smaller operating temperature approaches have to be considered in heat recovery schemes, in which the additional heat recovered with the lower permissible heating fluid exit temperature may be sufficient by itself to justify use of an enhanced boiling tube, so that heat transfer augmentation becomes a secondary consideration.

11-3 HEAT EXCHANGER CONFIGURATIONS

Enhanced boiling tubes can be installed in nearly all the heat exchanger configurations typically used for conventional reboilers, evaporators, and the like that are in service in the petroleum and chemical processing industries. These exchangers may be vertical, horizontal, or kettle type and have either circulating or once-through flow. For vertically oriented bundles boiling is almost always on the tube side, so that the selection of boiling enhancements is limited to a High Flux porous coating, internal finning, spirally fluted tubes, or inserts. For horizontal or kettle type units boiling occurs on the shell side, so that there are many more boiling enhancements from which to choose.

Figure 11-1 (from Antonelli and O'Neill [1981, 1986]) depicts the three most commonly used enhancing boiling heat exchanger layouts. The top one is for a refrigerant-chilled kettle-type condenser, a configuration that is widely used in the refrigeration systems of olefin and gas processing plants. Boiling is augmented on the shell side of the tubes to evaporate a refrigerant, such as propane or ethane. The refrigerant typically arrives at the inlet nozzle after being throttled from a higher pressure. Therefore, the inlet flow is two-phase with the vapor quality ranging from about 10% to 15% by weight. The inlet for an enhanced bundle can be located near the top of the shell so that the entering vapor does not pass over the tube bundle, which will reduce the tendency toward vapor blanketing (a conventional design typically has the inlet at the

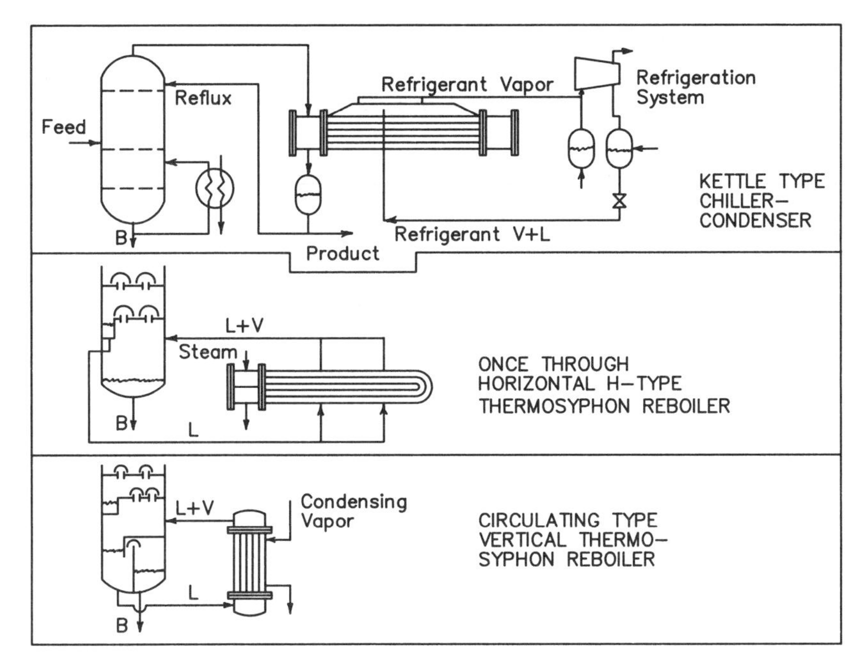

Figure 11-1 Three common enhanced boiling heat exchanger layouts.

bottom of the shell to increase two-phase convection in the bundle). An enlarged shell is utilized to remove entrained liquid from the outlet vapor flow.

The boiling process in these units is nearly isothermal. Condensation inside the tubes may occur over a temperature range, depending on the pressure drop and the dew points of the components. The minimum temperature approach has to be adjusted for the liquid subcooling on the shell side created by the hydrostatic pressure of the liquid pool, which decreases the temperature driving force for boiling in the lower part of the bundle and also tends to reduce the boiling heat transfer coefficient. Because the temperature approaches are typically quite close, as low as 3 K in some applications, the decrease in the dew point temperature of the condensing medium has to be taken into consideration in the design of an enhanced boiling exchanger to avoid a temperature cross.

Refrigerant-chilled condensers are usually large, having bundle diameters up to 1.6 m, shell diameters up to 3 m, and tube lengths up to 12 m. Because the operating temperatures in these units are as low as 170 K suitable tube materials must be chosen, such as fine-grained or aluminum-killed steel or 3.5% nickel steel. The most common TEMA head and shell designs used are BKM, CKU, and AKU. As in conventional design, the U-tube bundle geometry is favored for two-tube pass designs because this reduces the tendency for flow maldistribution to occur during condensation on the tube side compared to the M-type rear head, where phase separation can occur.

Enhanced boiling heat exchangers usually operate at higher heat fluxes and also transform a greater portion of the heat into vapor at the tube wall than conventional exchangers. This results in a greater boilup of vapor in the bundle. Thus, to prevent liquid entrainment in the exit flow, either larger shells or more exit nozzles may be required. Antonelli and O'Neill (1981, 1986) have recommended that shell diameters be at least 600 mm larger than the tube bundle diameter and, for very large units, at least 1.6 to 1.8 times the bundle diameter. As for conventional units, they suggest a ratio of tube pitch to tube diameter of 1.25 on a triangular layout.

The second configuration shown in Fig. 11-1 is the horizontal thermosyphon reboiler with the boiling enhancement on the shell side of a U-tube bundle. This layout is used in petroleum refining facilities where access to the bundle is required or when a vertical reboiler is not feasible. The typical applications are to benzene–toluene–mixed xylene (BTX) column reboilers, depentanizer reboilers, and naptha stabilizer reboilers. These units are generally steam-heated. Liquid feed (L) enters through nozzles at the bottom, and the outlet stream returns to the distillation tower partially vaporized ($L + V$). This type of tube bundle is removable, so that replacement of an existing plain tube bundle with an enhanced tube bundle is simple because the existing shell, head, and other components can be reused. A circulating type reboiler may also be specified depending on the existing tower control system. The liquid head driving a thermosyphon type enhanced boiling tube reboiler should be held to the minimum required for good circulation to reduce the subcooling of the entering liquid flow. Only low-finned and Gewa-T types of tubing are effective for sensible heating of subcooled liquid at the bottom of the bundle, and excessive subcooling should be avoided in any case. The common TEMA types used are BHM, BJU, BJM, CEN, and AES.

The vertical thermosyphon reboiler is the last configuration depicted in Fig. 11-1. Typical enhanced boiling tube applications are to demethanizers, C_2 splitters, and propane-propylene splitter reboilers in ethylene refining facilities and multieffect evaporators and reboilers for distilling aqueous solutions. Boiling enhancement is applied inside the tubes, and condensation on the outside benefits greatly from longitudinal or fluted fins to augment the falling film coefficient. When replacing an existing plain tube bundle to increase the reboiler's heat duty or to reduce its temperature approach, an identical size shell and tube sheets should be specified so that the existing heads, piping, and support structure can be used with the new exchanger. Tube diameters typically range from 25 to 32 mm (larger than conventional designs to offset the higher resistance to flow), and normal tube lengths vary from 3.5 to 6 m.

Two other heat exchanger configurations are often used in process plants: the kettle reboiler with a removable U-tube bundle and weir, and the internal column reboiler. For the first configuration, a new enhanced boiling tube bundle replacing an existing plain tube bundle can use the existing shell, head, and piping, which minimizes the cost and reduces the installation time. If this is done to concur with a scheduled replacement of the plain tube bundle, the effective cost can be reduced even further. The new enhanced boiling tube bundle may be smaller in diameter than or the same size as the conventional bundle it replaces. An internal reboiler is easily replaced by an enhanced boiling tube bundle to increase the capacity of a tower or to revamp the plant. When a new tower is being built, this configuration can be used to advantage to reduce the bundle size for a given heat duty and thus to minimize the space it occupies inside the column.

The above configurations only represent several of those that can be used with enhanced boiling tubes. The boiling and heating stream pressures, the existing tower layout, the influence of the liquid head on subcooling, the allowable pressure drop of the heating medium, and the particular process engineer's preference determine the configuration used in much the same way as for conventional units. Generally speaking, any configuration used with plain tubes can also be built with enhanced boiling tubes (when thermally justifiable). Restrictions due to fouling are discussed in Sec. 11-7.

Plate heat exchangers can also be built with enhanced boiling surfaces. Boiling tests with units using porous coated surfaces have been conducted by Uehara et al. (1983) for R-22 and ammonia and by Panchal, Hillis, and Thomas (1983) for R-22.

11-4 PROCESS APPLICATIONS

Typical process applications of enhanced boiling tubes in the chemical and petrochemical industries are described in the next sections. These applications were chosen with the high-performance enhanced boiling tubes in mind because their range of application is narrower than that of low-finned tubes. Only low-finned tubes can be applied to medium and heavy fouling services.

Air Separation Plants

The objective of the development of the first commercially successful high-performance boiling tube (High Flux) was to improve the performance of the main condenser in air separation plants, according to Milton and Gottzmann (1972). The process streams are primarily oxygen and nitrogen. Utilizing enhanced boiling surfaces is advantageous in reducing the size of the heat exchangers and decreasing compressor size and operating costs. The temperature approaches between the fluid streams can be closer because of the smaller boiling nucleation superheat required to initiate boiling. Figure 11-2 shows enhanced boiling tubes being expanded into the tube sheet of one of these units.

Boiling data for oxygen are presented in Fig. 3-23 and for nitrogen in Fig. 7-9. The heat transfer coefficients for nitrogen were shown to be much larger than those for serrated plate-fin heat exchangers, which are a competing technology for these units. For example, the boiling heat transfer coefficients for the High Flux and Thermoexcel-E tubes were from 18 to 80 times those of serrated plate-fin channels at a wall superheat of 0.7 K, which is a typical operating condition. By applying a doubly enhanced tube to augment the condensing side as well, the surface area of these shell-and-tube units would be smaller than that of a comparable plate-fin unit, although the heat duty per unit volume of heat exchanger might still be less because of the higher surface area density in plate-fin assemblies.

Gottzmann, O'Neill, and Minton (1973) and O'Neill and Gottzmann (1980) described the successful application of the High Flux tube to these facilities. The other

Figure 11-2 Main condenser of an air separation plant (courtesy Union Carbide).

enhanced boiling tubes apparently have not been made in aluminum, so that aluminum versions would have to be developed and tested to be applied in this industry.

Separation of Light Hydrocarbons in Ethylene Plants

In olefins plants, enhanced boiling tubes are used primarily in the ethylene refining or purification section. The High Flux tube has been used extensively in heat exchangers of world class ethylene plants as documented by O'Neill, King, and Ragi (1980). In one plant that they described in detail 19 High Flux heat exchangers were utilized, representing both intube boiling and shell-side boiling. A schematic diagram of the plant layout is shown in Fig. 11-3, and Table 11-1 lists the specifications of the heat exchangers.

A cold box with plate-fin exchangers was operated in parallel with the demethanizer chillers to obtain maximum heat recovery in the tail gas stream, which is mostly methane and hydrogen. Exchanger heat duties as large as 14.7 MW (50×10^6 Btu/h) were attained and operating pressures on the boiling side ranged from 1 to 30 bar. When boiling was on the tube side, external fluted fins were used to augment condensation. With boiling on the shell side, condensing mass velocities were raised on the tube side to improve performance rather than resorting to extended surfaces. None of the heat exchangers was designed to operate at temperature approaches less than 3 K. Condensation pressure drops were designed in the range from 0.14 to 0.34 bar (2 to 5 psi), including the nozzles, heads, and returns. Actual ethylene production was reported to be 108% of design capacity. Additional information was also given by O'Neill,

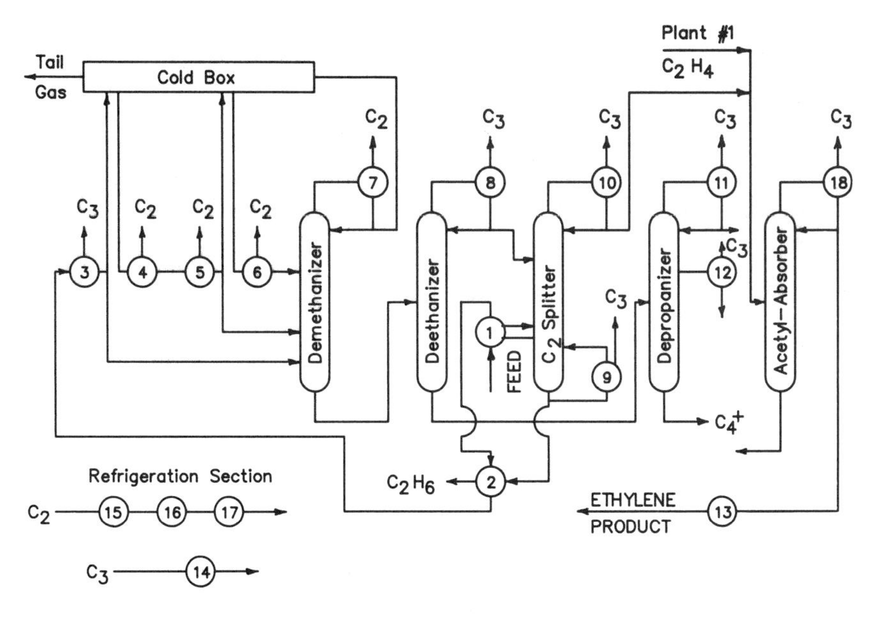

Figure 11-3 Enhanced boiling heat exchangers in an ethylene plant.

Table 11-1 Taft enhanced boiling heat exchangers

Exchanger number	Service	Heat duty, MW	U_o,W/(m^2·K)	Surface area, m^2	TEMA type	Size ($B/S/L^*$)	T_{sat},K
E-1	C_2 hip reboiler	4.3	1,556	224	CKU	76/157/4.86	255
E-2	Ethane vaporizer	1.5	1,437	93	CKU	71/132/2.43	246
E-3	Number 1 demethanizer feed chiller	3.4	1,022	368	CKU	86/175/6.08	233
E-4	Number 2 demethanizer feed chiller	0.7	1,544	93	CKU	58/127/3.65	221
E-5	Number 3 demethanizer feed chiller	1.9	1,272	163	CKU	64/127/5.47	200
E-6	Number 4 demethanizer feed chiller	0.7	1,397	69	CKU	41/102/6.08	172
E-7	Demethanizer condenser	1.3	1,499	205	CKU	64/122/6.99	172
E-8	Deethanizer condenser	3.0	1,579	226	CKN	56/122/9.12	250
E-9[†]	C_2 fraction reboiler	9.9	2,390	412	BEN	155/3.65	273
E-10	C_2 fraction condenser	14.7	1,772	1,032	CKN	122/251/10.94	241
E-11	Depropanizer condenser	5.5	1,204	328	CKU	84/173/6.08	272
E-12[†]	C_3 vent condenser	0.1	1,300	6	CEN	30/2.43	283
E-13	Ethylene product vaporizer	0.7	1,567	213	AKU	76/137/4.86	255
E-14	C_3 refrigeration subcooler	1.9	1,891	114	CKU	66/137/3.65	250
E-15	Ethylene 1 desuperheater	0.9	812	64	CKU	61/122/2.43	284
E-16	Ethylene 2 desuperheater	1.2	732	108	CKU	61/122/3.65	250
E-17	Ethylene refrigeration condenser	7.1	1,380	630	CKU	109/244/6.69	233
E-18	Acetone absorption overhead condenser	9.7	1,312	1,409	BKU	157/229/6.08	240
E-19	Vent column condenser	1.2	1,573	206	BKU	64/127/6.08	172

[*] B, bundle diameter (cm); S, shell diameter (cm); L, tube length (m).
[†] Intube boiling.

Fenner, and Williamson (1982) about the design of E-1 through E-6 and the cold box of the demethanizer feed chilling unit of the Taft plant.

Baseload Natural Gas Liquefaction

Enhanced boiling technology is especially adaptable to baseload liquid natural gas (LNG) liquefiers that use classic cascade refrigeration systems and also to propane forecooling systems for plants that use mixed refrigeration cycles. These systems have several large evaporator-condensers and a number of other evaporator coolers and partial condensers for the gas and thus can benefit significantly from heat transfer augmentation to reduce the size and first cost of these units and to minimize the compressor horsepower requirements by operating the heat exchangers at smaller temperature differences. This process application has been studied by O'Neill, Gottzmann, and Terbot (1971), Milton and Gottzmann (1972), and more extensively by O'Neill, Gottzmann, and Terbot (1972).

Figure 11-4 from the studies mentioned above depicts an LNG facility based on a classic cascade cycle (only the propane, ethylene, and methane refrigeration systems are shown). The simplified flow diagram is for a hypothetical plant producing 500 million standard cubic feet per day (scfd), or 14×10^6 m^3/d, of 1,020 Btu/lb (490 kJ/kg) high–heating value liquid natural gas. After passing through heavy hydrocarbon, carbon dioxide, and water removal units, the dried natural gas is assumed to enter the system at 41.7 bar (605 psig).

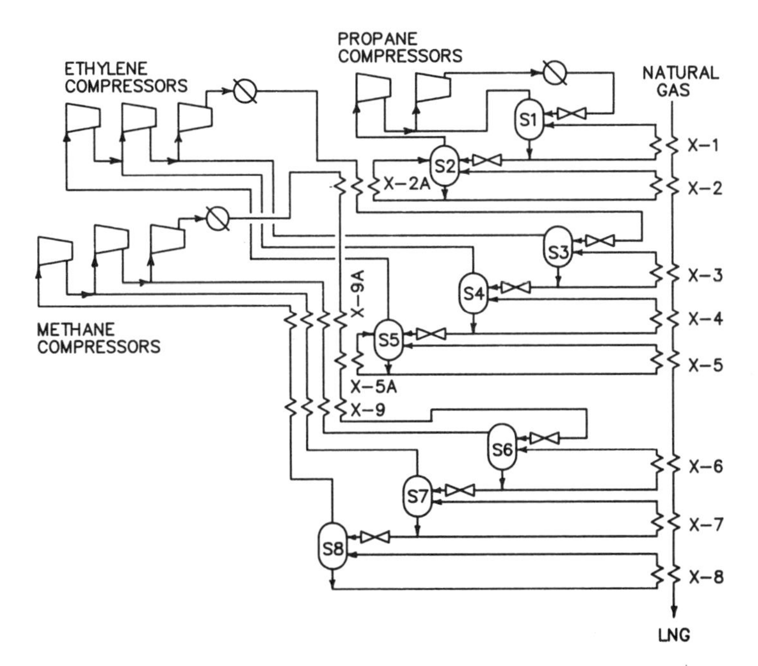

Figure 11-4 Flow diagram for a large LNG facility.

Applying enhanced boiling technology, each heat exchanger (X) in the diagram could be a kettle-type reboiler with enhanced boiling tubes except the two single-phase units X-9 and X-9A, which would be brazed aluminum plate-fin heat exchangers. The heat exchangers needed for the service were designed, and their specifications can be found in O'Neill, Gottzmann, and Terbot (1972). Assuming 75% adiabatic efficiencies for the compressors and tube-side pressure drops of 0.34 bar (5.0 psi) for all units, a LNG liquifier cycle was calculated. The results for a nonoptimized cycle are shown in Table 11-2 for two different temperature approaches among the three refrigeration streams in the two cascade condensers X-2A and X-5A. Lowering the temperature approach from 5.6 K to 2.8 K reduced the power required by 12,300 kW (16,500 hp). In addition, the total heat transfer areas for the exchangers depicted in Fig. 11-4 are reduced to less than one-third the conventional design values by using enhanced boiling technology, allowing for a much more compact design.

One operating problem with these types of refrigeration systems is the buildup of oil in the refrigerants. Figure 9-18 shows that the propylene boiling performance of High Flux tubing is adversely affected by the presence of only 1% oil by volume, as is also true to some extent for both conventional and plate-fin designs. Thus the buildup of oil in the refrigeration network must not be allowed to reach unreasonable levels (this would also raise the boiling point temperatures of the refrigerants, producing adverse effects on the overall cycle efficiency). Because enhancing the boiling side shifts most of the thermal resistance to the heating side, however, the effect of oil on the overall heat transfer coefficient U_o is usually minimal and within the bounds of normal design accuracy. (For further information about the effect of oil on shell-side boiling, refer to Sec. 9-5; for intube boiling effects, refer to Sec. 10-1.)

Refinery Light Ends Facilities

There are numerous potential heat exchanger applications for enhanced boiling tubes in refinery light ends separation facilities. Figure 11-5 (from Ragi, O'Neill, and Heck [1984]) illustrates locations of such enhancements on a simplified process flow sheet with all the entering streams combined at one single location for simplicity (not all the separations depicted would necessarily exist in any one plant). The diagram identifies reboilers that could be retubed with enhanced boiling tubes either (1) to operate with a lower-pressure steam or to be part of a heat integration network or (2) to be, in addition, potential units for heat pumping.

Table 11-2 Heat exchanger and power requirements for a LNG plant

X-2A/X-5A approach temperature, K	Power Hp/ million cmd (MW)	Plate-fin surface* area, m^2	Boiling area, m^2		Total area, m^2	
			High Flux	Conventional	High Flux	Conventional
5.6	11.9	7,440	13,940	57,630	21,400	65,070
2.8	11.0	13,940	25,100	134,800	39,040	148,740

*X-9 and X-9A.

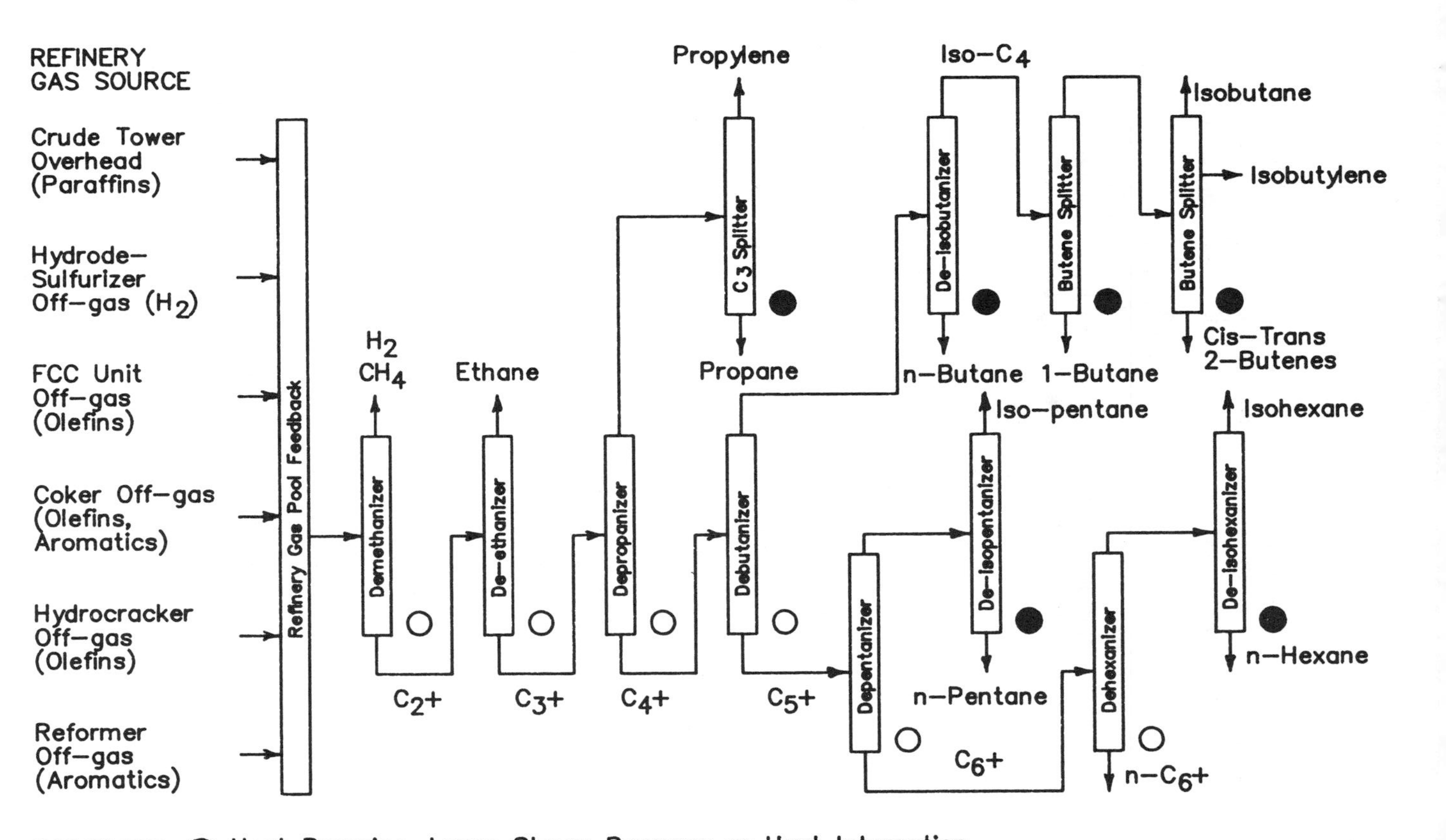

Figure 11-5 Enhanced boiling tube applications in refinery light ends separations.

When revamping or debottlenecking an existing plant, enhanced boiling tube reboilers can be installed so as to reuse as many of the existing heat exchanger shells as possible to reduce downtime and costs. The smaller allowable temperature approaches of enhanced boiling tubes make some heat integration and heat pump projects possible that would be too unwieldy with conventional tubing. In some plants, a hydrotreater may have to be added (if not already installed) to use enhanced boiling tubes without fouling. In heat integration networks that use a liquid heat transfer fluid, the smaller temperature approaches allow additional energy to be extracted from the fluid. A doubly enhanced tube may be suitable for these applications where space is limited by the existing plant layout.

Other prospective applications of enhanced boiling tubes are in the separation of xylene isomers in Benzene-Toluene-Xylene (BTX) units. Again, the use of lower-pressure steam, heat integration, and heat pumping are the principal alternative advantages to be gained. Kenney (1979) described Exxon's successful use of enhanced boiling tubes in a heat integration scheme between a xylene splitter tower and an adjacent debenzenizer. The project raised the pressure in the xylene splitter tower so that its overhead vapor could be used as the heat source in an adjacent debenzenizer reboiler (this amounts to the development of a cascade rectification system). Utilization of enhanced boiling tubing reduced the pressure-temperature increase in the overhead vapor to a minimum, and the project saved about 17 MW (58×10^6 Btu/h) of thermal energy.

Heat pumping or vapor recompression is another attractive way to save energy. The heat "pumped" can either be used to provide heat to the column's own reboiler or to provide heat to an adjacent tower's reboiler. The key parameters in analyzing the installation of a heat pump on an existing distillation tower include (1) the pressure difference between the top and the bottom of the tower; (2) the absolute pressure in the tower, which influences the relative volatility and the system's compression ratio; (3) the heat duty of the cycle; and (4) the reboiler's LMTD compared to the compressor horsepower required to provide it. According to Kenney (1979), a heat pump can decrease distillation tower energy consumption to about 10% to 15% of the energy normally consumed in the reboiler in a conventional unit, but at the expense of using an additional compressor.

Various heat pumping schemes are possible, and the most appropriate one depends on the existing facility (or on the laying out of a new facility), tower control systems, and so forth. A schematic diagram of a vapor recompression cycle for a propylene-propane splitter analyzed by Fiores, Castells, and Ferré (1984) is shown in Fig. 11-6. The overhead vapors were preheated in exchanger E-1 and then compressed in compressor K-1. The superheated vapor coming out of the compressor was split between the column reboiler E-2 and the trim cooler E-3. Part of the resulting condensate was refluxed to the column through exchanger E-1, and the remainder was removed as distillate. This scheme provided more effective energy usage compared to another cycle (also with enhanced boiling tubes) described by Quadri (1982), where the reflux exchanged heat with cooling water and thus was lost. Both schemes showed that enhanced boiling tubes are the most economic surface for the reboiler-condenser,

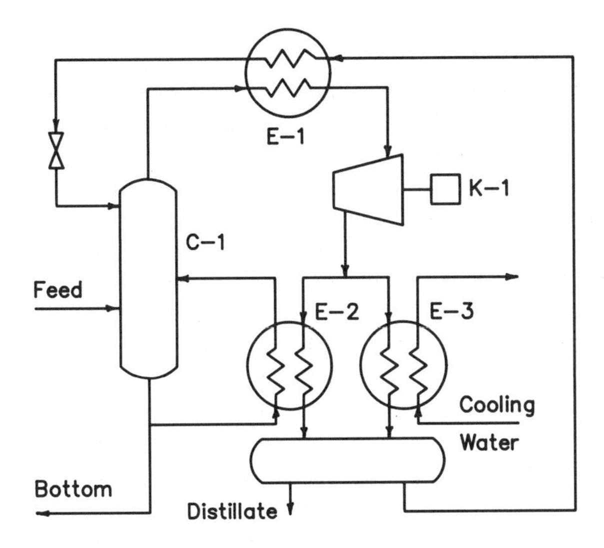

Figure 11-6 Heat pump cycle applied to a propylene-propane splitter.

primarily because they reduce the LMTD and hence the compressor power required for vapor compression.

Feed effluent heaters are yet another area of application that can benefit from the higher thermal performance of enhanced boiling tubes. Yampolsky and Pavlics (1983) made a parametric study of the application of spiral fluted tubes to these units. The process consisted of using a hot effluent leaving a fired heater to preheat and partially evaporate its feed into the heater. The primary advantages resulting were smaller heat exchangers and additional heat recovery made possible by the closer temperature approach.

Distillation of Aqueous Solutions

Enhanced boiling tubes can significantly reduce the size of evaporators in the staged multieffect concentration of aqueous solutions in large methanol plants, ethylene glycol facilities, and other alcohol and solvent facilities. Gottzmann, O'Neill, and Minton (1973) presented operating data for an ethanol kettle type vaporizer for ethanol drawn from a storage tank. Cupro-nickel High Flux tubes were used for this service. The investigators also provided data for several vertical multieffect ethylene glycol evaporators.

The operating cost savings are the principal factor favoring the use of enhanced boiling tubes in new ethylene glycol evaporator trains. The lower allowable mean temperature differences permit low-pressure steam to drive the train, thus avoiding more expensive high-pressure steam. Consequently, it may be possible to drive a five- or six-effect train with steam pressures less than 100 psig (7 bar) while concentrating the solution from 10% to 65% ethylene glycol.

Other Process Applications

Many other prospective applications of enhanced boiling tubes have not been mentioned above. One example is pipeline gas chilling. Here, externally enhanced boiling tubing has been used in the feed gas chiller of an enhanced oil recovery project, an application that was described by O'Neill and Ragi (1986). Another advantageous application is to seawater evaporators. The small concentration of salt in the water does not foul High Flux enhanced boiling tubing according to an experimental study by Gottzmann, O'Neill, and Minton (1973). Yet another application is to ammonia plant evaporators and syngas chillers. Some temperature-sensitive separations, such as styrene, could probably benefit in the form of better product quality from the lower operating temperatures that could be economically used in an enhanced boiling tube reboiler.

Another application area is to Rankine power recovery cycles. In these systems, heat is recovered from a waste heat process stream, such as a catalytic cracker overhead, and used to evaporate a pressurized fluorocarbon. The vapor generated is then utilized to drive a turbine and alternator to produce electricity. This cycle is shown schematically in Fig. 11-7. Yampolsky and Pavlics (1983) demonstrated that a dramatic decrease in evaporator size could be realized by using spiral fluted tubes for intube evaporation. Another similar application is to low-temperature steam generation from waste heat streams. Both these types of applications are dependent on the economic need for more electricity or steam in the facility.

11-5 SPECIAL THERMAL DESIGN CONSIDERATIONS

Enhanced boiling tube evaporators and reboilers perform like conventional units. A number of special thermal design considerations need to be taken into account, however.

Thermal Design Guidelines

Before initiating the thermal design of an enhanced boiling heat exchanger, several design constraints and features have to be considered. Some of the more important ones are reviewed below.

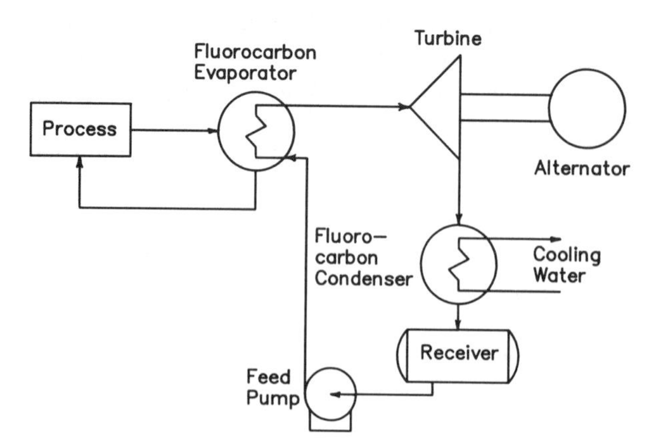

Figure 11-7 Low-temperature power recovery cycle.

Sensible heating and subcooled boiling. An enhanced boiling surface, whether external or on the tube side, may or may not be effective for augmenting the sensible heating of a subcooled liquid up to its boiling temperature or for superheating vapor. Internally enhanced tubes, with the exception of porous coated ones, are probably fairly efficient because they produce swirl flow. On the shell side boiling nucleation is adversely affected by subcooling, and the enhanced boiling heat transfer coefficient can be reduced substantially by small subcoolings because the wall superheats are typically small. For instance, for a unit operating with a mean temperature difference of 5 K, the boiling wall superheat will only be about 2 K. Thus 1 K of subcooling at the bottom of the shell reduces this by half. Using the effective superheat, which is the wall temperature minus the saturation temperature, to determine the boiling heat transfer coefficient does not yield a conservative value except at large heat fluxes (for more information, the reader is referred to Fig. 2-7 for subcooled boiling on a plain surface; subcooled boiling data for a low-finned tube is presented in Fig. 6-18).

Services with large sensible heating loads may be more economically handled by splitting the service into two exchangers, each of which best suits its heat transfer duty. When the subcooling is not large, however, its adverse effect can be overcome by introducing the feed above a horizontal bundle. For instance, Czikk, O'Neill, and Gottzmann (1981) noted that the introduction of a subcooled feed above a horizontal High Flux bundle had little adverse effect on performance, apparently because the subcooled feed mixes with the saturated liquid pool. Thus it is good design practice to use feed nozzles near the top of horizontal heat exchangers, either when the inlet feed is subcooled or when it is partially vaporized (to minimize the possibility of vapor binding).

Rise in bubble point temperature. When evaporating a large percentage of a liquid mixture feed, a moderate percentage of mixtures with wide boiling ranges, or a refrigerant contaminated with compressor oil, there is a possibility that a significant rise in the local bubble point temperature will occur along the fluid flow path because of the preferential evaporation of the lighter component (or components). In this instance, a sizeable portion of the heat absorbed by the fluid will serve to heat sensibly the liquid-phase from the inlet to the outlet temperature. This reduces the amount of vapor raised and can decrease the local boiling heat transfer coefficient by virtue of the smaller wall superheating. This factor has to be taken into account to avoid underdesign, especially at large pressures where the latent heat of vaporization is lower.

Process fluid composition. A nucleate pool boiling curve for a liquid mixture of one composition may not be a good approximation for another of only slightly differing composition. As is also true for conventional boiling, the boiling heat transfer coefficient can drop substantially with a change in composition, in particular with the addition of a small amount of a heavier component to a pure fluid (for instance, refer to Fig. 9-11 at small concentrations of acetone). For boiling inside tubes the composition change along the tube can be sizeable, and a heat transfer coefficient evaluated at the inlet composition may not be indicative of the actual overall performance. This problem can be identified and resolved by utilizing a fundamentally based, enhanced mixture boiling correlation to check for the effect of variations in the composition of the process fluid on performance. This effect is partly diminished because the heating-side thermal resistance becomes controlling with the application of an enhanced boiling tube.

Fouling resistances. Fouling resistances should be selected carefully. TEMA values are not recommended for enhanced boiling tubes because they were developed long ago for plain tube bundles under the assumption of a fixed boiling heat transfer coefficient for all reboilers as proscribed by the method of Kern (1956), which was popular at that time. The heating fluid's value should also be selectively chosen. Overspecifying fouling factors can result in discarding otherwise sound and beneficial energy conservation opportunities.

Pressure drop considerations. When replacement of an existing conventional tube bundle with an enhanced boiling bundle is being considered, the heating fluid's pressure drop often becomes one of the principal factors limiting design because the heating fluid side becomes the controlling thermal resistance. Thus the actual pressure drop available should be measured or estimated rather than just specifying a standard value.

Heat exchanger optimization. A new heat exchanger designed with enhanced boiling tubes should be correctly optimized for the particular operating service. One pitfall to avoid when comparing an enhanced boiling design to a conventional unit is to impose the same geometric limitations on both. Unnecessary, and sometimes unwitting, design restrictions should be avoided. For example, using a doubly enhanced tube with condensation on the tube side may benefit from a larger-diameter tube or more tubes in parallel of shorter length to meet the pressure drop restrictions. It is only logical that shifting the controlling thermal resistance from the boiling side to the heating side will result in a different optimum tube bundle geometry or even a different optimal heat exchanger configuration.

Inserting enhanced boiling data into conventional thermal design computer programs. Most thermal design codes available commercially for sizing and rating conventional reboilers and evaporators provide the user with the alternative choice of specifying (1) a specific value of the heat transfer coefficient, (2) a nucleate pool boiling curve defined by two data points or by a leading constant and slope, (3) a multiplying factor to apply to the program's calculated value, or (4) an externally low-finned tube. Consequently, it is fairly easy to make a thermal design comparison of an enhanced boiling design to a plain tube design if suitable boiling data or correlations are available. For shell-side boiling on steel and cupro-nickel tubes, the manufacturer should be consulted because copper tube data will probably not give a good indication of performance. For intube boiling, methods are discussed in Chapter Ten, Sec. 10-1.

These programs do not typically allow the user to modify the pressure drop calculation for the increased resistance to flow. Thus one must perform this calculation by hand. This is particularly crucial for exchangers with thermosyphon operation to predict the actual circulation rate and to avoid too large liquid heads that subcool the liquid feed. If the preliminary designs look reasonable, expert assistance should be used for the final design.

Temperature approaches. An important design specification in heat integration programs and for vapor recompression cycles is the minimum temperature approach. The better boiling nucleation characteristics of some enhanced boiling tubes allow high-performance operation to be attained at lower boiling wall superheats than those for plain tubes (refer to Secs. 5-3 and 9-3). Therefore, the most economic temperature approach when using an enhanced boiling tube is typically smaller. Hence the performance comparison will probably require optimizing the system separately with the various types of tubes under consideration to determine the most economically beneficial one. This typically includes the reduction in first cost by using a smaller-size compressor and its operating expense. For single-phase heat transfer fluids, the evaluation should include the additional heat recovered by the lower exit temperature approach.

Two-phase flow instabilities. For units operating as thermosyphons, sufficient liquid head or a valve in the inlet feed line should be utilized to avoid flow instabilities, as is common with conventional units. Enhanced boiling tubes are normally more stable hydrodynamically than conventional ones (see the discussion of mesh, brush, and spring inserts in Sec. 10-1).

Vapor disengagement space. For shell-side boiling, liquid entrainment in the outlet flow must be avoided in some applications. The larger heat fluxes of enhanced tube bundles compared to those of conventional designs (the first have larger values of U_o) can cause problems with liquid carryup, which results from higher liquid superficial velocities at the top of the bundle. When this presents a problem, such as for vapor going to a compressor, the local velocity and hence the amount of liquid entrainment can be reduced by increasing the shell disengagement space or the size and number of shell-side exit nozzles.

Special Thermal Constraints and Operational Limits

Several special constraints and limits apply to enhanced boiling heat exchanger bundle diameters for horizontal units. First of all, special attention has to be given to the influence of subcooling caused by the liquid head in the shell when boiling at low temperature approaches. For a tube bundle with a diameter and liquid head of 4 ft (1.2 m) of propylene, for example, there is an increase of 1.0 psi (0.07 bar) in the pressure at the bottom of the shell, which corresponds to about 1.0 K of subcooling. If the LMTD is only 3 K, then the outside wall temperature of the lower tubes may not be greater than the local boiling point temperature even though it is above the local bulk liquid temperature. In addition, the enhanced boiling heat transfer coefficient and its boiling nucleation characteristics are adversely affected by subcooling. Thus, as an alternative, the diameter of the bundle should be decreased and the tube length increased to reduce the effects of subcooling or a larger mean temperature difference must be specified.

The second important thermal limit on the maximum allowable diameter of horizontal reboilers and evaporators is the maximum bundle heat flux at which dryout of the top tube rows occurs. Enhanced tube bundles have higher heat fluxes than plain tube bundles operating at the same temperature approach, and thus the vapor carryup produced can be substantial even at low mean temperature differences. Allowing dryout, or vapor blanketing, to occur is not good design practice because it reduces the heat transfer effectiveness of those tubes affected and tends to promote fouling. Antonelli and O'Neill (1981) developed the following expression for a conservative estimate of the maximum bundle heat flux for High Flux tube bundles:

$$(Q/A)_B = 6\,(D_B L/A)(Q/A)_{st} \tag{11-1}$$

where the critical heat flux for a single tube is calculated using the Zuber/Kutateladze correlation:

$$(Q/A)_{st} = 0.18\rho_v^{1/4}\Delta h_v[gg_c(\rho_L - \rho_v)\sigma]^{1/4} \tag{11-2}$$

The first equation above was derived under the assumption that the superficial vapor velocity (based on the projected bundle surface area) can be at least six times the critical velocity calculated from Eq. (11-1), a point that was substantiated in O'Neill, Gottzmann, and Terbot (1972). Antonelli and O'Neill (1981) reported that Eqs. (11-1) and (11-2) gave a conservative estimate when compared to the safe operation of High Flux bundles functioning above this heat flux limit. The calculation procedure can also be used as a design guideline for determining the maximum permissible bundle diameter. It is not known whether this six-fold superficial velocity criterion applies to other enhanced boiling tubes.

For enhanced boiling inside tubes, no information is apparently available for the critical heat flux for porous coated tubes. For tube inserts and internally finned tubes, the limited information available (described in Chapter Ten's discussion of intube critical heat flux augmentation) suggests that the critical heat flux is increased substantially. For specific information, see, for instance, Cumo et al. (1974), Megerlin, Murphy, and Bergles (1974), and Kitto and Albrecht (1988). In vertical thermosyphon reboilers, however, the exit vapor qualities are typically substantially removed from these conditions.

Reboiler Circulation Rates

The circulation rate has a substantial effect on the thermal performance of conventional and enhanced boiling reboilers, both for horizontal and vertical units. An adequate liquid head has to be maintained to obtain predictable performance and to avoid two-phase flow oscillations. The higher the liquid head above the minimum required to ensure adequate circulation, however, the greater the degree of subcooling of the inlet liquid. For enhanced boiling tubes this point is especially important because some of the tube rows at the bottom of the bundle (or the tube length for intube boiling) may only be used for sensible heating, for which enhanced boiling tubes such as High Flux are no more effective than cheaper plain tubing. Failure to address this point can lead to overprediction of the reboiler performance and poor operational performance.

As an example, Table 11-3 from Antonelli and O'Neill (1981, 1986) demonstrates the large effect of circulation rate on the thermal performance of a horizontal High Flux BHU reboiler installed on a toluene distillation column. The overall heat transfer coefficient for the higher circulation rate was reduced to 54% of the design value by the 3 K of subcooling caused by the larger liquid driving head. The investigators estimated that about half the tube bundle was used for sensible heating, whereas operation achieved only 77% of the design heat duty. At the lower circulation rate the overall heat transfer coefficient was twice as high as the large circulation rate value, even with its smaller mean temperature difference (and hence smaller wall superheat to drive the boiling process). The heat duty matched its design value. Thus, in summary, the circulation rate is an important design and operational parameter when using enhanced boiling tubes in reboilers. These results are also indicative of good conventional design practice, in which excessive liquid sensible heating unnecessarily reduces reboiler performance.

For boiling inside vertical High Flux tubes, Antonelli and O'Neill (1981, 1986) recommended that liquid levels be maintained between 50% and 100% of the tube height, depending on liquid density and pressure. At elevated pressures, the liquid level is most commonly set at the top of the tube. At low pressures, where subcooling is more of a problem, lower levels are used to minimize the liquid head. For optimum performance, the investigators recommended maintaining exit vapor qualities less than 40% and the height of the sensible heating zone at less than 10% of the tube height. In calculating the circulation rate the friction factor in the boiling zone of a High Flux tube is roughly twice that of a plain tube, whereas in the sensible heating zone the friction factor is about 1.6 times that of an identical bare tube. Hence larger-diameter tubes may be required when an internally enhanced boiling tube is used.

The other internal enhancements discussed in Chapter Ten may also be successfully applied to vertical thermosyphon reboilers. Internally finned tubes with low fin heights appear to be attractive for new heat exchangers because of their low pressure drops; tube inserts may be the best alternative for existing units performing below expectations if sufficient liquid head is available. Conventional cleaning methods may be able to clean these low-finned tubes because fin heights are typically only about 0.3 to 0.6 mm and because some inserts are removable.

Enhanced Bundle Boiling Factor

For boiling on the outside of plain tube bundles, the shell-side heat transfer coefficient is augmented by the liquid circulation up through the bundle (this amounts to the two-phase convection process). Thus plain tube bundle boiling heat transfer coefficients are larger than those for nucleate pool boiling on the outside of a single, plain tube. The tube bundle boiling heat transfer coefficient will typically be from 1.5 to 3.0 times that of a single, plain tube under the same conditions (as described in Chapter Two).

Most studies on shell-side boiling with enhanced tubes have found the bundle performance to be equal to or better than the single tube performance. Suppression of

Table 11-3 Effect of circulation rate on a horizontal High Flux BHU toluene column reboiler

Circulation rate	Steam pressure, bar	Steam flow rate, kg/h	Steam saturation temperature, °C	Heat duty, MW	Surface area, m^2	Inlet boiling point, °C	Exit boiling point, °C	Reboiler flow rate, kg/h	Approximate vaporization, %	Mean temperature difference, K	U_o, $W/(m^2 \cdot K)$
Design Condition	7.93	36,165	175	20.3	767	163	165	688,360	33	9.9	2,680
High circulation	7.37	27,918	172	15.7	767	154	158	2,454,000	7	15.5	1,442
Low circulation	7.37	36,360	172	20.5	767	160	165	363,600	62	9.4	2,839

boiling by the induced flow therefore does not occur. Enhanced boiling tubes typically increase the nucleate pool boiling coefficient by a factor of 3 to 10, and thus the convection produced by circulation up through the bundle does not contribute much to the process, except for low-finned tubes.

The testing of prototype exchangers with High Flux tubes by Union Carbide has demonstrated that the single tube, nucleate pool boiling performance was essentially the same as its flow boiling performance, either for boiling on the outside of tube bundles or inside a tube, according to Czikk, O'Neill, and Gottzmann (1981). Hence the bundle boiling factor can be taken as 1.0, which simplifies the thermal design process immensely.

Arai et al. (1977) tested horizontal tube bundles with 225 tubes for boiling R-12 on both 16.4-mm diameter copper Thermoexcel-E tubes and 18.0-mm diameter low-finned (748 fins per meter) tubes. They found that both types of tubes had bundle boiling factors greater than 1.0 over most of the heat flux range studied. Other studies confirming the same trend are discussed in Chapter Ten, Sec. 10-2.

The principal data that appear to contradict this trend are some Heat Transfer Research, Inc. (HTRI) results. Yilmaz, Palen, and Taborek (1981) reported the bundle boiling factor for a 168-tube internal reboiler unit with 90-10 cupro-nickel Gewa-T tubes to be less than 1.0 (as low as 0.8) at low to medium heat fluxes. This, however, probably resulted from one of three different factors:

1. The single tube test was done with a copper test section, whose higher thermal conductivity would produce better boiling performance than the cupro-nickel version, over and above its effect on the fin efficiency.
2. There are unavoidable inaccuracies involved in determining the bundle boiling heat transfer coefficient from overall performance data, that is, using a calculated value for the steam condensation coefficient inside the tubes to back out the larger shell-side coefficient.
3. The circulation rate up through internal reboilers is not easily controlled and perhaps was too high (as shown in Table 11-3).

The liquid head above the bundle in the HTRI simulated column reservoir could have created sufficient subcooling to adversely affect the bundle's boiling performance. Taborek (personal communication 1987) has indicated that the liquid level was set to the top of the tube bundle, however.

In summary, the bundle boiling factor for a properly designed and operated enhanced boiling tube bundle can be assumed to be greater than or equal to 1.0. Its actual numerical value is not necessarily required for designing reboilers and evaporators in hydrocarbon services when the single tube values demonstrate high enhancement because the controlling thermal resistance will be shifted to the heating fluid inside the tubes. Hence even an enhanced bundle boiling factor of 1.5 will usually have only a marginal influence on the overall heat transfer coefficient, except for low-finned bundles, which have smaller single tube coefficients.

Tube Pitch

The spacing between tubes in horizontal bundles is not as critical for enhanced boiling tubes as for plain tubes because the enhanced boiling heat transfer coefficient for a bundle is affected less by the two-phase convection process. Arai et al. (1977) reported experimental results for the Thermoexcel-E bundle described above for two different staggered tube pitches: 25 mm horizontal spacing and 20 mm vertical spacing, and 21 mm and 17.5 mm spacings, respectively; these give vertical and horizontal ratios of tube pitch to tube diameter of 1.52 and 1.22, respectively, for the first bundle and 1.28 and 1.07 for the second. The boiling performances of the two bundles were identical.

The primary guidelines to follow in setting tube pitches are:

1. the larger the tube pitch, the smaller the tendency for vapor blanketing to occur at the top of the bundle;
2. the greater the tube pitch, the larger the bundle diameter and the adverse effect of subcooling; and
3. within limits, the smaller the tube pitch, the cheaper the exchanger.

Operating heat fluxes of enhanced boiling tube bundles are normally comfortably below the values predicted by Eq. (11-2). For this reason, conventional tube pitch ratios of 1.25 are recommended. Antonelli and O'Neill (1981, 1986) noted that a triangular pitch is preferred, but the use of a square, diamond, or rotated square pitch does not affect the boiling performance of the High Flux tube. If a low-finned tube bundle is installed in a fouling service, then a square pitch is superior because it is more easily cleaned.

11-6 MECHANICAL DESIGN CONSIDERATIONS

The following are some special mechanical design considerations related to enhanced boiling tube bundles:

1. Enhanced boiling tubes can be bent into U bends without buckling. The minimum recommended bend radii should be obtained from the tube manufacturer, however.
2. Enhanced boiling tubes have plain lengths at both ends to facilitate rolling or welding into tube sheets and may have additional plain sections at the baffles if desired.
3. The wall thickness under the boiling section is thinner than at the plain ends of mechanically worked tubes but is sufficiently thick for normal operating pressures. Extra-thick walls can be supplied for services in which stress and erosion conditions are severe (tube manufacturers typically recommend using the wall thickness at the root of the fin in stress calculation equations, but a stress concentration factor may be desirable).

4. The smaller temperature approaches and hence lower operating temperatures and pressures utilized for enhanced boiling tube exchangers reduce thermal stresses on the tubes, baffles, and tube sheets and thus can diminish their thickness and cost.
5. The more compact design made possible by the higher heat transfer performance decreases the exchanger's weight and volume and hence also the weight of its liquid contents. This reduces the size of the support structure and also decreases product losses during shutdown.
6. When tubes with external fluted fins are used to augment vertical shell-side condensation, the higher condensing stream inlet velocity will normally require the placement of an impingement plate at the inlet nozzle to prevent tube vibration problems, per TEMA standards.
7. External boiling enhancements can withstand normal wear and tear during tube bundle fabrication. As a rule low-finned tubes and modified low-finned tubes are easily pulled through the tube sheets and battles during assembly, whereas porous coated tubes may need some "persuasion" because of their large surface roughness.
8. Proper handling and storage of enhanced boiling tubes and bundles may be necessary with some materials to prevent rusting. Rust inhibitors sometimes applied to tubes during fabrication should be rinsed off with a suitable solvent before installation in the plant.

Other common mechanical design considerations that apply to conventional reboilers and evaporators also apply to enhanced boiling ones.

11-7 FOULING

Fouling is one of the foremost concerns of process heat transfer engineers, whether for boiling, condensation, or single-phase services. As pointed out by Gilmour (1965) more than 20 years ago, the most effective way to combat fouling is through proper design of the heat exchanger rather than the addition of large fouling factors, which produce oversized bundles with poor flow characteristics. Gilmour pointedly reminds us that poor performance occurs more often than not from poor thermal design rather than from fouling, but fouling always gets the blame. For this reason, it is rather difficult to obtain good operating data on fouling factors in the field even for well-instrumented units because it is not easy to separate poor thermal design from fouling.

Practical Concerns

The adverse effect of fouling on enhanced boiling performance is a particularly troublesome issue. Process heat transfer engineers bring it up as a reason for not using new technology in their plants more often than any other factor. Their concern is justifiable considering that a processing facility producing perhaps millions of dollars of product per day can be shut down or bottlenecked by one faulty heat exchanger. Nevertheless, allowing hundreds of thousands of dollars of energy to be wasted every year by not using enhanced boiling surfaces in reliable and effective services is not acceptable practice.

The most typically voiced anxieties are as follows:

1. Predicting the level of fouling beforehand for any boiling service, plain or enhanced, is difficult. The occurrence of a large fouling resistance would negate the benefit gained by spending extra money to augment the boiling heat transfer coefficient.
2. Particulate matter, such as rust, may enter the boiling enhancement and clog its passageways, rendering the enhanced boiling tube ineffective.
3. Oil or a heavy hydrocarbon component of mixture with a wide boiling range may become concentrated in the enhancement and decrease the expected improvement in thermal performance.

The first two concerns can be partly put to rest for situations in which a plain tube bundle is to be retubed with an enhanced tube bundle. If the original bundle did not foul, then the enhanced boiling tubes will probably not foul either. If the plain tube bundle did foul for boiling on the shell side, then low-finned tubes should be used, as is discussed below, to reduce the rate of fouling and to make cleaning easier. Whether or not a service is fouling should be determined from actual visual observation of the particular bundle because poor performance may be the result of poor thermal design rather than fouling.

A continuous buildup of heavy components or oils in the enhancement matrix does not occur, as judged from the available evidence. The boiling process is affected by the heavy component, which tends to reduce the enhanced boiling heat transfer coefficient (as discussed in Chapter Nine); this is similar to the degradation experienced by plain tubes for the same circumstances. The boiling performance is stable at this level over a long period of time, however, which demonstrates that the heavier components do not build up with time. In fact, Thome (1987) demonstrated that enhanced boiling tubes experience less degradation, percentage-wise, in their heat transfer coefficients in mixtures than plain tubes, especially for aqueous mixtures. As another example, in Fig. 9-12(*a*) the binary mixture of 95% *n*-pentane and 5% 1-tetradecene boiling on the Gewa-TX tube has a boiling range of 94 K, yet it still provides much better heat transfer performance than a plain tube, even with the small concentration of a C_{14} hydrocarbon. In addition, water–ethylene glycol mixtures have large boiling heat transfer coefficients on the High Flux tube according to data given by Czikk, O'Neill, and Gottzmann (1981), even though these mixtures have large boiling ranges.

From a physical standpoint, it is thought that the bubble pumping action, the two-phase flow inside the passageways, and the recirculation of the liquid in the passageways with the bulk prevents a buildup of the less volatile components.

Liquid impurities do not pose problems for enhanced boiling surfaces up to a certain concentration; this is similar to plain tube behavior. Compressor lubricating oils are known to build up in the refrigeration systems of ethylene and LNG plants, in which some of the common refrigerants are methane, ethane, ethylene, propylene, and propane. Gottzmann, O'Neill, and Minton (1973) reported that the nucleate pool boiling performance of the High Flux tube begins to deteriorate with as little as 1% (by volume) oil when boiling propylene at 1.0 bar. They pointed out, however, that the condensing side is the controlling thermal resistance in these exchangers, so that the overall heat transfer coefficient is only marginally affected. For example, they showed that the overall heat transfer coefficient decreased by only 3% for a 4% oil concentration in propylene. O'Neill, Gottzmann, and Terbot (1971) cited one prototype ethylene condenser in which an oil concentration as high as 9% in the evaporating propylene stream still operated at a steady heat duty after 30 months of operation.

Factors Working Against Fouling Buildup

Low-finned tubes. Kern (1956) reported that fouling on low-finned tubes in reboilers occurred at a reduced rate compared to plain tubes on the basis of a comprehensive survey that he and his associates performed. He reported that on low-finned tubing the hard scale tends to flake off in a platelike form, as though the fins acted as knife edges. It was also thought that the thermal expansion and contraction of the fins during startup or plant excursions loosened the scale.

There are several other advantages of low-finned tubes over plain tubes in fouling services:

1. As noted by Kern (1956), the fin efficiency increases as the effective heat transfer coefficient on a finned tube decreases with the buildup of scale, thus providing some compensation. This apparently explained why some bundles that had operated for long periods without a decrease in performance still appeared to be fouled on visual inspection,
2. The heat flux through the extended surface of a low-finned tube is smaller than that of a plain tube under the same conditions, which results in a smaller temperature drop across the scale,
3. Putting fins on the outside of a tube shifts the controlling thermal resistance to the inside, and thus more fouling on the shell side can be tolerated before the unit must be shut down for cleaning,
4. Scale formation on low-finned tubes is a relatively slow process in which the deposition of the particles follows the contour of the fin, producing a similar amount of wetted surface area compared to the clean fin (this can be observed in the photographs shown below).

There are apparently no operating data on the fouling of internally finned tubes in vertical thermosyphon evaporators, probably because little use has been made of them so far. They would seem to offer advantages at low pressures, where plain tube boiling heat transfer coefficients can be rather low compared to those for condensing steam or where a viscous fluid is being boiled.

Porous enhancements. Ragi, O'Neill, and Heck (1984) reported that the High Flux tube has been remarkably resistant to fouling during evaporation of many different petroleum refinery streams, light hydrocarbons, aqueous solutions, and cryogenic liquids. They attributed this to several special factors:

1. high liquid circulation rates through the porous matrix tend to flush out any particular matter, such as magnetite and iron oxides (rust);
2. capillary forces adequately wet the internal surfaces in the enhancement, thus preventing dry, hot spots from forming; and
3. higher heat transfer performance lowers the wall temperature, which reduces the polymerization or decomposition of heat-sensitive fluids.

Besides these factors, there is another key distinction to be made. Evaporation in enhanced boiling passageways occurs at the free vapor-liquid interface of the liquid film, not at the heated wall. In contrast, boiling on a plain surface is characterized by microlayer evaporation of a liquid layer trapped between the bubble and the heated wall. As this layer evaporates it dries out, which facilitates the deposition of solids on the wall. For example, microscopic observation of a plain surface after boiling undistilled water will disclose small "craters" where calcium and other minerals have been deposited around active boiling sites. Nonporous type enhanced boiling tubes have larger-diameter reentrant passageways and openings, which further reduces the likelihood of their trapping particular matter.

Fouling Experience in Practice

Low-finned tubes. Plant operating experience with low-finned tubes has been described in the literature from time to time by veterans in the field. Several episodes are recounted below.

Webber (1960) described the decision-making process leading to the application of low-finned tubes in reboilers at Humble Oil & Refining Company. Because the applications were to fouling conditions for which the investigators had no previous experience, they started out with some laboratory tests, which showed that finned tubes would not foul as quickly as plain tubes and, once dirty, were easier to clean. In their field studies, one of the first reboilers was retubed with finned tubes and operated in a deisopentanizing service, charging sweetened plant pentanes. The existing triangular pitch tube bundle fouled severely in this environment in only 6 months. This bundle was then replaced with a low-finned tube bundle with a square pitch. After 6 months of service, the column was brought down for reboiler cleaning. It was observed that the fouling layer formed in wavy shapes over the fins, following to some extent the fin contours. The outside tubes could be cleaned completely, but the inner ones could not. The fins were damaged slightly by cleaning with sandblasting, although not seriously. It was concluded that the finned tube bundle could tolerate fouling for a longer service time before its thermal resistance became a production limitation (photographs of the bundle before and after cleaning and of the effect of sandblasting on the fins can be found in Webber's [1960] report).

On the basis of this experience, Hunble Oil installed low-finned tubes on other column reboilers. After about 1 year of operation, the investigators obtained heat transfer data on seven different bundles, three finned and four plain, to try to make a performance comparison for Light Ends Fractionation Unit (LEFU) reboilers, which apparently were steam heated. Their plant data are reproduced in Table 11-4. The overall heat transfer coefficients U_o are based on the total wetted outside area of the finned tube bundles. The dimensions of the finned tubes were not noted; no attempt was made to back out the actual boiling-side coefficients themselves. The rightmost column gives Webber's (1960) comparison of the heat duty per unit length per unit degree to compare data obtained at different LMTD values.

Only one direct comparison can be made from these data: the deisobutanizing alkylate services, where two finned tube bundle data sets and one plain bundle set were available. These data demonstrated two points: (1) the overall heat transfer coefficients based on the total outside area of the finned tube bundles were about 20% higher for the finned tube bundles, and (2) the heat duty per unit length of tubing per degree showed a substantial increase (by a factor of about 2.5) even though the LMTD values were smaller for the finned tube bundles (and hence had smaller wall superheats for boiling). The heat duties per unit length (Q/L) for the three services are 3,009, 2,892, and 1,632 W/m, respectively, which represents an increase of 80% relative to the plain tube. The pentane splitting column finned tube reboiler had a lower overall heat transfer coefficient than the other two finned tube bundles, indicating that this bundle may be dirty, according to Webber (1960). With 3 years of operating data accumulated with finned tube reboilers, the use of finned tubes was reported to have become accepted practice for steam-heated reboilers at Humble Oil's Baytown plant in Texas.

Gilmour (1965) and Moore (1974) described some of their experience at Union Carbide's Texas City Plant with low-finned tube bundles. One particularly notable

Table 11-4 LEFU reboiler data

Service (tubing)	Heat duty, MW	LMTD, K	Total surface area, m^2	Total tube length, m	U_o, $W/(m^2 \cdot K)$	Q/L/LMTD*, $W/(m \cdot K)$
Deisobutanizing alkylate (finned)	9.81	25	376	3,262	1,045	120
Deisobutanizing alkylate (finned)	7.33	29	254	2,533	1,000	100
Deisobutanizing alkylate (plain)	12.89	38	394	7,899	852	43
Butane splitting (plain)	6.04	75	124	2,479	608	32
Butane splitting (plain)	8.97	46	138	2,758	1,369	71
Pentane splitting (finned)	7.85	44	434	3,957	415	45
Depentanizing alkylate (plain)	2.18	26	130	2,609	653	32

*Heat duty per unit length per unit degree.

application was a U-tube kettle reboiler of a hydrocarbon distillation column. The original plain tube bundle contained 15.9-mm (5/8-inch) diameter tubes on a 23.8-mm (15/16-inch) triangular pitch with a bundle diameter of about 76 cm (30 inches). The bundle had to be replaced with a spare every 6 to 8 weeks. On average, it took 300 to 400 manhours to clean the dirty bundle with saws, wedges, picks, and other hardware. This reboiler was retubed with 15.9-mm diameter low-finned tubes with a larger pitch of 31.8 mm (1.25 inches) on a rotated square layout. Approximately half the tubes were removed, but the outside wetted surface area remained about the same as that of the original bundle. The new coil was reported to operate as long as and sometimes longer than the original bundle before cleaning was necessary, but the cleaning time was reduced to only 4 to 8 manhours with a high-pressure water jet. The plain tube bundle became coated with a uniform deposit of a rubberlike polymer material that bridged from tube to tube; the finned tube bundle formed a similar coating that followed the contour of the fins, but the larger spacing prevented it from bridging between the tubes.

The finned tube bundle described above is shown in Figs. 11-8 through 11-11 (the prints are courtesy J. O'Donnell, Jr., of High Performance Tube, Inc., who obtained copies from the authors). The first photograph (Fig. 11-8) shows the entire tube bundle in a fouled condition after being drawn from the shell. The fouling is observed to be quite uniform along and around the bundle. The second photograph (Fig. 11-9) shows

Figure 11-8 Fouled low-finned tube bundle (courtesy High Performance Tube, Inc.).

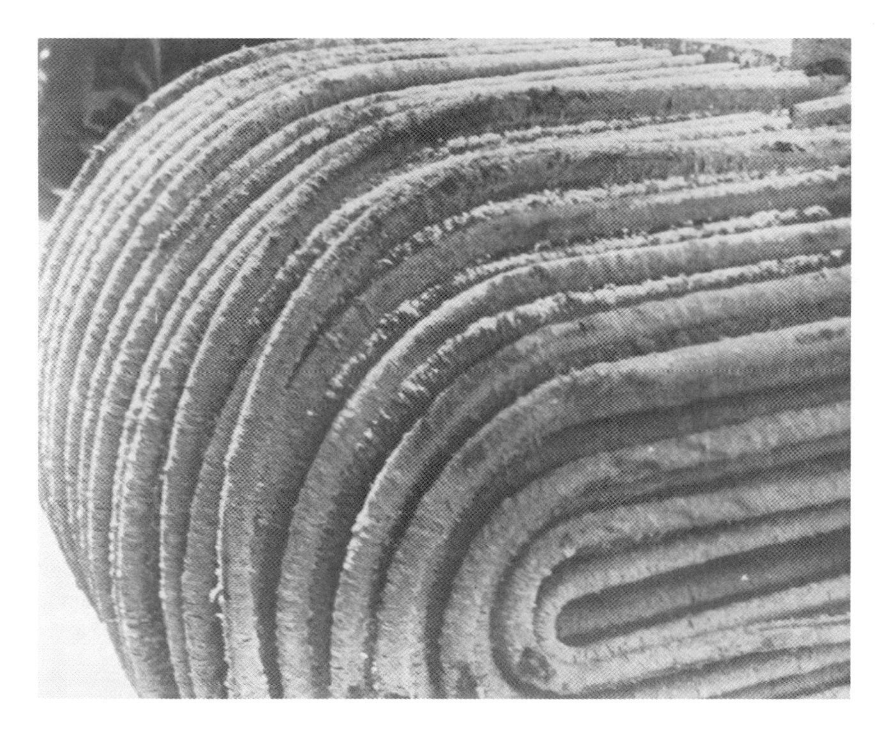

Figure 11-9 Close-up view of the low-finned tube bundle (courtesy High Performance Tube, Inc.).

Figure 11-10 Cleaning the low-finned tube bundle (courtesy High Performance Tube, Inc.).

Figure 11-11 Totally cleaned low-finned tube bundle (courtesy High Performance Tube, Inc.).

a close-up view of the U bend, where the contour of the fins in the fouling layer can be seen. Figure 11-10 depicts the bundle being cleaned with a water jet. Finally, Fig. 11-11 shows the completely cleaned bundle. Thus Gilmour (1965) and Moore (1974) reported that, even though the finned tube bundle fouled, the heat transfer duty was not reduced to less than the operating limits of the system, and a quick cleaning could be done while the process unit was shut down for other reasons.

In summary, low-finned tubes are not only suitable for nonfouling applications but are superior to plain tubes for some medium and heavy fouling services. Also, it was demonstrated that the use of larger tube pitches not only makes eventual tube bundle cleaning easier but also reduces the caking of foulants from one tube to the next that would cause flow maldistribution.

Porous enhancements. Long-term performance data for porous coatings are evidently only available for the High Flux tube. Figure 3-17, for instance, shows test data obtained for a prototype ethylene-propylene reboiler-condenser (boiling propylene on the shell side, condensing ethylene on the tube side with internal fins). Data spanning about 4 years of operating experience demonstrated that the heat transfer performance had been stable since startup. Gottzmann, O'Neill, and Minton (1973) reported that the operating environment was about as severe as can occur in this service because reciprocating rather than centrifugal compressors were used in the plant, which made oil contamination a frequent occurrence. According to Milton and Gottzmann (1972), the only cleaning during this time period was a hot methane thaw.

Other High Flux operating experiences have been reviewed by Gottzmann, O'Neill, and Minton (1973). For instance, a vertical ethylene column reboiler with the High Flux coating on the inside and fluted fins on the outside was monitored for a 1-year period. The 12-MW heat duty reboiler evaporated ethane and condensed propylene. No variation in performance was observed over this period. An ethanol kettle vaporizer with a 4.7-MW heat duty and two prototype vertical ethylene glycol evaporators also demonstrated consistent operation without fouling for test periods of 1 year or more. A list of successful operation of other units in other services was also given in this report.

In summary, extended field test data have demonstrated that a porous enhancement in prototype and full-size exchangers can maintain high heat transfer performance in services considered mildly fouling. Pilot plant and operating data for the other types of boiling enhancements are apparently not yet available.

Fouling Factors for Design

The value of the fouling factor specified for design has a large effect on the range of practical applications of enhanced boiling tubes. For plain tube bundles evaporating light normal hydrocarbons (C_1 to C_8), Palen and Small (1964) recommended the use of fouling factors in the range from 0 to 0.00018 K m^2/W (0 to 0.001 h·°F·ft^2/Btu). This range is probably reasonable for enhanced boiling tubes as well, depending on the particular service.

11-8 OPERATING CONSIDERATIONS

One important operational limit that must not be exceeded is the maximum allowable boiling wall superheat, which divides the nucleate boiling regime from the film boiling regime. This can be a problem with conventional reboilers when a high-temperature steam reaches the unit during startup. Once the tube bundle goes into the film boiling regime, the heat flux must be lowered to less than the minimum heat flux of film boiling to bring the bundle back into the nucleate boiling regime. Because the tube wall is heated to a higher temperature in film boiling, and total evaporation near the tube wall occurs, fouling tends to be worse than normal. Enhanced boiling reboilers and evaporators should be less prone to this operating problem because temperature approaches are typically low in most of their applications. The critical heat flux for an enhanced boiling tube, however, is reached at a smaller wall superheat than for a plain tube unit (see Chapter Six's discussion of miscellaneous effects on peak nucleate heat flux). Thus startup wall superheats should be monitored to ensure proper operation. (This restraint is actually the same as the maximum bundle heat flux discussed above but viewed from a perspective that is perhaps easier to implement in a plant.) The maximum wall superheat is determined by dividing the maximum bundle heat flux in Eqs. (11-1) and (11-2) by the boiling heat transfer coefficient.

11-9 ECONOMIC CONSIDERATIONS AND SURFACE SELECTION

Financial Incentives and Tube Costs

The financial incentives for reducing temperature approaches and increasing thermal efficiency are impressive. For example, evaporator duties range from about 1 to 15 MW in a typical olefin plant producing 10^9 lb/year (0.45×10^9 kg/year) of ethylene. Thus reducing the energy consumption in the compressors (rated at about 30,000 kW) that drive the plant's refrigeration system by only 5% represents an annual savings on the order of $500,000.

An economic comparison between enhanced boiling tubes and plain tubes should (at the minimum) include the value of the energy saved; the value of an increase in production capacity; the cost of the tubing, shell, tube sheet, support structure, fabrication, and installation; and the extended time before cleaning with a low-finned tube bundle. Some of these factors are not easily estimated.

As a rule of thumb, low-finned tubing costs about half as much as plain tubing per unit heat transfer surface area. The mechanically worked surfaces, such as Turbo-B, Thermoexcel-E, and Gewa-TX, cost more to manufacture than integral low-finned tubes, with the additional expense depending primarily on the ductility of the material. The porous coated tubes would normally cost the most because of their higher production costs. The mechanically worked doubly enhanced tubes may or may not cost more than a single enhanced tube because the doubly enhanced tube may be the manufacturer's standard production tube. (It is not possible to prepare a table with actual unit costs of these tubes because they vary depending on too many factors, such as tube material and size of the order.)

In soliciting bids, the process heat transfer engineer should keep in mind that integral low-finned tubes are more or less sold as a bulk commodity because the manufacturer is not including expensive boiling data or heat exchanger thermal design as part of the contract. The high-performance enhanced boiling tubes may be sold on the basis of specific boiling data for the fluid or with thermal design services and thermal guarantee for the particular application (or both). Thus these tubes have higher prices to cover their research and development costs and any technical services provided. They may also be priced partly on the value of the energy to be saved in the particular application; this is similar in many respects to how special distillation tower packings are marketed, for example.

Notwithstanding their higher costs per unit length, enhanced boiling tubes can have payback periods of only several months, as demonstrated in three applications discussed by Ragi, O'Neill, and Heck (1984) for the High Flux tube. According to Webber (1960), retubing a plain tube bundle with low-finned tubes to increase capacity and reusing the existing shell will cost between one-fifth and one-half as much as a new larger plain tube bundle requiring a new shell. For completely new installations, Webber estimated the cost ratio of finned tube bundle to plain tube bundle to be from 0.5 to 0.9.

Tube Selection

At first glance, selecting the most advantageous enhanced boiling tube appears to be a complex task involving the comparison of many different optimized thermal designs and the analysis of their economic advantages, especially for refrigeration, air conditioning, and vapor recompression services, in which an entire system must be designed and optimized. Therefore, the first step is to eliminate those tubes for which little or no boiling data are available for the fluid to be evaporated. The manufacturer may be willing to obtain or supply specific data on request. The following simple shortcut method presented here can then be used to reduce the remaining field to several tubes before a more in-depth analysis is attempted.

Figure 11-12 and Table 11-5 depict the overall heat transfer coefficient as a function of the boiling-side and heating-side heat transfer coefficients. The boiling-side coefficient may be for either intube boiling or shell-side boiling depending on the case at hand. The heating-side heat transfer coefficient includes all the other thermal resistances: the heating fluid's film coefficient, the fouling factors, and the thermal resistance of the tube wall. The procedure then proceeds as follows:

1. The heating-side coefficient is estimated by simple hand calculations or from the computer output of a plain tube thermal design (its value will not change appreciably

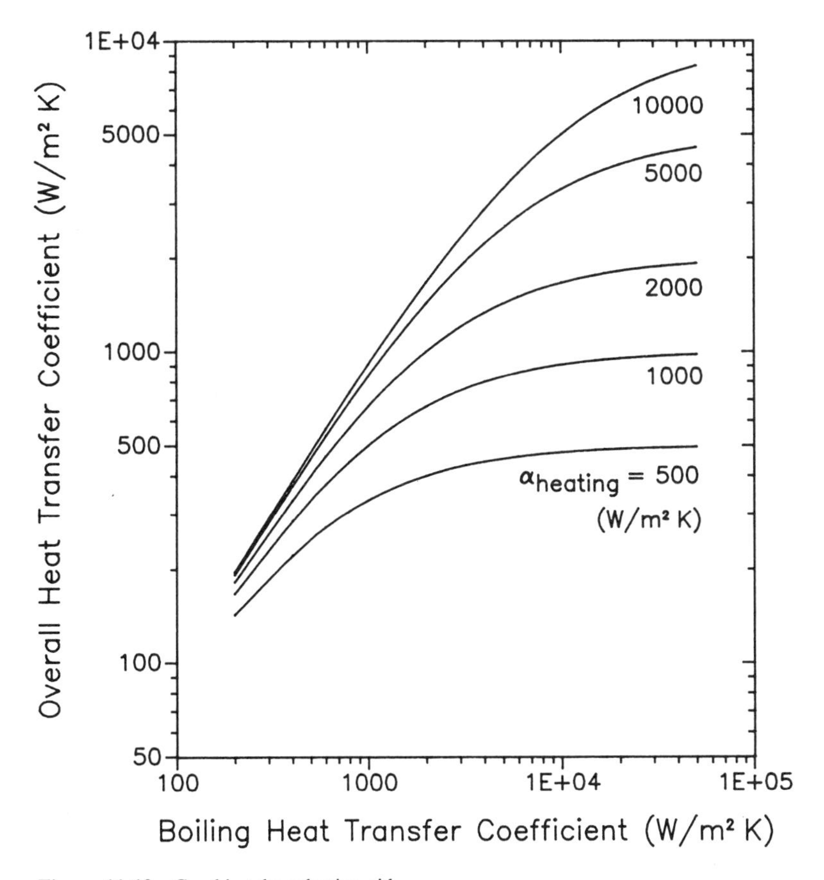

Figure 11-12 Graphic tube selection aid.

Table 11-5 Boiling augmentation evaluation table (data are in watts per square meter per Kelvin)

Boiling heat transfer coefficient	Heating-side heat transfer coefficients				
	500	1,000	2,000	5,000	10,000
	U_o	U_o	U_o	U_o	U_o
200	143	167	182	192	196
400	222	286	333	370	385
600	273	375	462	536	566
800	308	444	571	690	741
1000	333	500	667	833	909
2000	400	667	1,000	1,429	1,667
4000	444	800	1,333	2,222	2,857
6000	461	857	1,500	2,727	3,750
8000	471	889	1,600	3,077	4,444
10000	476	909	1,667	3,333	5,000
15000	484	937	1,765	3,750	6,000
20000	488	952	1,818	4,000	6,667
30000	492	968	1,875	4,286	7,500
40000	494	976	1,905	4,444	8,000
50000	495	980	1,923	4,545	8,333
Infinity	500	1,000	2,000	5,000	10,000

with the application of an enhanced boiling tube as a first approximation but can be appropriately modified for any doubly enhanced tubes under consideration),

2. The plain tube and the enhanced boiling tube heat transfer coefficients are then obtained from a product bulletin or the boiling curve in Chapters Seven, Nine, and Ten or are calculated by the methods presented in Chapters Two, Eight, and Ten (performance with tubes in materials other than copper should be obtained from the manufacturer),
3. From Table 11-5 or Fig. 11-12, the overall heat transfer coefficient is determined for each of the different tubes,
4. The ratios of the overall heat transfer coefficients obtained relative to the plain tube value give an approximate but reasonable estimate of the heat transfer augmentations possible and determine whether further investigation is called for,
5. If the augmentation indicated is sufficiently large, the overall heat transfer coefficient can then be used to perform a hand calculation on the size of a new heat exchanger or the new heat duty or energy saved when replacing an existing plain tube bundle,
6. From the unit costs of the tubes, an economic analysis can be performed to determine the first cost of a new heat exchanger or the payback period for retubing or replacing an existing one.

When actual values cannot be easily found for a low-finned tube, multiplying a single plain tube's boiling heat transfer coefficient (calculated by one of the methods given in Chapter Two's discussion of heat transfer correlations) by the ratio of the finned tube's surface area to that for a plain tube of the same size gives an approximate estimate. For the high-performance tubes, the curves in Fig. 9-8 cover many applications as a first approximation to determine whether the investigation is worth continuing.

Although the above method is unsophisticated, it is a simple first step to determine whether the application of an enhanced boiling tube is warranted or which tubes are the most advantageous without an undue investment of time. Analyzing the values in Table 11-5, one notes that the best heat transfer augmentation is obtained when the heating-side heat transfer coefficient is large, which means that the boiling side is controlling. The benefit of replacing or choosing a high-performance boiling tube over a low-finned tube is also easily determined. In addition, it can be noted that, as the boiling heat transfer coefficient becomes higher than a certain threshold value in Fig. 11-12, the additional gain in the overall heat transfer coefficient U_o becomes less significant.

As an example, consider a reboiler evaporating a light hydrocarbon stream with condensing steam whose composite heating-side heat transfer coefficient is fixed at 5,000 W/(m^2·K) and for which the following bundle boiling heat transfer coefficients (in watts per square meter per Kelvin) are possible: 2,000 for a plain tube, 5,000 for a low-finned tube, 15,000 for one high-performance tube, and 20,000 for a second high-performance tube. Finally, consider use of fins to augment the condensing heat transfer coefficient by a factor of 2.5 that increases the composite heating-side coefficient to 10,000 W/(m^2·K).

From Table 11-5 or Fig. 11-12, the results obtained are shown in Table 11-6. Estimates of the increase in the heat duty Q and the decrease in the total surface area (or length of tubing) possible are also shown for a fixed mean temperature difference. Application of a low-finned tube bundle appears to be quite reasonable. Using enhanced tube 1 or 2 provides still better performance (the difference in their boiling heat transfer coefficients has little effect in this example). Selection of a doubly enhanced tube increases the overall heat transfer coefficient to 4.67 times the plain tube bundle's value and decreases the amount of surface area required by 79%. Now that a preliminary estimate has been made and the apparent advantages of several of the tubes demonstrated, optimized thermal designs could be completed to select the most economical tube.

Table 11-6 Example reboiler results

Tube	U_o, W/(m^2·K)	Increase in U_o (or Q)*, %	Decrease in surface area†, %
Plain	1,427		
Finned	2,500	75	43
Enhanced 1	3,750	163	62
Enhanced 2	4,000	180	64
Doubly enhanced	6,667	367	79

*Assuming a fixed nominal surface area and mean temperature difference (MTD).
†Assuming a fixed heat duty and MTD.

11-10 SUMMARY

The practical considerations of interest to process heat transfer design engineers for the application of enhanced boiling tubes to reboilers and evaporators in the petroleum and chemical processing industries have been discussed. The more important points are summarized below:

1. Important economic advantages can be gained by utilizing enhanced boiling tubes, such as a notable reduction in the initial cost of a heat exchanger, a substantial decrease in its energy-related operating cost, and additional heat recovery.
2. Enhanced boiling tubes, some doubly enhanced, are available commercially in materials suitable for refinery and chemical plant services.
3. Overall heat transfer coefficients for reboilers and evaporators with enhanced boiling tubes can often be from two to five times higher than those of conventional, plain tube designs in appropriate applications.
4. All common reboiler configurations can be used with enhanced boiling tubes.
5. Common process applications are to ethylene plants, refinery light ends facilities, natural gas liquefiers, air separation plants, ethylene glycol facilities, ammonia plants, methanol plants, seawater distillation facilities, and other alcohol-solvent distillation units.
6. Special thermal design considerations for enhanced boiling tube reboilers include their proper thermal optimization, the maximum bundle heat flux and maximum bundle diameter, the maximum boiling wall superheat, and optimum liquid circulation rates.
7. High-performance enhanced boiling tubes, such as Turbo-B, Gewa-TX, High Flux, and Thermoexcel-E, should only be applied to nonfouling and light fouling services; integral low-finned tubes can be used to best advantage in high fouling services.
8. A tube selection and evaluation guide was developed for quick determination of the advantages of using an enhanced boiling tube in a particular application.

11-11 NOMENCLATURE

A	heat transfer surface area (m^2)
D_B	bundle diameter (m)
g	gravitational acceleration (m/s^2)
g_c	gravitational constant
Δh_v	heat of vaporization (J/kg)
L	tube length (m)
Q	heat duty (W or MW)
$(Q/A)_B$	maximum bundle heat flux (W/m^2)
$(Q/A)_{st}$	single tube maximum heat flux (W/m^2)
T_{sat}	saturation temperature
U_o	overall heat transfer coefficient [$W/(m^2 \cdot K)$]

α_{heating} heat transfer coefficient of heating fluid [W/(m^2·K)]
ρ_L density of liquid (kg/m^3)
ρ_v density of vapor (kg/m^3)
σ surface tension (N/m)

REFERENCES

Antonelli, R., and P. S. O'Neill. 1981. Design and application considerations for heat exchangers with enhanced boiling surfaces. Paper read at International Conference on Advances in Heat Exchangers, September, Dubrovnik, Yugoslavia.

———. 1986. Design and application considerations for heat exchangers with enhanced boiling surfaces. In *Heat Exchanger Sourcebook,* ed. J. W. Palen, 645–61. Washington, D.C.: Hemisphere.

Arai, N., T. Fukushima, A. Arai, T. Nakajima, K. Fujie, and Y. Nakayama. 1977. Heat transfer tubes enhancing boiling and condensation in heat exchangers of a refrigerating machine. *ASHRAE Trans.* 83 (part 2): 58–70.

Cumo, M., G. E. Farello, G. Ferrari, and G. Palazzi. 1974. The influence of twisted tapes in subcritical, once-through vapor generators in counter flow. *J. Heat Transfer* 96:365–70.

Czikk, A. M., P. S. O'Neill, and C. F. Gottzmann. 1981. Nucleate pool boiling from porous metal films: Effect of primary variables. In *Advances in Enhanced Heat Transfer*, HTD Vol. 18, 109–22. New York: American Society of Mechanical Engineers.

Fiores, J., F. Castells, and J. A. Ferré. 1984. Recompression saves energy. *Hydrocarbon Process.* 63(7):59–62.

Gilmour, C. H. 1965. No fooling—no fouling. *Chem. Eng. Sci.* 61(7):49–54.

Gottzmann, C. F., P. S. O'Neill, and P. E. Minton. 1973. High efficiency heat exchangers. *Chem. Eng. Progr.* 69(7):69–75.

Kenney, W. F. 1979. Reducing the energy demand of separation processes. *Chem. Eng. Progr.* 75(3):68–71.

Kern, D. Q. 1956. A new process tool: low-finned tubing. *Petrol. Refiner* 8:128–32.

Kitto, J. B., Jr., and M. J. Albrecht. 1988. Elements of two-phase flow in fossil boilers. In *Two-phase flow heat exchangers: Thermal-hydraulics fundamentals and design,* ed. S. Kakac, A. E. Bergles, and E. O. Fernandes, NATO ASI series E, vol. 143, 495–551. Dordrecht: Kluwer Academic Publishers.

Megerlin, F. E., R. W. Murphy, and A. E. Bergles. 1974. Augmentation of heat transfer by use of mesh and brush inserts. *J. Heat Transfer* 96:145–51.

Milton, R. M., and C. F. Gottzmann. 1972. High efficiency reboilers and condensers. *Chem. Eng. Progr.,* 68(9):56–61.

Moore, J. A. 1974. Fintubes foil fouling for scaling services. *Chem. Process.* (7):8–10.

O'Neill, P. S., G. W. Fenner, and E. F. Williamson. 1982. Advanced heat transfer technology in cracked gas separation. Paper read at AIChE Annual Meeting, November, Los Angeles.

O'Neill, P. S., and C. F. Gottzmann. 1980. Improved air plant main condenser. Paper read at ASME Century 2 Emerging Technology Conferences (cryogenic processes and equipment session), August, San Francisco.

———. and J. W. Terbot. 1971. Heat exchanger for NGL. *Chem. Eng. Progr.* 67(7):80–82.

———. 1972. Novel heat exchanger increases cascade cycle efficiency for natural gas liquefaction. *Adv. Cryog. Eng.* 17:420–37.

O'Neill, P. S., R. C. King, and E. G. Ragi. 1980. Application of high performance evaporator tubing in refrigeration systems of large olefins plants. *AIChE Symp. Ser.* 76(199):289–300.

O'Neill, P. S., and E. G. Ragi. 1986. Recent application trends for enhanced boiling surface tubing. Paper read at AIChE Winter Annual Meeting, November 2–7, Miami Beach (paper 26c).

Palen, J. W., and W. M. Small. 1964. Kettle and internal reboilers. *Hydrocarbon Process.* 43(11):199–208.

Panchal, C. B., D. L. Hillis, and A. Thomas. 1983. Convective boiling of ammonia and Freon 22 in plate heat exchangers. *Proc. ASME-JSME Therm. Eng. Joint Conf.* 2:261–68.

Quadri, G. P. 1982. Use heat pump for P-P splitter. *Hydrocarbon Process.* 60(3):147–51.

Ragi, E. G., P. S. O'Neill, and J. L. Heck. 1984. Retrofit refinery reboilers for energy cost savings. Paper read at Pacific Energy Association Heat Transfer Symposium, May 9–10, San Francisco.

Thome, J. R. 1987. Enhanced boiling of mixtures. *Chem. Eng. Sci.* 42:909–17.

Uehara, H., H. Kusuda, M. Monde, and T. Nakaoka. 1983. Shell-and-plate heat exchangers for an OTEC plant. *Proc. ASME-JSME Therm. Eng. Joint Conf.* 2:253–60.

Webber, W. D. 1960. Under fouling conditions—Finned tubes can save money. *Chem. Eng.* (3):149–52.

Yampolsky, J. S., and P. Pavlics. 1983. *Spiral fluted tubing for augmented heat transfer in process industries, final report, Idaho operations office*. Washington, D.C.: Department of Energy.

Yilmaz, S., J. W. Palen, and J. Taborek. 1981. Enhanced boiling surfaces as single tubes and tube bundles. In *Advances in Enhanced Heat Transfer*, HTD Vol. 18, 123–29. New York: American Society of Mechanical Engineers.

CHAPTER

TWELVE

DESIGN CASE HISTORIES

12-1 INTRODUCTION

The application of enhanced boiling tubes, whether low-finned tubes or high-performance versions for shell-side or tube-side boiling, to evaporators and reboilers requires special thermal design expertise to obtain the maximum benefit of the technology. In addition, it is not necessarily sufficient to prove that enhanced boiling tubes function with superior performance as single tubes or in small laboratory test bundles to guarantee that their performance in practice will also be exemplary.

As in the thermal design of conventional evaporators, the geometry of the bundle, shell, baffles, and the like can have a large influence on the actual performance of the unit and hence on the effectiveness of the enhanced boiling tubes themselves. Thus there is a "learning curve" associated with understanding the effect of these geometric parameters on performance that is vital for competitive applications such as fluorocarbon refrigeration systems. Secondary factors affecting the tube bundle performance typically can only be determined from careful study of large-scale laboratory test data or by monitoring the performances of actual units in the field. For instance, several such factors are the optimum locations for the inlet and outlet nozzles, the vapor distributor design for water chillers, the tube pitch, and the tube sheet layout.

In the present chapter, the results of large-scale tests and field operating performances are presented for the few cases in which these have been documented in the open literature. The case histories are organized by industrial sector: fluorocarbon refrigeration services, refinery and chemical plant refrigeration services, and reboilers. The objectives are to demonstrate some successful applications in the field and to discuss the effect of the secondary factors mentioned above on overall heat exchanger performance. Nearly all the laboratory test information available is for shell-side boiling because intube boiling tests are invariably run with just one tube. Some results of several parametric studies that used actual boiling data are also included where appropriate. For complete information of any one study, the reader is referred to the original report.

12-2 FLUOROCARBON REFRIGERATION SERVICES

Enhanced boiling tubes have been used extensively in fluorocarbon refrigeration applications for many years, both for intube boiling and for boiling on the outside of tubes.

Low-Finned Tube Bundle Prototype

Nakajima and Shiozawa (1975) carried out an extensive performance study on a flooded type evaporator with low-finned tubes boiling refrigerant-11 on the shell side with cooling water flowing inside the tubes. The rated heat duty of the evaporator was 352 kW (100 tons of refrigeration) at full capacity operating at the following conditions: evaporation temperature, 1°C; inlet and outlet temperatures of the chilled water, 10° and 6°C, respectively; chilled water flow rate, 75,450 kg/h (166,000 lb/h); and refrigerant circulation rate, 8,884 kg/h (19,545 lb/h). The horizontal tube bundle was made with 185 copper integral low-finned tubes 2.4 m long with 750 fins per meter (19 fins per inch [fpi]). The finned tubes had a ratio of outside to inside surface area of 3.47. Only the inside tube diameter (14 mm) was cited. The wall thickness at the plain ends was 1.4 mm. The bundle had a total outside surface area (including the finned surface) of 67 m^2. The tube pattern was triangular, with a layout angle of 30° and a tube pitch ratio of 1.32. The bundle height was 260 mm inside an oversized shell of 772 mm inside diameter. The bundle had a floating head with four tube passes. A weir was also installed behind the floating head, as in a kettle reboiler.

Three different inlet nozzle locations were tested, one at the bottom near the fixed head, another at the bottom near the floating head, and one in the end of the shell behind the weir (the last was to allow some tests to be done without the inlet flash vapor passing over the tube bundle). Fourteen tubes in the bundle were instrumented with thermocouples to obtain local water temperatures inside the tubes at about one-third the length of the bundle from the fixed head (a sectional drawing of the evaporator was shown in the original report).

Initial heat transfer tests showed that the heat transfer coefficient of the chilled water was well predicted by the Dittus and Boelter (1930) correlation when the leading empirical coefficient was changed from 0.023 to 0.027 and the Prandtl number exponent was changed to 0.25. Thus for water at 6°C the small ripples on the inner surface of the tubing resulting from the external finning process only reflected about a 5% increase compared to a plain tube's performance with a smooth interior.

With the use of a distributor to obtain uniform allocation of the inlet vapor under the bundle (determined from air-water tests), it was found that the average boiling side heat transfer coefficient for tube passes 1 and 2 (the bottom half of the bundle) and for passes 3 and 4 (the upper half of the bundle) increased appreciably with increasing heat flux. The heat transfer coefficients for passes 1 and 2 were 1.1 to 2 times those for boiling on a single finned tube over the same heat flux range (4.65 to 8.14 kW/m^2). Passes 3 and 4 performed about 10% to 50% higher than the lower two passes, depending on the operating conditions. The inlet vapor quality of the refrigerant was not measured, but it was stated that the ratio of the vapor generated in the bundle to that in the inlet flow was approximately 4:1, which translates into about a 20% inlet vapor quality for the test conditions.

The effect of the evaporation pressure was tested over the narrow range of 0.45 to 0.55 bar. The average bundle boiling heat transfer coefficient was found to increase with increasing pressure, although the data only deviated about 5% from the values obtained at the intermediate test pressure of 0.5 bar.

The effect of the nozzle location and type of inlet vapor distributor were tested by using the three different inlet nozzles and two different vapor distributors. Vapor distributor 1 was purposely made to deflect only the vapor at the inlet nozzles, and distributor 2 was specially made to obtain uniform distribution of vapor under the bundle. When distributor 1 and either the inlet nozzle near the fixed head end of the shell or the inlet nozzle near the floating head (or both) were used, no substantial variation in the mean bundle boiling heat transfer coefficient occurred over a wide range of heat fluxes. For distributor 2 the inlet near the fixed head gave much better performance than the inlet behind the weir alone. Thus it was concluded that, with poor inlet vapor distribution, even an eccentric location of the inlet nozzle relative to the outlet nozzle (which was located above the floating head) did not have a strong influence on the bundle's boiling performance. Uniform vapor distribution underneath the bundle, however, improved performance relative to vapor totally bypassing the bundle, as in the case of the nozzle behind the weir. In addition, using the inlet near the fixed head with distributor 2 produced improvements of about 10% to 80% in the mean bundle boiling heat transfer coefficient compared to those obtained when distributor 1 was used, pointing to the crucial role of the inlet vapor distributor on the boiling performance of a low-finned tube bundle.

Performance Comparison of Thermoexcel-E and Low-Finned Tube Bundles

Arai et al. (1977) built two identical evaporators for a new Hitachi R-12 refrigeration unit, one with Thermoexcel-E tubes with pore sizes of 0.1 mm and 2,000 reentrant tunnels per meter of length and the second with low-finned tubes with 748 fins per meter (19 fpi) of 1.4 mm height. The rated duty of the unit was 703 kW (200 tons of refrigeration). The lengths of the copper tubes were 2.5 m in each case, but their diameters were not given. The tube spacings in the bundles were reported to be more compact than for the existing low-finned tube model. Because of the tighter packing, simple baffle plates were used to diverge the flow under the bundle from three inlet orifices rather than an elaborate inlet vapor distributor. The Thermoexcel-E tubes were externally rolled every 50 mm to provide internal ridges to enhance the chilled water flow. Water flow rates were in the range of 2 to 3 m/s. Also, the condenser shell was located inside the oversized evaporator shell, above the evaporator tube bundle. This geometry achieved a more compact unit and allowed both bundles to have one common end plate at either end. The evaporator's vapor outlet was at the center of its oversized shell above the condenser's shell (the locations of the vapor inlet orifice were not cited). The evaporator tube layout apparently was rectangular in cross-section in view of the shape of the chilled water inlet manifold shown in the photograph of the unit.

Extensive tests were run on the refrigeration unit after about 200 hours of operation on a test stand. The doubly enhanced Thermoexcel-E evaporator bundle's overall heat transfer coefficients were observed to be consistently 45% to 55% higher than those of the low-finned unit at refrigeration loads ranging from 60% to 120% design capacity. About one-third of the augmentation was attributed to the water-side enhancement and two-thirds to the boiling-side enhancement. Varying the evaporation temperature from 0° to 2°C had no effect on the level of augmentation. Measurements of oil concentrations in the evaporator revealed concentrations of 1% to 2% by weight. Poor vapor distribution underneath the finned tube bundle may have occurred and could have been responsible for part of the increase observed in the Thermoexcel-E's performance relative to the finned tube bundle (tubes such as the Thermoexcel-E have higher single tube boiling performances than low-finned tubes and therefore gain less from the two-phase convection process caused by vapor circulation up through the bundle). As a consequence, uniform vapor distribution is felt not to be an important design factor when tubes such as the Thermoexcel-E are used.

High Flux Tube Bundle Prototypes

In an earlier enhanced boiling study by Czikk et al. (1970), the High Flux tube was tested in two different 70-kW (20-ton) flooded chiller prototypes. In these two chillers, the porous matrix had a void fraction of approximately 60%. The working fluid in both prototypes was refrigerant-11.

The first prototype consisted of a 20-cm internal diameter shell and a four-tube pass bundle with a total of 20 plain copper tubes externally coated with a 0.3-mm thick porous High Flux layer. The tubes were 19.9 mm in outside diameter, 16.8 mm in internal diameter, and 1.52 m long and were laid out in a triangular pattern with a 23.8-mm (15⁄16-inch) pitch. The ratio of pitch to outside tube diameter was therefore 1.20. The top row had eight tubes, the second seven tubes, the third four tubes, and the bottom row one tube, all symmetrically spaced. The top tube row was at the horizontal center line of the shell. The liquid feed was provided from the bottom. Steam jackets were welded along the bottom exterior of the shell to generate vapor to simulate incoming vapor from an expansion valve in an actual refrigeration unit. The water velocity was held fixed at 2.90 m/s in the tubes, and the vapor generated by the steam jackets simulated an inlet vapor quality of 20%. The R-11's evaporation temperature was fixed at 1°C. The log mean temperature difference was varied between 5.8 and 7.7 K for a heat flux range from 31.4 to 45.6 kW/m^2, respectively. For these conditions, the following results were obtained:

1. the overall heat transfer coefficient was typically about 5700 W/(m^2·K);
2. the water-side heat transfer coefficient in the tubes was 8,500 W/(m^2·K);
3. the shell-side boiling heat transfer coefficient varied from 27,400 to 41,500 W/(m^2·K) from low to high LMTD values, respectively, and were slightly better than

those measured for the same porous enhancement sprayed on a flat plate and tested under pool boiling conditions; and

4. the refrigeration performance obtained range from 1.28 m/ton at large LMTD values to 1.8 at the lowest LMTD value.

It should be noted, however, that the simulated vapor injection rate cited was not the actual quantity of vapor raised by the steam jackets because only part of the entering heat would be transformed into vapor and the rest into sensible heat. The percentage apparently represents the fraction of the heat load entering through the steam jacket relative to the total refrigeration duty of the steam jacket and the chilled water in the tubes.

Several additional tests were performed to determine the effect of some special operating conditions.

1. A simulated vapor injection rate of 46% was found to have no effect on the evaporator's boiling performance, indicating that the porous coated bundle was not susceptible to vapor binding and that there was no appreciable effect of two-phase convection induced in the bundle.
2. Introducing an oil concentration of 2% by weight of the R-11 charge had no appreciable effect on boiling performance.
3. Lowering the liquid level in the shell to the third tube row from the top of the four rows had no effect on boiling performance.
4. A nonstop 2,000-hour test with 1% oil in the refrigerant charge demonstrated that no degradation in the ton of refrigeration per meter of tubing occurred relative to the initial data obtained at 0% oil content.
5. Descaling the water side with a solution of inhibited sulfamic acid in water immediately after the above test increased the evaporator's performance briefly, but it returned to the previous value after only 1 hour of operation. No fouling of the porous layer from the oil was thought to have occurred.
6. Varying the evaporation temperature to 0°, 2.2°, 3.3°, and 5.6°C caused the boiling-side performance to vary with the corresponding change in the boiling-side temperature difference.

Czikk and co-workers then built a second prototype using a Wolverine corrugated tube coated with a 0.25-mm thick porous layer of copper alloy particles to determine the augmentation that could be attained from a doubly enhanced tube. A 20-ton evaporator was built inside a 30-cm internal diameter shell consisting of 18 corrugated tubes of 23.7 mm outside diameter, 21.3 mm internal diameter, and 0.70 m length on a 28.1-mm triangular pitch. The ratio of pitch to tube diameter was thus 1.195. The tubes were made of a copper alloy. The tube bundle had six tube passes with alternate rows of three and two tubes per row, making the bundle seven rows deep. The sides of the bundle were flanked by machined, contoured plates to provide half the clearance that would have existed from other adjacent tubes in a larger bundle. This geometry, depicted schematically in Fig. 12-1, was meant to simulate a wide bundle of a 400-ton

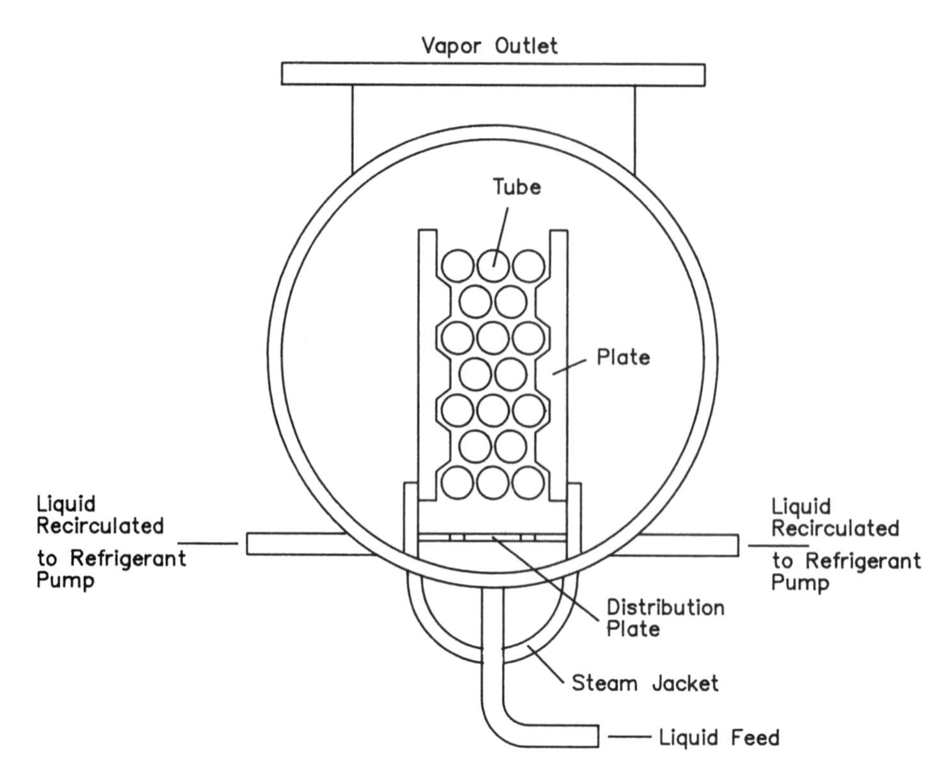

Figure 12-1 Cross-sectional view of a corrugated tube chiller prototype.

chiller. A vapor distribution plate was placed below the bundle as shown, and a steam jacket was welded to the bottom of the shell exterior.

Extensive performance data were obtained for conditions similar to those of the other test described above and were given in tabular form in the report. The following conclusions were made from these tests:

1. Varying the liquid level from 6.4 cm above the top of the bundle to 7.1 cm below it had no measurable effect on the shell-side boiling performance.
2. The refrigeration duty ranged from 0.55 to 1.40 m/ton, which was an improvement over the smaller-diameter, plain High Flux tube in the other prototype described above.
3. The boiling performance of the bundle matched that obtained in tests run for the same coating on a disk under pool boiling conditions.
4. The boiling performance was not affected by varying the vapor injection rate between 0% and 17.4% or by vapor binding, the latter even under conditions of high heat flux (up to 86 kW/m^2) and low liquid recirculation into the shell.

Comparison of High Flux and Low-Finned Tube Water Chillers

Starner and Cromis (1977) described some of their extensive experience in testing and applying enhanced boiling tubes to flooded chillers. Using in-house experimental data, they performed a parametric study on a full-size refrigerant-22 system with a 0.14-bar (2-psi) pressure drop limitation on the water side and a chiller water cooling temperature drop of 5.6 K from 12.2° to 5.6°C (and a second option of 8.3 K); they assumed a fouling factor of 0.00009 $m^2 \cdot K/W$ (0.0005 $h \cdot ft^2 \cdot °F/Btu$) for the water side. They determined the ratios of the meters of tubing per ton of refrigeration of a low-finned tube unit to those of a High Flux tube unit as a function of the "small temperature difference," which they defined as the difference between the exit chiller water temperature and the evaporation temperature. The tube outside diameters used in the study were 19 mm. The low-finned tube had 1,026 fins per meter (26 fpi) with an external wetted surface area per meter of tube length of 0.181 m^2/m (0.64 ft^2/ft). The High Flux tube had no internal enhancement for the water.

Figure 12-2 depicts the results. The use of the High Flux tubing reduced the required tube length per unit of refrigeration duty by about half compared to the industry standard low-finned tube. The benefit of the higher boiling performance of the High Flux tube increased as the small temperature difference decreased. For two-pass designs, the investigators stated that the optimum shell lengths ranged from 3.7 to 9.1 m, with the High Flux designs always giving optimum lengths slightly shorter than those of the finned tube for the same small temperature difference. The High Flux evaporator design

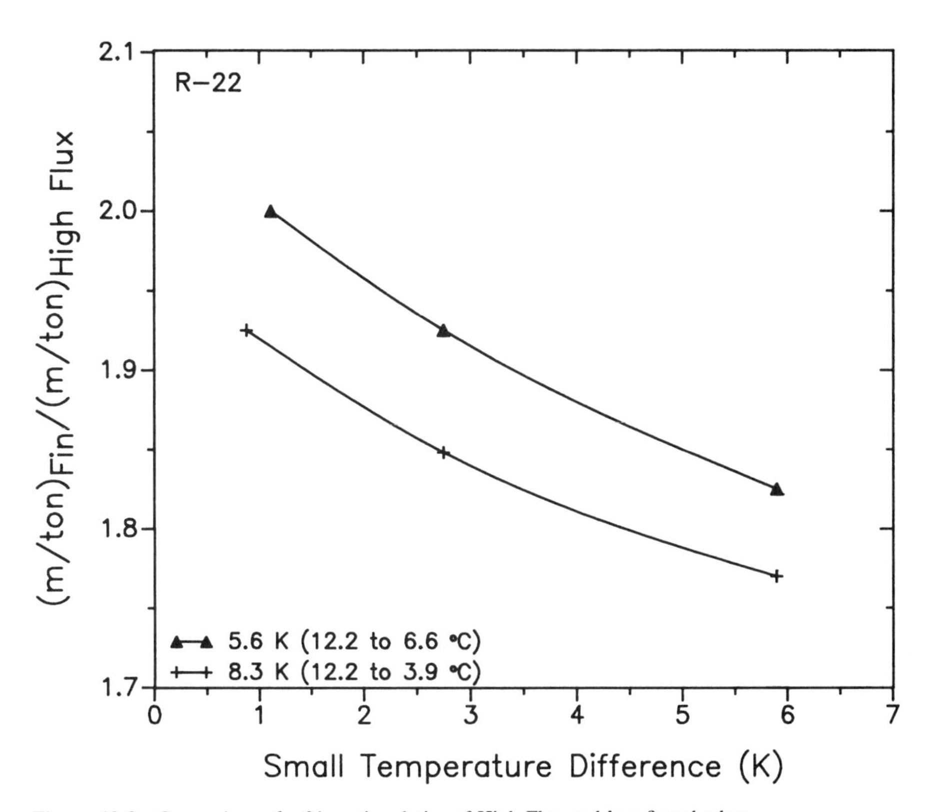

Figure 12-2 Comparison of refrigeration duties of High Flux and low-finned tubes.

gave a compressor power savings of from 3% to 10% compared to the finned tube design for bundles of the same length.

Copper Flame Sprayed Tube Evaporator

Mochida, Takahata, and Miyoshi (1983) tested a 4-MW evaporator for use in an ocean thermal energy conversion (OTEC) demonstration plant with a rated electrical power output of 120 kW. Titanium tubes were flame sprayed with copper on their external surface to produce a porous coating 0.15 mm thick. A complex shell-and-tube design was developed that divided the evaporator into two compartments, one above the other, to reduce the effect of subcooling produced by the liquid head inside the 1.9-m diameter shell. The two-pass unit had 870 tubes of 25.4 mm outside diameter and 0.7 mm wall thickness with a porous coated length of 5.66 m. Refrigerant-22 was the working fluid, entering at 13.1°C after being pumped from a hot-well tank in the line before the evaporator and was boiled at a saturation temperature of 24.8°C (10.4 bar).

The investigators obtained a handful of operating data for water velocities in the tubes ranging from 1.2 to 2 m/s, for which the measured overall heat transfer coefficient ranged from 3,100 to 4,300 W/(m^2·K). A sponge-ball cleaning system was used once daily to clean the seawater biofouling from the inside of the tubes. Perhaps the most interesting part of this study was the evaporator design to reduce the liquid head and hence the subcooling in the liquid pool in the shell.

Air Conditioning Evaporator with Microfinned Tubes

A thermal design computer program was developed by Huang and Pate (1988) to compare the performances of microfinned and smooth tubes in air conditioning evaporators and condensers. The computational model featured a numerical solution scheme to solve the momentum and energy equations together with the equations of state along incremental lengths of the heat exchanger tube. The boiling and condensation heat transfer coefficients for the microfinned tube were calculated from experimental data obtained by Pate and co-workers (see Chapter Ten's discussion of experiments with internally finned tubes).

The evaporator coil analyzed was a typical plate-and-tube version similar to that shown in Fig. 4-14 and had 551 plates per meter. The tubes were 10.16 mm in outside diameter. The refrigerant was R-22 and entered at a temperature of 1°C. Dry air entering at 21°C was the fluid to be cooled. The unit had a heat duty of about 30 kW.

The simulation results showed that for the same inlet conditions the microfinned unit required about 12% less tubing to evaporate all the refrigerant compared to the plain tube unit. Also, the exit saturation temperature was about 1°C lower than that of the plain tube unit because of the microfinned tube's larger pressure drop. For units of identical dimensions the microfinned tube provided a superheating of 5 K at the exit, whereas the smooth tube unit produced saturated vapor only. The performance augmentation obtained was not that substantial considering the 80% (approximately) increase in the boiling heat transfer coefficient and may not be economically justifiable because of the microfinned tube's higher unit cost. Using higher-performance louvered fins on the plates would improve the results. The comparison was done with just one tube circuiting arrangement, however, and no attempt was made to optimize either unit.

For further information about the use of enhanced boiling tubes in air conditioning systems, the reader is referred to Tatsumi et al. (1982), Tojo et al. (1984), and Pate (1988).

12-3 REFINERY AND CHEMICAL PLANT REFRIGERATION SERVICES

Another area of application of enhanced boiling tubes is refrigeration services in ethylene plants, gas plants, and air separation units. The only well-documented studies found for this class of services are for porous coated (High Flux) tubes. Plate-fin heat exchangers (described in Chapter Thirteen) are another widely used enhanced boiling geometry applied to these services.

Ethylene Column C_2 Splitter

Milton and Gottzmann (1972) and Gottzmann, O'Neill, and Minton (1973) described one of the first applications of High Flux tubing to a large refrigeration service, a vertical thermosyphon reboiler on an ethylene column C_2 splitter in Ponce, Puerto Rico with ethane evaporating at 21.7 bar (315 psia) on the tube side and propylene condensing on the outside. The porous coating was sprayed on the inside of the tube, and the outside had longitudinal fluted fins to augment condensation. The tubes were 50.4 mm in outside diameter and 3.05 m long. The exchanger had an effective surface area of 253 m^2 and a design thermal duty of 12 MW (41×10^6 Btu/h). Horizontal baffles were used to interrupt the condensate film falling down the outside of the vertical fluted tubes.

The heat transfer coefficient on the vertical fluted tubes was calculated with a method presented by Antonelli and O'Neill (1981, 1986) to handle three different flute profiles. The investigators gave the following easy to apply expression for predicting the condensing heat transfer coefficient on vertical fluted fins:

$$\alpha_{\text{cond}} = 0.925\, a \left(\frac{\lambda_L^3 \rho_L^2 g}{\mu_L \Gamma} \right)^{1/3} \tag{12-1}$$

where the parameter a depends on the flute geometry and fluid type. The following are their experimentally obtained values: profile A, $a = 8.7$ (steam); profile B, $a = 12.7$ (light hydrocarbons); and profile C, $a = 16.1$ (nitrogen and methane). Profile A had flutes 0.9 mm high, profile B had flutes 1.7 mm high, and profile C had flutes 2.0 mm high. The correlation was valid for tube lengths up to 6 m and at conditions for which the condensing liquid film was turbulent. The investigators reported condensation heat transfer augmentations to be a factor of 8 to 12 relative to those of a plain vertical tube bundle.

Figure 12-3 depicts the field test data measured 2, 9, and 12 months after startup. Also depicted are the original High Flux design curve and a conventional, plain tube bundle design curve. A breakdown of a measured overall High Flux heat transfer coefficient of 5,850 W/(m^2·K) showed the boiling side to have a value of 56,780 W/(m^2·K) and the condensing side to have a value of 7,670 W/(m^2·K), representing a thermal resistance ratio of condensing side to boiling side of 7.4:1. Therefore, the additional surface area provided by the fluted fins and their contours' effect of thinning the liquid film were important to obtaining the maximum benefit of the high boiling heat transfer coefficient. The large tube diameter was utilized to reduce the two-phase pressure drop inside the tubes to provide sufficient liquid recirculation at high heat fluxes so as not to approach the dryout point. In addition, the large tube diameter increased the flow area between adjacent tubes on the shell side, thus minimizing the shell-side pressure drop and its adverse effect on the LMTD. The performance level of the doubly enhanced unit was about five times that of a simulated bare tube unit. Notice

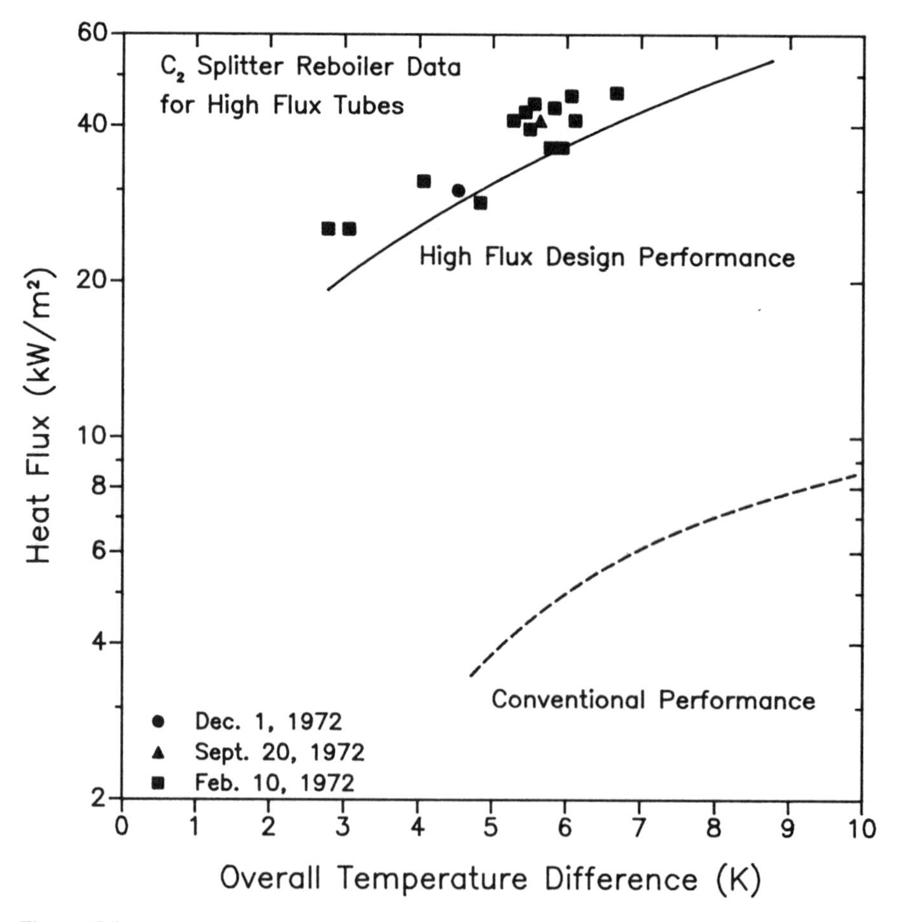

Figure 12-3 Performance of a C_2 splitter condenser.

should also be taken of the fact that the enhanced unit operated effectively at LMTD values as low as 2.8 K.

Other Applications

Further examples of large-scale studies applying High Flux tubing to refrigeration services are available. These include an ethane-ethylene splitter reboiler described in Antonelli and O'Neill (1981, 1986), an ethylene column condenser prototype (see Fig. 3-17) discussed by Gottzmann, O'Neill, and Minton (1973), and a large natural gas chiller (3,700 m^2 in surface area) described by O'Neill and Ragi (1986).

12-4 REBOILERS

Low-Finned Tube Column Internal Reboiler

Yilmaz, Palen, and Taborek (1981) and Yilmaz and Palen (1984) completed a comprehensive experimental study on internal column reboilers with plain, low-finned, and high-performance enhanced boiling tubes. *p*-Xylene was tested at two different pressures, 0.34 and 1.03 bar (5 and 15 psia). Table 12-1 lists the specifications of the four nearly identical U-tube bundles. The outside surface areas of the bundles were based on the outside diameter of the tubes, except for the finned tube, where nominal and total wetted areas are shown. The low-finned tubes were the Wieland-Werke Gewa-K type with 736 fins per meter (18.7 fpi) and fin heights of 1.50 mm. The fin thickness was 0.49 mm at the base and 0.34 mm at the tip. The ratio of outside wetted surface area to inside surface area was 4.11; the ratio of the outside wetted surface area of the finned tube to that of a plain tube was 2.69. The Thermoexcel-E and Gewa-T tube specifications in 90-10 cupro-nickel were not cited and may not be the same as those for the copper versions tested as single tubes. The geometries apparently were not optimized for boiling *p*-xylene.

The tube bundles were heated by condensing steam. In the vapor shear flow regime, the smaller internal diameter of the finned tube and the Gewa-T tube would yield higher intube condensation coefficients than the plain tube at the same heat flux, but only about 4% higher. In addition, the ripples on the internal tube surface created by the exterior finning process would provide a small degree of augmentation. The smaller internal diameters and rippled internal surfaces would also increase the three enhanced boiling tube bundles' pressure drops, which would slightly reduce their mean temperature differences relative to the plain tube bundle.

Figure 12-4 shows the overall performances of the four bundles at 1.03 bar. Calculating the finned tube bundle heat flux with the outside wetted surface area of the tubes gave a finned bundle performance less than that of the plain tube; calculations based on the nominal outside area of the tubes (with the diameter at the tips of the fins) showed that the finned tube bundle was superior to the plain tube bundle by about a factor of 2. The Gewa-T bundle slightly out-performed both the Thermoexcel-E and the finned tube bundles over most of the heat flux range when compared in terms of

Table 12-1 Tube bundle specifications

Tube type (material)	Surface area*, m^2	Bundle diameter, cm	Bundle length, m	Tube outside diameter, mm	Tube inside diameter, mm	Tube pitch, mm	Ratio of tube pitch to tube outside diameter	Number of tubes	Layout angle, degrees
Plain (90-10 cupro-nickel)	19.2	37.8	2.38	15.88	12.60	23.83	1.50	168	90
Gewa-K-finned (90-10 cupro-nickel)	19.2	37.8	2.38	15.85	10.36	23.83	1.50	168	90
Thermoexcel-E (90-10 cupro-nickel)	21.0	40.1	2.39	17.63	14.33	25.40	1.44	168	90
Gewa-T (90-10 cupro-nickel)	17.8	37.6	2.38	14.73	10.16	23.83	1.62	168	90

*The total wetted surface area of the finned tube was 51.6 m^2.

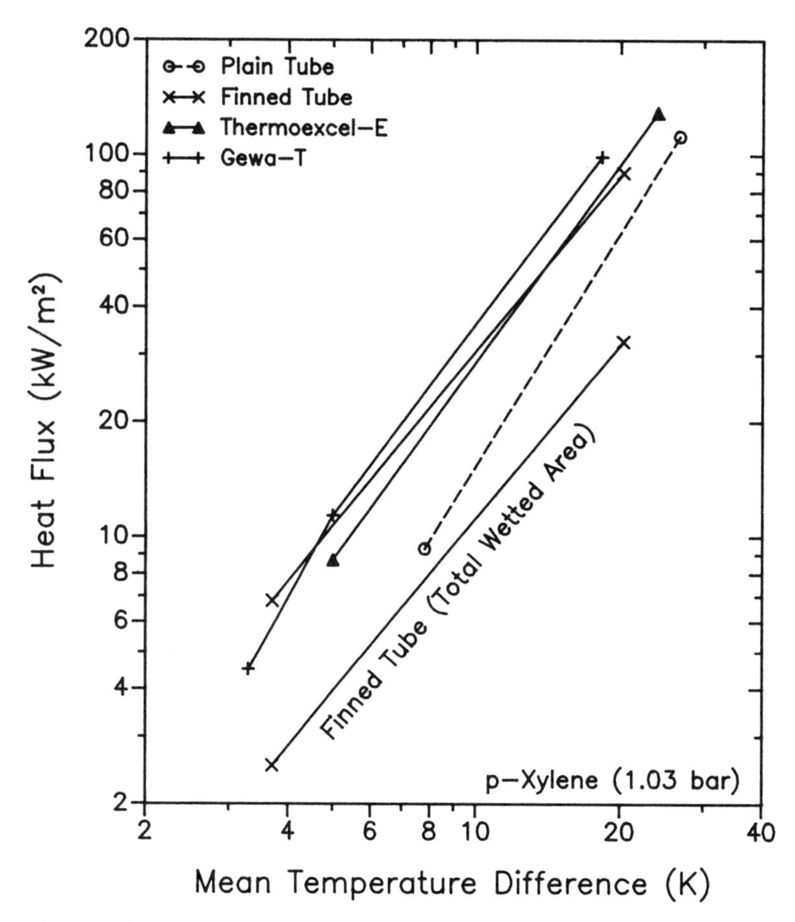

Figure 12-4 Internal reboiler performance comparison.

their outside nominal surface areas. Similar results were obtained for the 0.34-bar tests. These results are, of course, not general but specific to the test conditions.

It is somewhat surprising that the high-performance enhanced boiling tube bundles provided augmentation only equal to that of a low-finned tube bundle. The thermal resistances of the boiling and condensing streams were not cited in the study. In Yilmaz, Palen, and Taborek (1981), however, the heat flux was plotted against the shell-side wall superheat for the plain and Gewa-T tube bundles, which allows one approximately to back out the thermal resistances for these two bundles with the data in Fig. 12-4.

As an example, at a heat flux of 30 kW/m^2 the ratio of boiling-side to condensing-side thermal resistances for the plain tube bundle was determined to be 2:1, whereas for the Gewa-T bundle it was 0.8:1. Thus the maximum performance to be gained by utilizing externally enhanced boiling tubes under the cited conditions was not all that high because the condensation heat transfer coefficient was rather low. For an actual application in which the tube bundle was to be replaced, using doubly enhanced versions of these tubes would provide much more augmentation. In a new installation, the reboiler would be thermally optimized differently for the enhanced boiling tubes to increase the tube-side vapor velocity and hence the condensation-side heat transfer coefficient, perhaps by use of a doubly enhanced tube.

Other low-finned reboiler data have been given by Gilmour (1965) for several hydrocarbon services (see Table 11-4).

Porous Coated Tube Reboilers

Quite a few case histories describing the application of the High Flux tube to reboilers in refineries have been published by Union Carbide. Five different ones are presented in Ragi, O'Neill, and Heck (1984) and O'Neill and Ragi (1986). The depentanizer reboiler application (discussed in both reports) is reviewed here because this appears to be quite a favorable area for enhanced boiling tubes in light ends units.

In this application the revamp of a petroleum refinery in the Netherlands was necessary to increase the output of an existing plain tube depentanizer reboiler, a situation that was created by below-design performance occurring after a heat integration program was implemented in 1969. The heat integration scheme utilized heat in a top circulating reflux (TCR) stream from a crude tower to drive reboilers on other distillation columns in the light ends recovery area of the plant. The crude distillation section was composed of a super fractionation system including depentanizer, deisohexanizer, and deisopentanizer columns and their reboilers. Figure 12-5 depicts a simplified flow sheet of the system showing just the crude and the depentanizer towers.

The refinery was designed to operate with a range of crude oils from Kuwait, Saudi Arabia, and Nigeria, each with different cut-point requirements. The products of the crude tower were straight-run gasoline, heavy and light naphtha, gas oil, C_1 and C_2

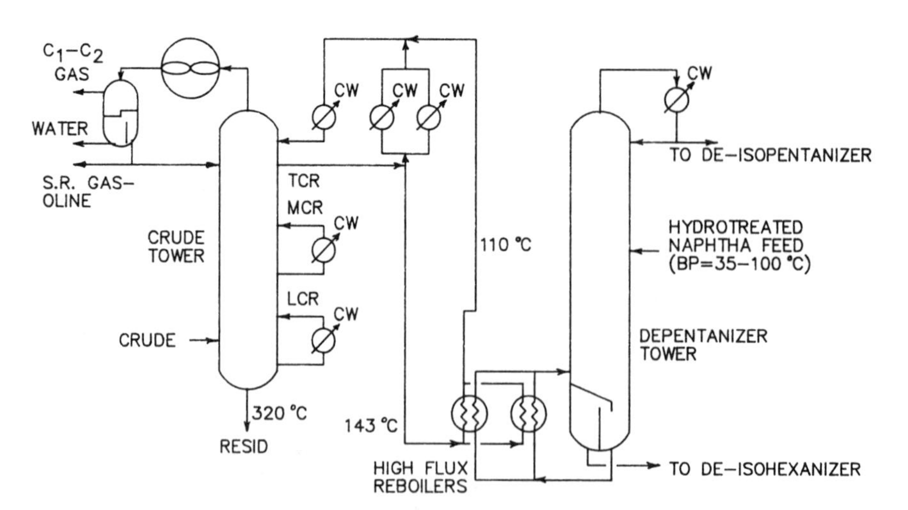

Figure 12-5 Simplified flow diagram for crude tower and depentanizer heat integration scheme. TCR, top circulating reflux.

gases, and atmospheric tower residuals. Top circulating reflux (primarily naphtha) was withdrawn from the crude tower; part of this was used to drive the two dependentanizer reboilers. For tower control purposes, the extra heat in the reflux not recovered in the reboilers was removed by the water trim coolers. The dependentanizer naphtha feed came from a hydrotreater. The feed components had boiling points ranging from 35° to 100°C. The overhead product stream was sent to a deisopentanizer, and the bottoms, consisting mostly of C_6 and heavier hydrocarbons, were sent to the deisopropanizer.

In the original plain tube reboiler design, the circulating reflux stream was to be cooled from 149° to 126°C in the reboilers and then down to 98°C in the crude tower overhead coolers. Its design flow rate in 1969 was originally 488,000 kg/h, but with other modifications to the plant and changes in refinery crude, test data in 1980 showed that the top circulating reflux stream flow rate eventually dropped to only about 350,000 kg/h and that its inlet temperature to the reboilers was only 143°C. The reduced TCR flow rate and temperature lowered the two dependentanizer reboilers' heat duties to about 71% of the original design, Table 12-2 summarizes these bare tube reboiler data.

Analysis of the refinery showed that the reboiler output could be improved by decreasing the outlet temperature of the TCR stream from the reboilers and thus rejecting less heat to the tower trim coolers. Rather than adding additional heat exchangers with plain tube surface area to recover this heat, a high-performance boiling surface was installed in the existing reboilers to increase the overall heat transfer coefficient. Thus the existing TEMA type BES floating head reboilers with plain tubes were replaced by new "E" sections (i.e., a bundle of High Flux tubes and two tube sheets for each reboiler). This solution allowed the existing shells, bonnets, floating head assembly, piping, reboiler foundations, and controls to be reused and minimized the capital expense and the downtime for the changeover. In addition, the existing TCR pressure drop was sufficient for the new bundles because the heating fluid's flow length through the new bundles was the same as that in the old ones.

Table 12-2 shows the thermal design specifications for the High Flux tubes. The outlet temperature of the TCR stream was lowered by 16°C compared to the existing plain tube operating conditions. The reboiler shell-side inlet and outlet temperatures remained the same. Thus the overall heat transfer coefficient had to be increased by about 80% to raise the existing reboiler performance to its original design value while simultaneously reducing the wall superheating available for boiling. The maximum expected performance, which was 12% higher than design, is also shown for a yet lower exit temperature of the TCR heating stream, which took advantage of the enhanced boiling tube's ability to operate at small wall superheats. This design therefore provided a margin of safety in the heat duty of the reboilers.

The new reboilers were started up in 1981, and the expected reboiler heat duties were realized. No reboiler fouling was observed, nor was there a reduction in heat duty with time. The refinery reportedly was able to recover about $500,000 annually from the increased dependentanizer column capacity. On the basis of this dollar value, the new reboiler tube bundles had a payback period of about 6 months.

This application demonstrated that enhanced boiling tubes can be applied successfully to dependentanizers if the gum type compounds are reduced to paraffins in a hydrotreater. Also, this example illustrates that sometimes an increase in the overall heat

Table 12-2 Dependanizer reboiler specifications (parameters represent totals for two reboilers)

Reboiler type	Number of tubes	Surface area, m^2	TCR* flow rate, kg/h	TCR inlet temperature, °C	TCR outlet temperature, °C	Reboiler inlet temperature, °C	Reboiler outlet temperature, °C	LMTD, °C	U_o, W/(m^2 · K)	Heat duty, MW
Bare tube										
Original design	1,274	484	488,000	149	126	103	104	33.1	429	7.99
Actual values	1,274	484	349,250	143	120	97	98	33.2	305	5.71
High Flux										
Design duty	1,274	484	349,250	143	110	97	98	25.8	552	7.99
Maximum duty	1,274	484	349,250	143	106	97	98	22.3	713	8.96

*Top circulating reflux.

transfer coefficient of only 80% is necessary to make an enhanced boiling tube economically justifiable because the lower possible temperature approach increases heat recovery. In addition, using enhanced boiling tubes in this case was the cheapest solution to the problem.

In summary, combining good thermal engineering practice with the application of enhanced boiling tubes can save energy, reduce capital expenditures, and provide for more reliable plant operation. Because low thermal performance of bare tube reboilers is usually only part of the problem causing a bottleneck in existing plants, the application of enhanced boiling tube units should be weighed together with other process changes to obtain the optimum performance of the process unit.

12-5 NOMENCLATURE

a	empirical parameter
g	acceleration due to gravity (m/s^2)
U_o	overall heat transfer coefficient [$W/(m^2 \cdot K)$]
α_{cond}	condensing heat transfer coefficient [$W/(m^2 \cdot K)$]
Γ	condensate loading [$kg/(m \cdot s)$]
λ_L	liquid thermal conductivity [$W/(m \cdot K)$]
μ_L	liquid viscosity (cp)
ρ_L	liquid density (kg/m^3)

REFERENCES

Antonelli, R., and P. S. O'Neill. 1981. Design and application considerations for heat exchangers with enhanced boiling surfaces. Paper read at International Conference on Advances in Heat Exchangers, September, Dubrovnik, Yugoslavia.

———. 1986. Design and application considerations for heat exchangers with enhanced boiling surfaces. *Heat Exchanger Sourcebook,* ed. J. W. Palen, 645–61. Washington, D.C.: Hemisphere.

Arai, N., T. Fukushima, A. Arai, T. Nakajima, K. Fujie, and Y. Nakayama. Heat transfer tubes enhancing boiling and condensation in heat exchangers of a refrigerating machine. *ASHRAE Trans.* 83(part 2):58–70.

Czikk, A. M., C. F. Gottzmann, E. G. Ragi, J. G. Withers, and E. P. Habdas. 1970. Performance of advanced heat transfer tubes in refrigerant-flooded liquid coolers. *ASHRAE Trans.* 76(part 1):96–109.

Dittus, F. W., and L. M. K. Boelter. 1930. *Univ. Calif. (Berkeley) Publ. Eng.* 2:443.

Gilmour, C. H. 1965. No fooling—No fouling. *Chem. Eng. Sci.* 61(7):49–54.

Gottzmann, C. F., P. S. O'Neill, and P. E. Minton. 1973. High efficiency heat exchangers. *Chem. Eng. Progr.* 69(7):69–75.

Huang, K., and M. B. Pate. 1988. A model of an air-conditioning condenser and evaporator with emphasis on intube enhancement. IIR Conference, July 18–21, Purdue.

Milton, R. M., and C. F. Gottzmann. 1972. High efficiency reboilers and condensers. *Chem. Eng. Progr.* 68(9):56–61.

Mochida, Y., T. Takahata, and M. Miyoshi. 1983. Performance tests of an evaporator for a 100-kW (gross) OTEC plant. *Proc. ASME-JSME Therm. Eng. Joint Conf.* 2:241–45.

Nakajima, K., and A. Shiozawa. 1975. An experimental study on the performance of a flooded type evaporator. *Heat Transfer Jap. Res.* 4(3):49–66.

O'Neill, P. S., and E. G. Ragi. 1986. Recent application trends for enhanced boiling surface tubing. Paper read at AIChE Winter Annual Meeting, November 2–7, Miami Beach (paper 26c).

Pate, M. B. 1988. Design considerations for air-conditioning evaporator and condenser coils. In *Two-phase flow heat exchangers: Thermal-hydraulic fundamentals and design,* ed. S. Kakac, A. E. Bergles, and E. O. Fernandes, NATO ASI series E, vol. 143, 849–84. Dordrecht: Kluwer Academic Publishers.

Ragi, E. G., P. S. O'Neill, and J. L. Heck. 1984. Retrofit refinery reboilers for energy cost savings. Paper read at Pacific Energy Association Heat Transfer Symposium, May 9–10, San Francisco.

Starner, K. E., and R. A. Cromis. 1977. Energy savings using High Flux evaporator surface in centrifugal chillers. *ASHRAE J.* 19(12):24–27.

Tatsumi, A., K. Oizumi, M. Hayashi, and M. Ito. 1982. Application of inner grooved tubes to air conditioners. *Hitachi Rev.* 32(1):55–60.

Tojo, S., K. Hosokawa, T. Arimoto, H. Yamada, and Y. Ohta. 1984. Performance characteristics of multigrooved tubes for air conditioners. *Aust. Refrig. Air Cond. Heat.* 38(8):45–61.

Yilmaz, S., and J. W. Palen. 1984. *Performance of finned tube reboilers in hydrocarbon service* (ASME paper 84-HT-91).

———. and J. Taborek. 1981. Enhanced boiling surfaces as single tubes and tube bundles. In *Advances in Enhanced Heat Transfer*, HTD Vol. 18, 123–29. New York: American Society of Mechanical Engineers.

CHAPTER

THIRTEEN

BOILING IN PLATE-FIN HEAT EXCHANGERS

13-1 INTRODUCTION

Plate-fin heat exchangers are an important alternative to enhanced boiling tubes for augmenting boiling heat transfer. In addition, large heat transfer surface areas per unit volume can be attained compared to those of conventional shell-and-tube units with plain tubes. Thus they not only augment the heat transfer coefficient but also increase the thermal performance on a per unit volume basis.

The original immersion brazing process was invented by Alcoa in 1939, and the Trane Company pioneered its use for constructing plate-fin heat exchangers in 1949. Initially, plate-fin designs were utilized primarily for single-phase applications, especially gaseous flows, in which large heat transfer surface areas result from the low heat transfer coefficients. More recently, relevant experimental data for boiling (and condensation) have been obtained for plate-fin geometries, and proprietary thermal design procedures have been developed for plate-fin evaporators. Thus plate-fin heat exchangers are increasingly applied to boiling services, taking over some of the traditional applications formerly provided by relatively more expensive shell-and-tube heat exchangers made with plain tubes.

Figure 13-1 depicts a diagram of a typical plate-fin heat exchanger. Fluid stream A enters through one header, from which it is distributed to every third layer. Streams B and C are countercurrent to stream A in the other layers. Although only three streams are shown in the diagram, a single plate-fin heat exchanger can handle up to ten or more process streams, yielding a compact design compared to separate shell-and-tube units for each pair of process fluids.

Plate-fin heat exchangers typically have from 700 to 1,500 m^2/m^3 of heat transfer

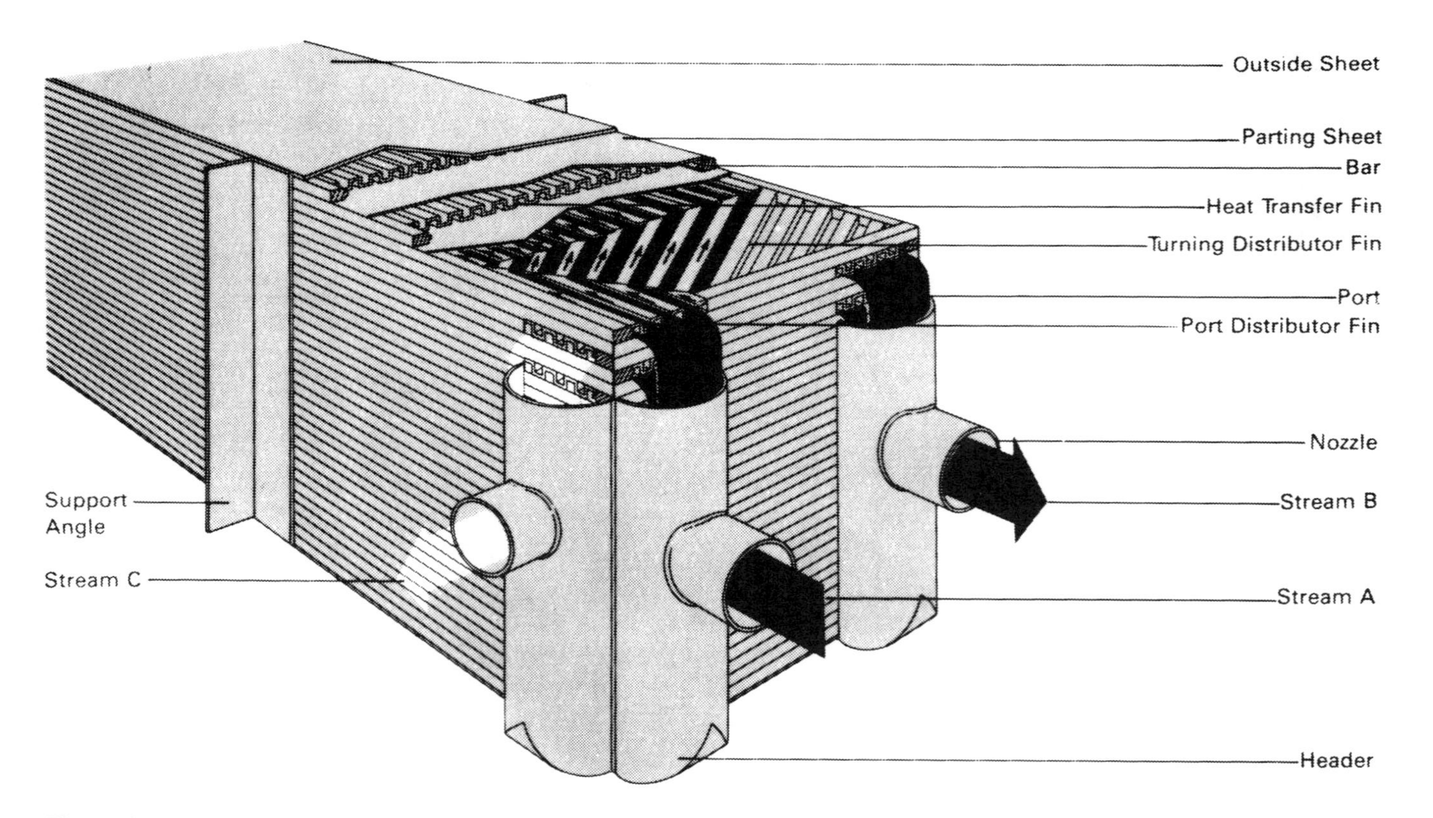

Figure 13-1 Plate-fin heat exchanger (courtesy ALTEC International, Inc.).

surface per unit volume, which is approximately six to ten times that of a comparable shell-and-tube heat exchanger with 19-mm plain tubes and about two to four times that of a low-finned tube bundle. The brazing of the tips of the corrugations to the flat parting sheets provides local mechanical support throughout the unit. Thus light-weight sheet metal, typically from 0.15 to 0.6 mm thick, can be used for the finned corrugations. This is considerably thinner than standard tube wall thicknesses, which are about 1.65 to 2.11 mm. Consequently, plate-fin heat exchangers normally weigh about one-third as much as conventional shell-and-tube units but provide the same heat transfer performance in terms of UA. Plate-fin units can be utilized at temperatures ranging from 4 K to about 1,070 K, depending on the materials of construction. Aluminum alloy is the primary construction material for temperatures less than about 500 K, and stainless steel is used for higher temperatures. Plate-fin units are usually employed for service pressures less than 20 bar, but special designs can be built to permit operation at pressures as high as 96.5 bar according to ALTEC International (1986).

The principal disadvantage of plate-fin heat exchangers is the susceptibility of their small passageways to clogging. In cryogenic units, the principal foulants are water vapor, carbon dioxide gas, and compressor oil, which freeze in the passageways if present. Modern operating units have special compressor designs and other features to minimize these problems, however. Frozen foulants can be removed by allowing the exchanger to "thaw" by bringing its temperature back to ambient or by circulating a warm gas. Mechanical cleaning is not practical.

The most important boiling services of plate-fin heat exchangers are in air separation plants, in which the principal application is to evaporate a liquid oxygen thermosyphon stream with a condensing nitrogen stream. Plate-fin heat exchangers are also used in natural gas processing plants to evaporate both pure fluids and liquid mixtures, such as liquid natural gas. In refinery and petrochemical heat recovery and heat integration schemes, plate-fin heat exchangers are used to advantage in the light end of ethylene facilities, for instance to transfer heat efficiently between various distillation streams (refer to Chapter Eleven's discussion of separation of light hydrocarbons in ethylene plants for a description of one such application). Plate-fin units are also used in refrigeration systems with working fluids such as ammonia, methane, propane, and R-11. Because the fluid flows can be designed to be nearly countercurrent, as shown in Fig. 13-1, temperature approaches of as little as 1 K can be used. This reduces expensive compressor horsepower consumption in refrigeration and heat pumping applications. Plate-fin heat exchangers are also used in aircraft refrigeration and air conditioning services and on oil platforms, where size and weight are the foremost design constraints.

The limited access to well-proven design technology for boiling (and condensation) at the present time inhibits the more widespread application of plate-fin heat exchangers, especially for nontraditional applications. In recent years, however, the progress in plate-fin boiling research has been impressive, and fundamentally based correlations are now becoming available. The objective of the present chapter is to describe the recent advances in two-phase flow and boiling in plate-fin heat exchangers and to review the various predictive methods proposed. The interested reader is also referred to several other reviews by Robertson (1980), Westwater (1983), and Carey and Shah (1988).

13-2 TYPES OF PLATE FINS

A wide variety of fin geometries have been developed by the manufacturers of plate-fin heat exchangers. The simplest geometry is the plain fin shown in Fig. 13-2(*a*). This is the cheapest geometry to produce but also gives the lowest thermal performance. Figure 13-2(*b*) depicts a serrated fin geometry, also known industrially by names such as lanced, offset, and multientry. The corrugations are stamped so as to cut the fins into short lengths, as shown in Fig. 13-3. Perforated fins, Fig. 13-2(*c*), are another type of common geometry. Typical fin spacings are from 6 to 25 fins per inch (fpi) (237 to 987 fins per meter). Available fin heights range from about 2.5 to 16 mm, although for boiling the smaller sizes are used. Choice of fin type, fin size and spacing, and so forth is dependent on the service application and can be different for each different process fluid stream in the same unit.

13-3 TWO-PHASE FLOW REGIMES

The principal two-phase flow patterns expected to occur in plate-fin heat exchanger channels are slug, churn, and annular flow. Bubbly flow may also be present for a short length until the bubbles grow to the size of the channel. At high vapor qualities, liquid entrainment would also be expected to occur from the shedding of liquid at the ends of serrated fins. The flow patterns occuring in the small, nonuniform flow channels affect the heat transfer process and the two-phase pressure drop. Thus study of flow patterns is of fundamental importance to understanding the physics of the process. Nevertheless, only a few studies on two-phase flow patterns in plate-fin heat exchangers have been performed.

The first visualization study on two-phase flows in plate-fin geometries was that of Carey (1985). He investigated two-phase flows of water-air mixtures in a horizontal serrated plate-fin test unit. The test section was made by installing a corrugated finned sheet inside a clear Plexiglass channel. The geometry of the fins was as follows: 1.59 mm serration length, 3.86 mm fin height, 0.2 mm fin thickness, and 276 fins per meter width. Stationary air bubbles were observed in the flow at the narrowest points in the channels at low flow rates. On increasing flow rate, however, these bubbles were washed downstream. Carey attributed this phenomenon to the effects of buoyancy, surface tension, and frictional forces acting on the bubbles. When a bubble was stationary, the analysis showed that the frictional and buoyancy forces were insufficient to overcome the surface tension holding the bubble in place. Then, by incrementally increasing the flow rate, the drag on the bubble was increased until the surface tension was overcome and the bubble broke away. Thus Carey concluded that slug flow does not occur at low qualities and flow rates in serrated fin designs unless specific flow criteria are satisfied.

Carey and Mandrusiak (1986) continued this flow visualization study with another test section to investigate two-phase flow patterns during actual boiling. Their serrated fin type of test section is shown in Fig. 13-4. The channel was mounted vertically so that upward vertical flow could be studied. The front face of the flow channel comprised several glass cover plates to allow observation of the boiling process and thus to identify the flow patterns. The back "plate" was an instrumented copper slab. The fins tested were longer and thicker than standard plate-fin designs: 12.7 mm in serration length, 3.8 mm in height, 1.59 mm thick, and spaced 7.94 mm apart. Two-phase flow patterns were obtained from photographs taken during flow boiling of water, methanol, and *n*-butanol in the channel. Bubbly, slug, churn, and annular flow patterns were observed over the following range of vapor qualities (x) and mass fluxes (G): $0.05 < x < 0.7$ and $3 < G < 100$ kg/(m^2·s). Using the flow regime map proposed by Hewitt and Roberts (1969) for upward cocurrent flow in round tubes, the investigators identified a transition between churn flow and annular flow in their data. They made the following observations about the flow process:

1. Bubbly and slug flow existed for only short lengths after the point of onset of nucleate boiling.
2. Separated flows in the form of churn flow and annular flow were the predominant flow regimes observed.
3. The liquid films flowing along the walls appeared to "pile up" at the front of the fins. This created waves that periodically traveled along the fins and apparently was one of the causes for liquid shedding from the downstream end of the fins, which was accompanied by a local fluctuation in the flow.
4. The liquid films in the corners were thicker than on the flat portions of the test section.
5. Dry patches were observed under some flow conditions near the exit of the test section. As total dryout was approached liquid still partially wetted the unheated front plate, whereas the heated back plate was seen to be dry between the fins but wetted in the corners formed by the fins. In general, no liquid entrainment in the vapor core was evident.

Mandrusiak, Carey, and Xu (1986) also tested the same section in a horizontal position with the heated back plate facing either upward or downward. Methanol and water were the test fluids. Stratified, wavy, and annular flow regimes were observed to occur shortly downstream from the point of initiation of saturated boiling.

Plotting their data on a Taitel and Dukler (1976) type of flow regime map for horizontal cocurrent flow in plain tubes, the investigators were able to delineate transition boundaries between stratified and wavy flows and between wavy and annular flows, including data for both top and bottom heating. These transition boundaries between the various flow regimes were not distinct, but their general locations were fairly well predicted by the original Taitel and Dukler (1976) curves.

Virtually no nucleate boiling was observed in these tests, even for wall superheats of 4 K. Small individual bubbles were occasionally observed in the corners, where the liquid films were thickest. For identical flow and heat flux conditions, the investigators noted that partial dryout generally occurred at lower vapor qualities for top heating than for bottom heating because the liquid film was thicker on the bottom plate, so that the top plate was more difficult to wet.

The two-phase flow for the single serrated fin geometry in the above studies has

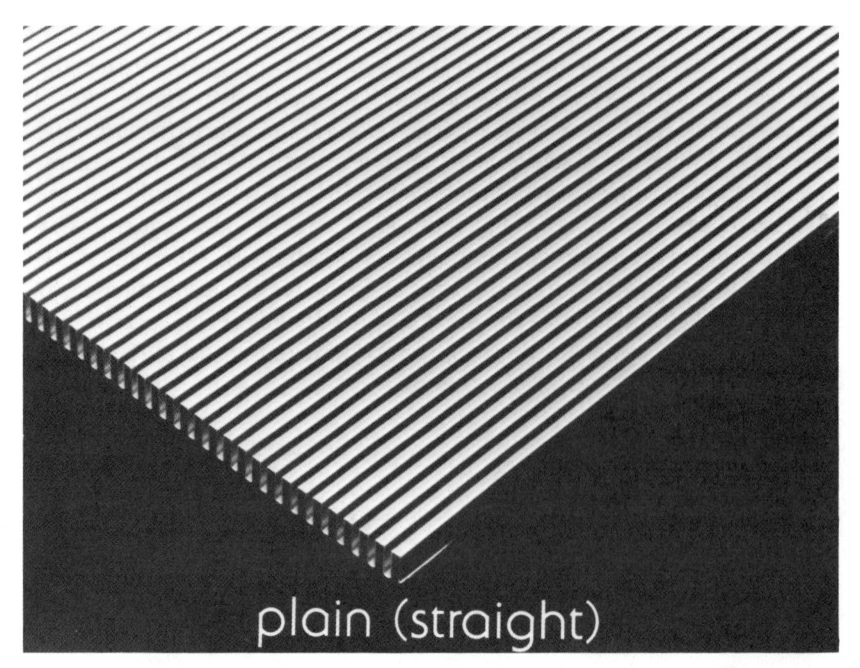

(*a*) plain

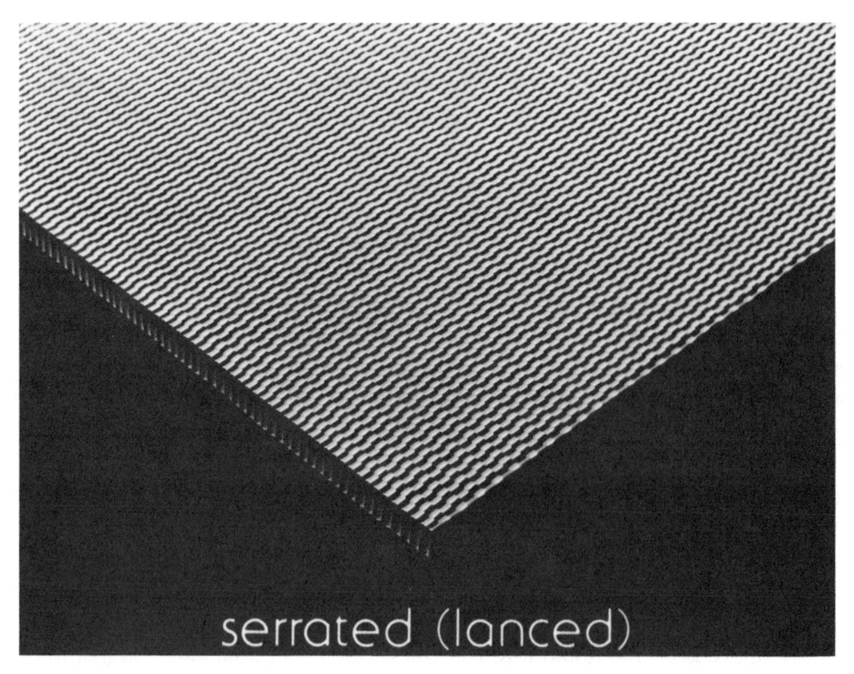

(*b*) serrated

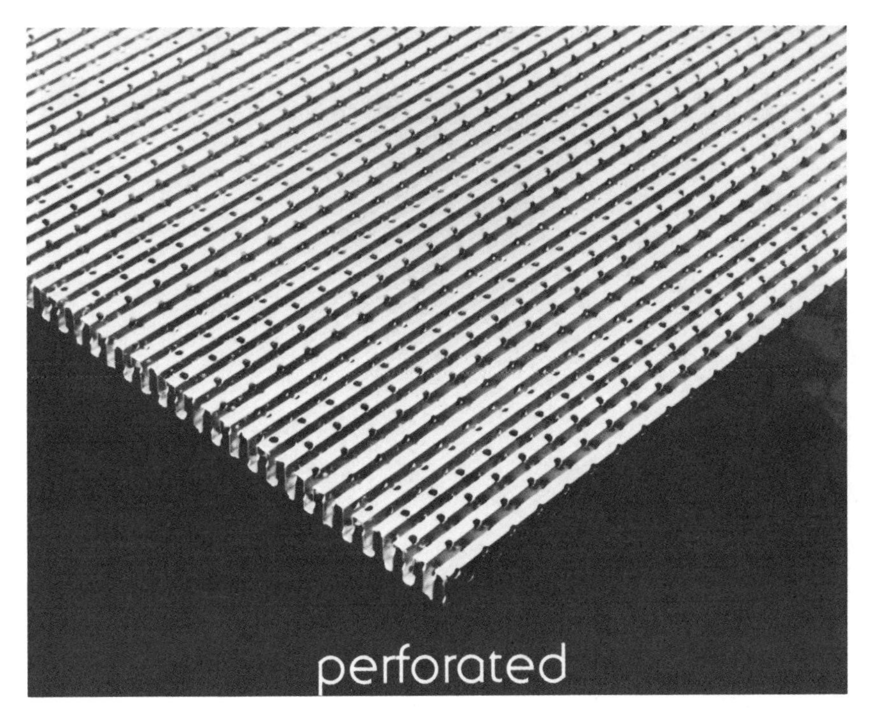

(*c*) perforated.

Figure 13-2 Plate-fin geometries (photographs courtesy ALTEC International, Inc.).

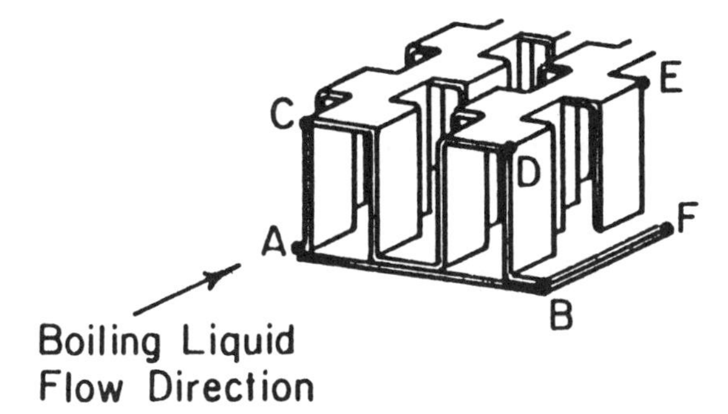

Figure 13-3 Schematic diagram of a serrated fin.

many similarities to boiling in larger-diameter, round tubes. More research on two-phase flow patterns in other fin geometries and fin sizes is needed, however, together with measurement of the corresponding two-phase pressure drops.

13-4 ONSET OF NUCLEATE BOILING

The onset of nucleate boiling in plate-fin heat exchangers is the point at which boiling nucleation occurs in the channel. For partially vaporized liquid feeds vapor is introduced

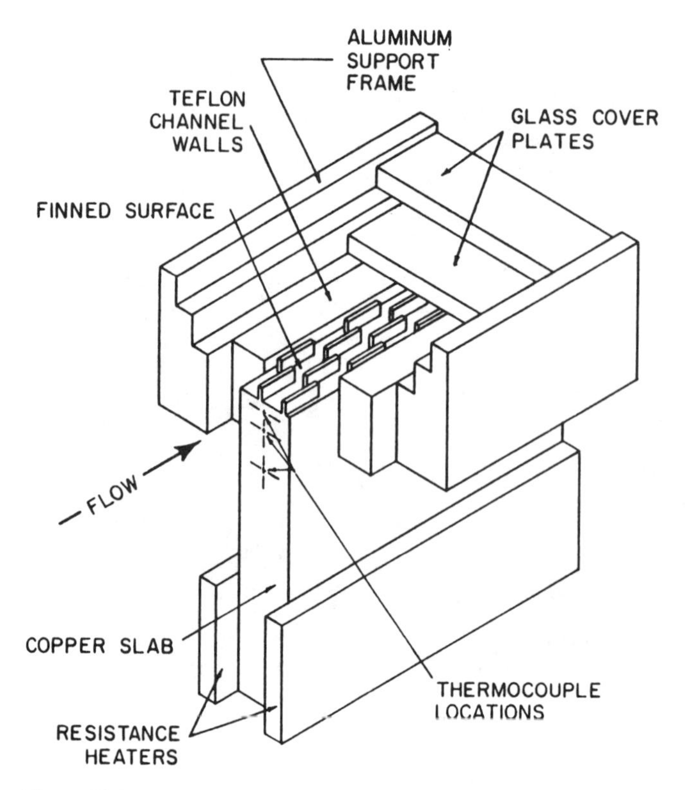

Figure 13-4 Carey and Mandrusiak (1986) test section.

at the entrance, and boiling nucleation is not necessary. Yet, even when the inlet stream is partially vaporized, separation of the vapor and liquid phases in the inlet header can lead to some passageways receiving no vapor. In these channels nucleation must occur for the fluid to evaporate. For subcooled entrance conditions, the liquid feed must first be heated to its saturation temperature before nucleation can occur. At this location the wall temperature has to be superheated to induce boiling nucleation. In serrated and perforated channels, however, vapor from a neighboring channel may migrate to a completely wetted channel to provide the initial vapor. Plain fin designs thus have the disadvantage that vapor from one channel cannot pass into a neighboring channel to minimize the effects of vapor maldistribution.

Robertson and co-workers investigated the onset of nucleate boiling phenomena in plate-fin heat exchangers. Robertson (1979), for instance, measured the temperature profiles of the parting sheet and the bulk fluid for vertical upward flow boiling in a long serrated fin test unit with a subcooled feed of liquid nitrogen. Because it was not practical to mount a large number of closely spaced thermocouples along the parting sheet, Robertson varied either the inlet subcooling or the heat flux in small increments to move the point of onset of nucleate boiling back and forth along the test section so that a composite of the temperature profile at the fixed thermocouple locations could be developed. Using this approach, Robertson and Lovegrove (1980, 1983) later obtained data for R-11 on the same test unit, and Robertson (1983) obtained results for a perforated fin test unit with liquid nitrogen.

Robertson and Clarke (1981, 1984) analyzed these data and observed a number of interesting trends. Figure 13-5 depicts several temperature profiles that they obtained with liquid nitrogen in a vertical serrated fin test section 3.4 m long. The fins were 3.18 mm long, 6.35 mm high, and 0.2 mm thick. The wall and bulk liquid temperatures were found to increase linearly with length up the test section in the subcooled liquid region, as would be expected. Under certain conditions, however, the liquid temperature "overshot" the local saturation temperature before nucleation occurred. Then, at the point of onset of nucleate boiling, both the wall temperature and the fluid temperature fell sharply, apparently as a result of flashing of the superheated liquid. This type of profile was observed to occur when the inlet subcooling was decreased or when the heat flux was increased and the point of onset moved lower down the test section to a location at which boiling was not previously present. Boiling nucleation probably occurred in a corner or in the crevice formed at the brazed joint between the parting sheet and the fin, where the wall is hottest and the liquid film is thickest.

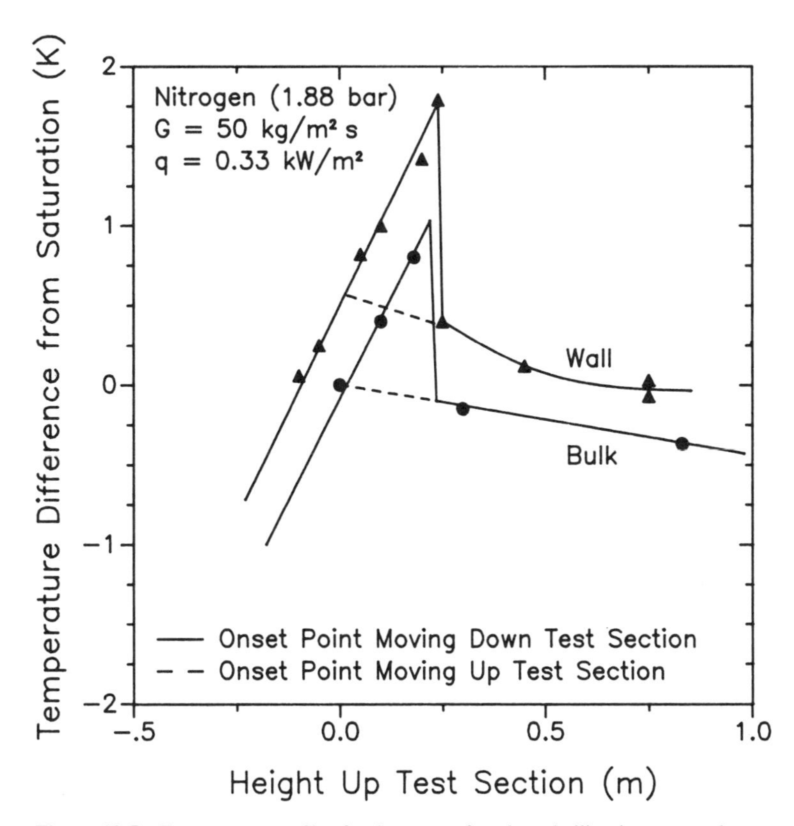

Figure 13-5 Temperature profiles for the onset of nucleate boiling in a serrated test section.

When the point of onset of boiling was moved up the test section by increasing the inlet subcooling or decreasing the heat flux, the temperature overshoots were no longer observed (as noted by the short dashed lines in Fig. 13-5). Therefore, the bubbles in the short nucleate boiling region after the onset of boiling observed by Carey and co-workers (described in Sec. 13-3) probably acted to "seed" new boiling nucleation sites along the test section as the point of onset moved higher up the channel.

Robertson and Clarke (1981) also reported a strong influence of heat flux and pressure on the wall superheat necessary for the onset of nucleate boiling in serrated and perforated fin geometries. The nucleation superheats (the difference between the peak in Fig. 13-5 and the saturation temperature) were found to be roughly proportional to the square root of the heat flux. At a given heat flux and liquid Reynolds number, the investigators observed that the superheat peaks were smaller at 5 bar than at 1.9 bar. Various plain tube nucleation expressions were found to be inadequate for describing the nucleation superheats in these tests.

13-5 FLOW BOILING STUDIES

There have been surprisingly few experimental studies on flow boiling in plate-fin heat exchangers compared to the large number of investigations on boiling inside enhanced boiling tubes (described in Chapter Ten). In flow boiling, it is important to distinguish between convective boiling and nucleate boiling. In convective boiling, evaporation occurs at the free vapor-liquid interface of a liquid film; in nucleate boiling, evaporation occurs at bubbles attached to the heated wall. These two processes are quite different, and the heat transfer coefficients in one regime are quite different from those in the other. Yet most test facilities do not provide for flow visualization, and thus the boiling regime is usually inferred from trends in the experimental heat transfer data themselves.

Table 13-1 gives a composite summary of the flow boiling investigations carried out with plate-fin geometries. Results of several of the more extensive studies are described below.

Boiling in Short Narrow Channels

Johannes and Mollard (1972) investigated boiling of liquid helium-I in individual short narrow channels and also in geometries with several short channels in parallel. Channel lengths varied from 10 to 20 mm. Flow in the vertical channels was induced by the thermosyphon effect of heating the channel walls. The investigators observed substantial effects of channel size and mass velocity on heat transfer. At a fixed mass velocity, a pronounced maximum in the heat dissipated to the fluid occurred at a channel size of 1.65 mm. In addition, thermal performance was found to decrease when both the front and back walls were heated compared to heating only one side. For instance, reductions of up to 50% occurred for the four different channel geometries relative to the same geometry with only one side heated. One other interesting result was that the variation in the Nusselt number with the product of the single-phase liquid inlet Reynolds number and the local vapor quality produced parallel lines for different channel heights. Thus the investigators were able to correlate their data to within 20% by use of the following two expressions:

For $L/D < 50$

$$\mathrm{Nu}_D = \frac{\alpha D}{\lambda_L} = 5.5\mathrm{Re}_{Di}^{0.8} x^{0.8} (Y/D)^{-0.8} \tag{13-1}$$

For $L/D > 50$

$$\mathrm{Nu}_D = \frac{\alpha D}{\lambda_L} = 0.24\mathrm{Re}_{Di}^{0.8} x^{0.8} \tag{13-2}$$

These expressions for the local boiling heat transfer coefficient are valid for vapor qualities from about 0.01 to 0.70 for helium-I at 1 atm. The hydraulic diameter was used for defining both the Nusselt and inlet Reynolds numbers. L is the length of the vertically heated channel, and Y is the distance from the entrance.

Boiling in Plate-Fin Geometries

An extensive flow boiling study on plate-fin heat exchangers was performed by Panitsidis, Gresham, and Westwater (1975). They tested a serrated fin design (see Table 13-1 for the specifications). Their test unit was composed of 22 parallel plates with a core size of 825 × 825 × 787 mm. Saturated liquid entered under thermosyphon action from the bottom, and steam entered from the side and condensed during two horizontal passes. Because no local wall temperatures were measured, total heat duties were determined over a range of condensing temperatures. The test fluids were isopropanol and R-113. The variation in the total heat duty with mean temperature difference between the condensing fluid and the evaporating fluid was found to be nearly linear over a wide range of mean temperature differences. The investigators reported extremely high heat duties per unit volume, up to 101,000 and 76,000 kW/m^3 for isopropanol and R-113, respectively, which are much larger than those for conventional shell-and-tube units under similar operating conditions.

Galezha, Usyukin, and Kan (1976) completed a comprehensive study on the boiling of R-12 and R-22 in vertical plate-fin heat exchangers with four different serrated fin geometries and one plain fin geometry. Complete specifications of the fins are given in Table 13-2. The test units were made of corrugated aluminum elements soldered between two 1-mm thick aluminum parting sheets. The units varied from 200 to 220 mm in height and were all 40 mm wide. Liquid was supplied at the bottom from a condenser, and the unit operated as a thermosyphon. Heating was applied by electric heaters on both parting sheets, and thermocouples were located in the parting sheets to obtain five local temperature measurements along the flow channel.

Figure 13-6 shows the data for boiling on the five different fin geometries with R-22 and R-12 and also on one surface with R-22 at four different saturation temperatures. The heat flux densities were determined from the total wetted surface area of each unit, and the heat transfer coefficients shown are mean values for the entire unit. For R-22 at 0°C, fin geometry 4 gave considerably better performance

Table 13-1 Plate-fin boiling studies

Study	Plate-fin geometry and $L/H/t$/fpm*	Fluid(s)	Pressure or temperature	Mass flux, kg/(m^2·s)	Superheat, K	Orientation (heating)	Type of data
Johannes and Mollard (1972)	Single channels, 100 to 200/0.37 to 4.34/40;	Helium-I	1.01 bar	10 to 30	0.02 to 1.0	Vertical (one or two sides heated)	Local Values
	plain, 100 to 200/0.35, 0.6, and 1.65†	Helium-I	Same	Same	Same	Same	Same
Collier, Kennedy, and Ward (1974)	Plain, 1115/6.35/0.2/590	Nitrogen	2.5 to 4.2 bar	14 to 130	0.06 to 1.0	Vertical	Pressure drops only
Shorin et al. (1974)	Plain, 1625/6/0.22/222	Oxygen	1.2 to 1.6 bar	0.03 to 0.35 (m/s)	600 to 4,500 (W/m^2)	Vertical (one side heated)	Correlated data only
Panitsidis, Gresham, and Westwater (1975)	Serrated, 4.67/3.88/0.1/641	R-113 Isopropanol	1.01 bar Same	0 to 72 Same	0 to 80 Same	Vertical (two sides heated)	Average values
Galezha, Usyukin, and Kan (1976)	Serrated, 10/6/0.25/351, 1.5/6/0.25/206,1.0/4/0.25/351, and 1.5/6/0.25/351;	R-12 R-22	0°C -20°, -10°, 0°, and +10°C	0 to 16 Same	0.6 to 5 Same	Vertical (two sides heated)	Average values
	Plain, —/6/0.25/206	R-12 R-22	Same Same	Same Same	Same Same	Same Same	Same Same
Robertson (1979)	Serrated, 3.18/6.35/0.2/591	Nitrogen	2 to 7 bar	11 to 110	0.3 to 2	Vertical (two sides heated)	Local values
Robertson and Lovegrove (1980, 1983)	Serrated, 3.18/6.35/0.2/591	R-11	3 to 7.2 bar	34 to 159	0.3 to 2	Vertical (two sides heated)	Local values

Yung, Lorenz, and Panchal (1980)	Serrated, 3.18/4.48/0.4/590	Ammonia		12 to 28	0 to 3	Vertical (two sides heated)	Local values
Robertson (1983)	Perforated (5%), —/6.35 /0.2/591	Nitrogen	2 to 7 bar	11 to 110	0.3 to 2	Vertical (two sides heated)	Local values
Chen and Westwater (1984)	Serrated, 4.7/3.9/0.1/629, 3.2/2.5 /0.2/709, and 3.2/2.5/0.2 /629	R-113	1.01 bar	14 to 750	15 to 120	Vertical (two sides heated)	Average values
Carey and Mandrusiak (1986)	Serrated, 12.7/3.8/1.6/105	Water	1.01 bar	6 to 26.5	0 to 15	Vertical (one side heated)	Local values
		Methanol	Same	9 to 35	Same		
		n-Butanol	Same	10 to 36	Same		
Mandrusiak, Carey, and Xu (1986)	Serrated, 12.7/3.8/1.6/105	Water	1.01 bar	4.5 to 32	0 to 15	Horizontal (top or bottom heated)	Local values
		Methanol	Same	8.7 to 34	Same		
Mandrusiak, Carey, and Xu (1988)	Serrated, 12.7/9.5/1.9/98 and 12.7 /1.9/1.6/105	Water	1.01		0 to 15	Vertical (one side heated)	Local values
		Methanol	Same	14 to 62	Same		
		n-Butanol	Same	50 to 320	Same		
		R-113	Same		Same		
Carey, Mandrusiak and Roddy (1987)	Serrated, 3.4/5.1/0.15/969, 3.2/7.9 /0.15/401, and 1.6/3.9 /0.20/280	Water-air		30 to 60		Vertical	Correlated pressure drops only
Robertson and Wadekar (1988)	Perforated (5%), 3000 /6.35/0.2/591	Cyclohexane	1.5 bar	50 to 290	1,000 to 10,000 (W/m^2)	Vertical (two sides heated)	Local values

[*] L, serration length (mm) or length of nonserrated test sections; H, fin height (mm); t, fin thickness (mm); and fpm, fins per meter.
[†] t and fpm not cited.

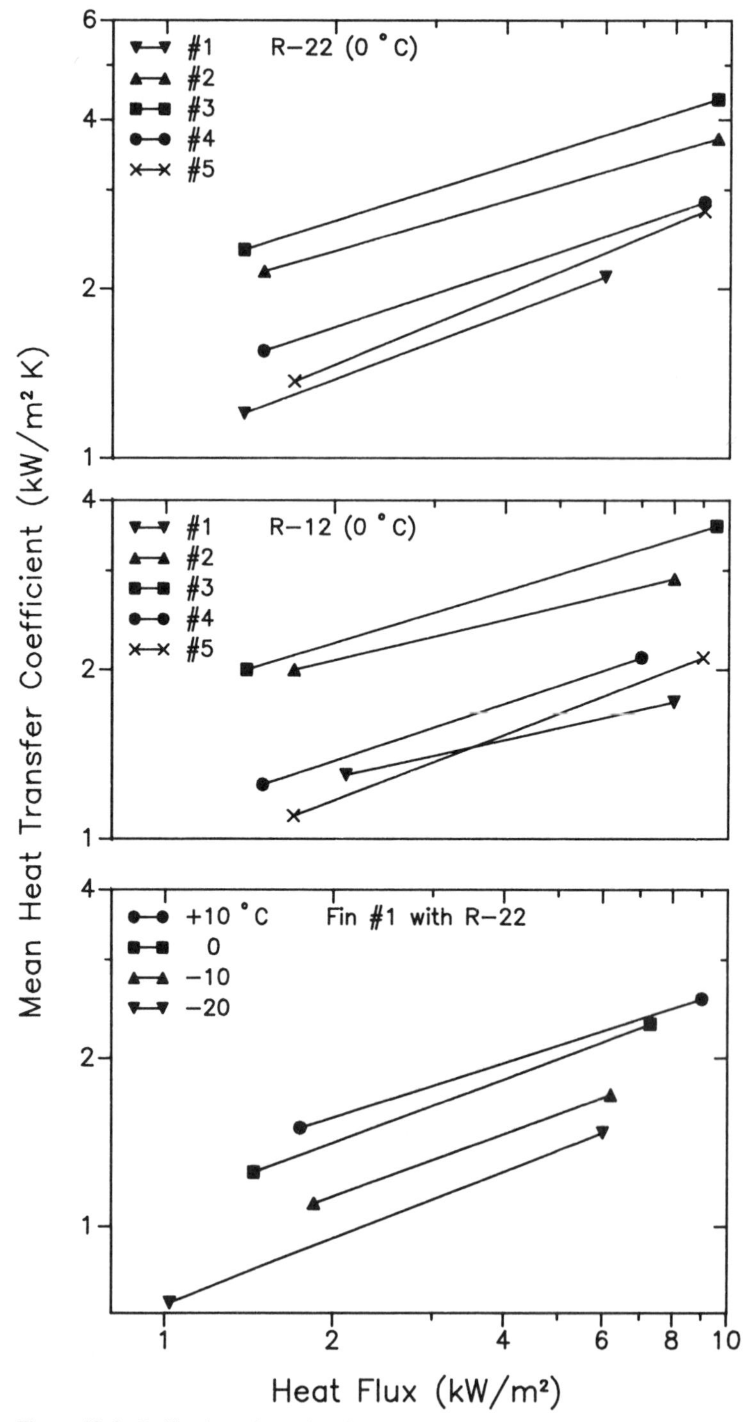

Figure 13-6 Boiling in various plate-fin geometries by Galezha, Usyukin, and Kan (1976).

Table 13-2 Plate-fin geometries tested by Galezha, Usyukin, and Kan (1976)

Fin number	Fin type	Height, mm	Length, mm	Pitch, mm	Thickness, mm	Hydraulic diameter, mm	Area/ volume, m^2/m^3
1	Serrated	6	10	2.6	0.25	3.1	1,170
2	Serrated	6	1.5	4.6	0.25	4.6	781
3	Serrated	4	1.0	2.6	0.25	2.5	1,387
4	Serrated	6	1.5	2.6	0.25	3.1	1,084
5	Plain	6		4.6	0.25	4.6	791

than the other geometries. Fin 4 had the shortest serration length and the smallest fin height of those tested. Serrated fin 2 out-performed its identical plain fin (5) by about 30% to 50%. Serrated fin 1, which had long fins and a smaller channel width than the plain fin design, actually performed worse than the plain fin. The results for R-12 at 0°C followed the same trend as those of R-22. The boiling curves for the different serrated fin geometries were essentially all parallel to one another, but the slope for the plain fin was slightly higher for both test fluids.

The bottom plot in Fig. 13-6 demonstrates that there was a strong effect of pressure on the R-22 boiling performance of fin 1 (the pressures corresponding to the saturation temperatures shown were 6.8, 5.0, 3.5, and 2.4 bar, respectively). Boiling performance improved with increasing pressure. All the data shown in Fig. 13-6 were taken for mass fluxes less than 16 kg/(m^2·s), however, so that the effect of mass flux on heat transfer was not considered.

Robertson (1979) and Robertson and Lovegrove (1980, 1983) measured local boiling heat transfer coefficients for liquid nitrogen and R-11 in vertical upward flow in a serrated fin test unit over a range of vapor qualities, heat and mass fluxes, and pressures (their test unit is described in Sec. 13-4, and the fin geometry is cited in Table 13-1). Figure 13-7 (from Robertson [1979]) depicts a cross-plot of their data for liquid nitrogen boiling at a nominal pressure of 3 bar, showing the local heat transfer coefficient as a function of mass flux for various fixed vapor qualities. Their single-phase liquid heat transfer data curve for the same test section is also shown.

The boiling heat transfer coefficient did not vary with mass flux at low vapor qualities in this study. For each curve for a fixed vapor quality, there was an inflection point at which the slope increased to about 0.8. The investigators hypothesized that the flow pattern was characterized by a thin liquid film flowing along the periphery of the narrow channels, which could be either laminar or turbulent depending on the local conditions. The inflection point was thus explained as a transition from laminar to turbulent flow in the liquid film.

The curve for vapor quality approaching zero does not tend to the single-phase liquid curve. According to Carey (1985), this unexpected result can be explained by the existence of stationary bubbles in the flow passages. Robertson and Wadekar (1988) presented a different physical explanation. They hypothesized that at vapor qualities between 0 and 0.05 the bubbles divide the flow into liquid slugs that undergo a thermal "entrance" effect each time a new serration is passed. Thus the

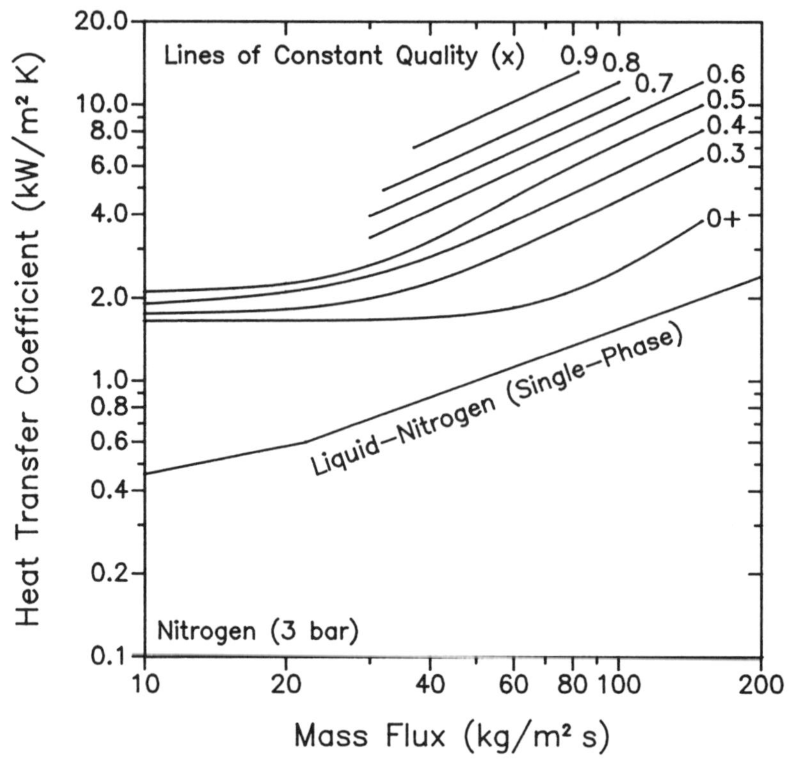

Figure 13-7 Flow boiling in a serrated plate-fin heat exchanger.

local heat transfer coefficients are similar to those at the leading edge of a plate and much larger than those for single-phase flow conditions.

A strong effect of vapor quality on the heat transfer coefficient is evident in Fig. 13-7 when the curves are compared at a fixed mass flux. Therefore, the heat transfer coefficient tended to increase along the length of an exchanger. Performance improved with increasing vapor quality up to qualities as high as 0.9.

Chen and Westwater (1984) compared the boiling performances of three plate-fin cores, each with a similar but unique serrated fin geometry obtained from three different manufacturers (see Table 13-1 for their specifications). Using condensing steam for heating, the investigators obtained overall heat duties for R-113 together with pressure drop measurements. Running tests with boiling wall superheats varying between 15 to 120 K, they observed that the two designs with shorter and thicker fins provided the best thermal performance at superheats greater than about 30 K; the other fin design gave better results at lower superheats. At high heat duties the flow became oscillatory in the first two fin designs, apparently as dryout began to occur. Pressure drops increased with increasing heat duty and exit velocity, ranging from about 3 to 14 kPa for the test conditions.

Carey and co-workers ran boiling tests on serrated fin test sections in both vertical and horizontal positions with only one side of the flow channel heated. Figure 13-4 depicts their test unit (see Table 13-1 for the fin dimensions studied). Figure 13-8 (from Mandrusiak and Carey [1988]; also shown in Carey and Shah [1988]) depicts the variation in the local boiling heat transfer coefficient for methanol as a function of local vapor quality at several mass fluxes. At the two lowest mass fluxes vapor qualities as high as 0.9 were attained, and a maximum in the local heat transfer coefficient occurred along the test section at a vapor quality of about 0.7. Partial dryout was observed at the two highest vapor qualities. The influence of increasing mass flux was to increase the heat transfer coefficient, as was seen in the Robertson (1979) data for liquid nitrogen. The data indicate, however, that there may be an upper limit to performance with increasing mass flux at about 49 kg/($m^2 \cdot s$). For boiling in a horizontal position, heating only the bottom side produced heat transfer coefficients marginally higher than those for top heating at the same test conditions. The effect of surface orientation, vertical or horizontal, was also observed to be small for both water and methanol.

To estimate the boiling augmentation afforded by their horizontal serrated fin unit compared to that of a plain tube with the same hydraulic diameter, Mandrusiak, Carey, and Xu (1986) evaluated the Shah (1976) correlation for horizontal flow over the same

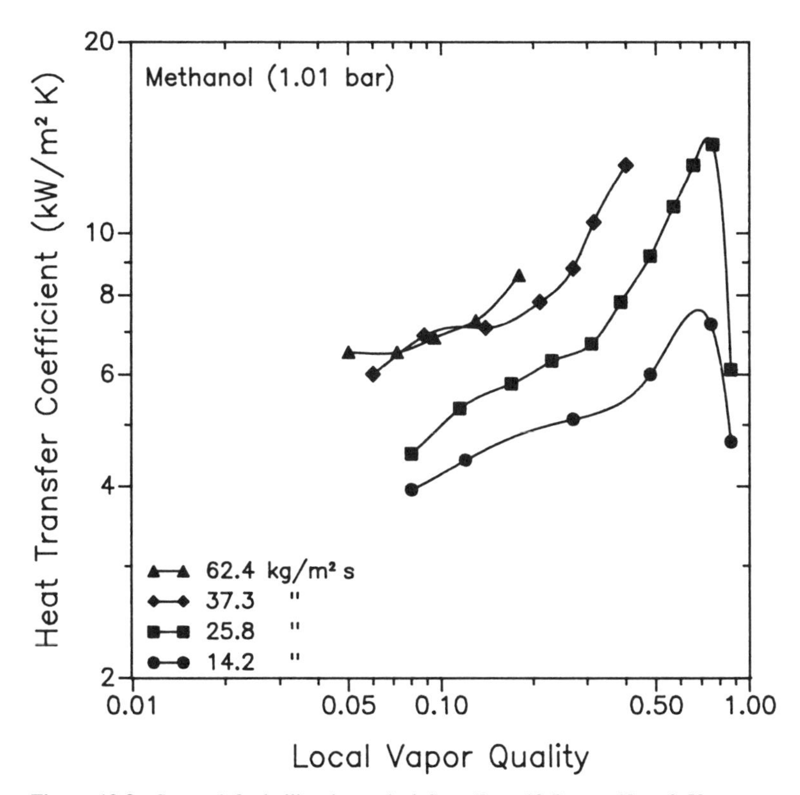

Figure 13-8 Serrated fin boiling in vertical flow (L = 12.7 mm, H = 9.52 mm, t = 1.91 mm, and fpm = 98).

heat flux and mass flux range as used for their methanol data. They observed that their test unit produced heat transfer coefficients about two to seven times those of the equivalent plain tube. This may not be indicative of the performance with smaller serrated fins with both sides heated, however. In general, the plate-fin heat transfer coefficients were substantially less than those obtainable with boiling on the outside of enhanced boiling tubes, as discussed in Chapter Seven.

Robertson and Wadekar (1988) ran a series of upflow boiling tests in a vertical plate-fin unit with perforated fins (see Table 13-1 for specifications) using cyclohexane at low heat fluxes. Test conditions covered mass fluxes of 50, 105, and 290 kg/(m^2·s) at a nominal pressure of 1.5 bar. For comparison purposes, similar tests were run with a vertical smooth tube 10.7 mm in diameter and 3.0 m long. The investigators observed distinct similarities between the results from the two different test sections. For instance, at the highest mass flux a Chen (1963) type of correlation fit both convective boiling data sets well. At low mass fluxes, however, the convective boiling heat transfer coefficients of the perforated fin test section were completely different, apparently because of the low Reynolds numbers produced by their small hydraulic diameters. Thus under these conditions the heat transfer process was characterized by bubble and slug flow patterns dominated by laminar flow. Therefore, the investigators concluded that in plate-fin passages the Reynolds number based on the total mass flux is of profound importance for determining the type of heat transfer occurring (laminar or turbulent).

In summary, a limited amount of boiling data is available in the open literature for plate-fin heat exchanger geometries. From these studies, it can be concluded that the local heat transfer coefficient under boiling conditions is a function of mass flux, vapor quality, pressure, and fin geometry. Further research is needed to determine the optimum fin configuration from a heat transfer–pressure loss–cost effectiveness standpoint. Work with mixtures is required to study the degradation in the boiling heat transfer coefficient with the addition of one or more components (refer to Chapter Nine for a review of the research on mixtures evaporating on enhanced boiling tubes) so that these units can be reliably evaluated as possible alternatives to shell-and-tube reboilers for clean operating conditions.

13-6 PLATE-FIN BOILING CORRELATIONS

Plain Fin Correlations

Collier, Kennedy, and Ward (1974) boiled liquid nitrogen in a vertical plate-fin heat exchanger 3.5 m long and 75 mm wide with plain fins (refer to Table 13-1 for the dimensions). They found that the Chen (1963) method described in Chapter Two for vertical round tubes gave reasonable results when their unit was divided into short segments and the total heat duty calculated by integration. This suggests that boiling performance is not augmented in plate-fin heat exchangers with plain fins but that only additional surface area is provided. The two-phase pressure drops in the test unit were adequately predicted by the following conventional method:

$$-\frac{dP}{dz} = \left(\frac{dP_f}{dz}\right)_L \phi_L^2 + G^2\frac{d}{dz}\left[\frac{(1-x)^2}{\rho_L(1-a)} + \frac{x^2}{\rho_v a}\right] + g\,[(1-a)\,\rho_L + a\rho_v] \quad (13\text{-}3)$$

Round tube expressions for the two-phase multiplier and void fraction were utilized and predicted the data to within about 25%. The investigation covered exit vapour qualities ranging from 0 to 0.7, mass fluxes varying from 14 to 62 kg/(m^2·s), and inlet pressures between 2.5 to 4.2 bar.

In another plain fin study, Shorin et al. (1974) presented results for liquid oxygen evaporating in a vertical plate-fin heat exchanger (see Table 13-1 for the fin dimensions). The length of the plate-fin channels was 1.625 m. Their correlation for boiling oxygen in 12-mm diameter, round vertical tubes 1.46 and 2.94 m long also adequately predicted their plate-fin data. Thus a plain fin geometry was again found not to augment heat transfer but only to increase the available surface area. The expression for the unit's heat duty Q was given as

$$\frac{Q}{\dot{m}\Delta H} = 1 + 0.15\mathrm{Re}_{Di}^{1.1}\mathrm{Re}_b^{-0.75}\left(\frac{L}{D}\right)^{-1.1} \quad (13\text{-}4)$$

where L is the length of their test unit, D is the hydraulic diameter, and the maximum enthalpy change is

$$\Delta H = c_{PL}\,(T_w - T_{sat}) + x\Delta h_v \quad (13\text{-}5)$$

where x is the outlet quality, T_w is the average wall temperature, and T_{sat} is the saturation temperature at the inlet. The single-phase liquid Reynolds number at the inlet Re_{Di} is defined as

$$\mathrm{Re}_{Di} \equiv \frac{u_i D}{\nu_L} \quad (13\text{-}6)$$

and the boiling Reynolds number Re_b is defined as

$$\mathrm{Re}_b \equiv \frac{qD}{\rho_v \Delta h_v \nu_L} \quad (13\text{-}7)$$

Equation (13-5) represents the maximum heat transfer rate thermodynamically possible (i.e., with an exit state at the wall temperature). Q is the actual heat duty obtained. Thus the ratio in Eq. (13-4) should always be less than or equal to 1.0.

In summary, two different studies have shown plain tube boiling correlations to describe successfully evaporation in vertical plate-fin heat exchangers with plain fins. Only two fluids in two fin sizes were tested, however, and the heat duty expression of Shorin et al. (1974) apparently is only applicable to their test unit.

Serrated Fin Correlations

Panitsidis, Gresham, and Westwater (1975) developed a method for predicting heat transfer in serrated fin type plate-fin heat exchangers with nucleate pool boiling data.

As a first approximation, they assumed that the flow in the channels had no effect on the heat transfer process. Hence they proposed to predict the boiling performance by using the pool boiling curve for the same fluid at the same operating conditions. This essentially assumes that the heat transfer coefficient for flow boiling in a plate-fin channel is defined by the local wall superheat alone. The investigators called this the "local assumption" method. To include the effect of fin efficiency each fin was divided into 26 nodes, and the local heat transfer coefficient at each node was then obtained with the wall superheat at that node. Summing the totals for all the fins provided the heat duty of the unit. This simple approach predicted the results for R-113 and isopropanol described above (Sec. 13-5) with considerable accuracy for wall superheats from about 10 to 50 K. At larger wall superheats, the method substantially unpredicted actual performance.

To improve this method's accuracy, Chen, Loh, and Westwater (1981) and Chen and Westwater (1984) included the effect of velocity on the local boiling heat transfer coefficient. This was accomplished by obtaining boiling curves for upward flow normal to a single horizontal heated cylinder. The necessary data for R-113 at atmospheric pressure for velocities from 0 to 10 m/s were taken by Yilmaz and Westwater (1980). These data were then utilized to predict the local heat transfer coefficients over a fin with the use of the local wall superheat and local two-phase velocity. The flow velocities in the plate-fin core were determined from a homogenous flow assumption for the vapor quality at the preceding row of fins. The inclusion of the velocity effect improved the accuracy of the local assumption method, especially at large wall superheats.

Although the local assumption method is advantageous because of its simplicity, there are several drawbacks. First of all, suitable single tube boiling data for the fluid of interest at the design pressure are not normally available. In addition, the effects of the particular serrated fin's geometry on boiling augmentation are ignored, except for the change in fin efficiency. Finally, the boiling process in serrated fin geometries is dominated by film flow evaporation rather than by nucleate boiling, especially at low superheats.

Galezha, Usyukin, and Kan (1976) also developed a serrated plate-fin correlation based on a nucleate pool boiling type of expression; they used their data for five different fin geometries for R-12 and R-22 (described above). The functional dependency of the boiling heat transfer coefficient on heat flux was determined by fitting the data by least squares to the expression

$$\overline{\alpha} = Aq^n \tag{13-8}$$

for each different fin geometry. The exponent n was found to be approximately equal to 0.33 for all the geometries, which agreed with other data for these fluids boiling in a 1-mm wide flat channel (this slope is much smaller than for nucleate pool boiling on a single tube). The empirical constant A varied with refrigerant, saturation temperature, and fin size and type. Using the thermodynamic similitude considerations proposed by Borishanskiy (1969) to include fluid property effects, the investigators arrived at the following correlation:

$$\overline{\alpha} = C\left(\frac{P_{cr}^{1/3}}{T_{cr}^{5/6}M^{1/6}}\right)\left(0.65 + 3.3\frac{P}{P_{cr}}\right)q^{1/3} \tag{13-9}$$

where the constant C is equal to 8,300, 12,850, 16,200, 10,120, and 9,880 for each fin in Table 13-2, respectively. The expression also includes a pressure correction term. The investigators were able to predict their data to within about 15% by using this approach. The ratio of the coefficients C for the serrated fin geometries to the plain fin values gives the boiling augmentation factor relative to a plain fin. The correlation in practice, however, requires C to be determined experimentally for at least one fluid for the specific fin geometry under consideration. The method does not allow local heat transfer coefficients to be determined because it was developed from mean values, and it does not include the effect of mass velocity.

Robertson (1982) studied the problem of correlating boiling heat transfer coefficients in serrated plate-fin heat exchangers with the use of a film flow boiling model. Essentially, he assumed a separated flow model to describe the boiling process with a thin film of liquid flowing upward along the channel walls around a vapor core. From his boiling data (described above), Robertson concluded that the boiling process was controlled by the convection of heat across the liquid film to the free interface, where evaporation occurs, and in the absence of nucleate boiling effects. The problem then becomes similar to that of convective film condensation (i.e., predicting the local thickness of the evaporating liquid film under the influence of interfacial and wall shear stresses). To do this, Robertson utilized a universal velocity profile in the liquid film.

Robertson (1984) and Robertson and Wadekar (1988) later presented an improved version that utilized a "double velocity profile." They also divided the flow into laminar and turbulent flow regimes to differentiate the boiling performance on the basis of Reynolds number. Complete details of the methods were not given, however. Yung, Lorenz, and Panchal (1980) also proposed a film flow model for serrated fin geometries on the basis of their experimental work with ammonia.

A relatively simple method for predicting the local heat transfer coefficients in serrated fin geometries was developed by Carey (1985). Evaluating the boiling data of Robertson (1979) and Robertson and Lovegrove (1980, 1983) for R-11 and nitrogen, Carey made two observations. First, at high vapor qualities and moderate to large flow rates, the trends in the data were consistent with the two film flow models described above. Second, at low vapor qualities and low flow rates the boiling data did not tend toward the single-phase liquid data obtained in the same studies, as one would expect. As can be noted in Fig. 13-7, the values for the heat transfer curve with a vapor quality approaching zero (0+) are much larger than those of the single-phase flow curve. Carey explained this anomaly as being the result of a surface tension effect on slug flow, which holds up the bubbles at the smallest flow passage locations under certain conditions. Thus the heat transfer process should be analyzed by two different flow models, one for essentially single-phase flow interrupted by stationary bubbles at low flow rates and vapor qualities and a second one for film flow conditions.

On the basis of a surface tension controlled flow model for low flow rates and vapor qualities, Carey (1985) developed the flowing expression for the minimum heat transfer coefficient (which describes the flat portion of the curves in Fig. 13-7):

$$\alpha_{\min} = \alpha_{\mathrm{L}} \mathrm{Pr}_{\mathrm{L}}^{0.296} (1.0 + 0.34\Omega^{0.68}) \tag{13-10}$$

The liquid-only heat transfer coefficient is calculated with the Wieting (1975) correlation for single-phase turbulent flow in serrated fin geometries and is given as

$$\alpha_{\mathrm{L}} = 0.242 \left(\frac{L}{D_{\mathrm{h}}}\right)^{-0.322} \left(\frac{t}{D_{\mathrm{h}}}\right)^{0.089} \left(\frac{\lambda_{\mathrm{L}}}{D_{\mathrm{h}}}\right) \mathrm{Pr}_{\mathrm{L}}^{1/3} \mathrm{Re}^{0.632} \tag{13-11}$$

The Reynolds number Re was based on the total mass flux (liquid and vapor) and was thus defined as

$$\mathrm{Re} \equiv \frac{G D_{\mathrm{h}}}{\mu_{\mathrm{L}}} \tag{13-12}$$

The flow parameter Ω was obtained from a force balance on a stationary bubble held in place by surface tension and is given by the following expression:

$$\Omega = \left[\frac{2\mathrm{Re}^{0.198}}{\gamma_{\mathrm{I}} \mathrm{We}} \left(\frac{D_{\mathrm{h}}}{D_{\mathrm{hs}}}\right) (1 - 0.25\mathrm{Bo})\right]^{1/2} \tag{13-13}$$

where the Wieting parameter is

$$\gamma_{\mathrm{I}} = 1.136 \left(\frac{L}{D_{\mathrm{h}}}\right)^{-0.781} \left(\frac{t}{D_{\mathrm{h}}}\right)^{0.534} \tag{13-14}$$

The Weber number We is defined as

$$\mathrm{We} \equiv \frac{G^2 D_{\mathrm{hs}}}{\sigma \rho_{\mathrm{L}}} \tag{13-15}$$

The Bond number Bo is defined as

$$\mathrm{Bo} \equiv \left(\frac{\rho_{\mathrm{L}} - \rho_{\mathrm{v}}}{\sigma}\right) g D_{\mathrm{hs}}^2 \sin\theta \tag{13-16}$$

In these expressions, two different hydraulic diameters have been used. D_{h} refers to the hydraulic diameter of the passages between the fins, and D_{hs} is the hydraulic diameter at the smallest cross-section of the flow channel at the end of the fin. L is the length of the fin, and t is its thickness.

For the film flow regime, Carey (1985) developed another correlation. Starting with the pure fluid version of the Bennett and Chen (1980) correlation but neglecting the nucleate boiling contribution, the convective heat transfer coefficient is given as

$$\alpha = \alpha_{L} \mathrm{Pr}_{L}^{0.296} F\,(1/X_{tt}) \tag{13-17}$$

where the liquid-only heat transfer coefficient here is calculated from Eq. (13-11) with the local liquid mass flux rather than the total mass flux. The Martinelli parameter is that of Wieting (1975) for turbulent flow:

$$\frac{1}{X_{tt}} = \left(\frac{\nu_{v}}{\nu_{L}}\right)^{0.099} \left(\frac{\rho_{L}}{\rho_{v}}\right)^{0.401} \left(\frac{x}{1-x}\right)^{0.901} \tag{13-18}$$

The Chen parameter F was modified to fit the Robertson R-11 and nitrogen data for serrated fins. Its new expression is

$$F = 1.65\left(\frac{1}{X_{tt}} + 0.60\right)^{0.78} \quad \text{for} \quad \frac{1}{X_{tt}} > 0.1 \tag{13-19}$$

To predict the local heat transfer coefficient, one evaluates both correlations. The actual heat transfer coefficient is the larger of the two.

Figure 13-9 shows a comparison of this method to the Robertson (1979) nitrogen data at several mass fluxes. The inflection point in the heat transfer coefficient is correctly predicted to move to the left as the mass flux increases, which is characteristic of both the R-11 and the nitrogen data. This method was found to correlate the R-11 and nitrogen data with a mean absolute deviation of 14%.

On the basis of their own serrated (offset fin) data for a test unit with heating on only one side (described above), Carey and Mandrusiak (1986) developed a new film flow model similar to Robertson (1982) for serrated fins and that of Bennett and Chen (1980) for round tubes. Their general expression for the ratio of the boiling heat transfer coefficient to the liquid-only heat transfer coefficient is

$$\frac{\alpha}{\alpha_{L}} = \frac{\mathrm{Re}_{L} \mathrm{Pr}_{L}^{1/6} (D_{hp}/D_{h})^{n/2}}{4.74 A^{1/2} \tan^{-1}\left[0.149 \mathrm{Re}_{L}^{1/2} \mathrm{Pr}_{L}^{1/2} (D_{h}/D_{hp})^{1/2}\right]} \left(1 + \frac{20}{X_{tt}} + \frac{1}{X_{tt}^{2}}\right)^{1/2} \tag{13-20}$$

where the liquid-only Reynolds number Re_{L} was defined as

$$\mathrm{Re}_{L} \equiv \frac{G(1-x)D_{h}}{\mu_{L}} \tag{13-21}$$

Two different hydraulic diameters were utilized: D_{h} is based on the wetted perimeter, and D_{hp} is based on the heated perimeter. The Lockhart and Martinelli (1949) correlation for turbulent-turbulent flow in round tubes was used for the two-phase multiplier. The Martinelli parameter for serrated fin flow was derived to be

$$\frac{1}{X_{tt}} = \left(\frac{\rho_{L}}{\rho_{v}}\right)^{1/2} \left(\frac{\mu_{v}}{\mu_{L}}\right)^{n/2} \left(\frac{x}{1-x}\right)^{1-(n/2)} \tag{13-22}$$

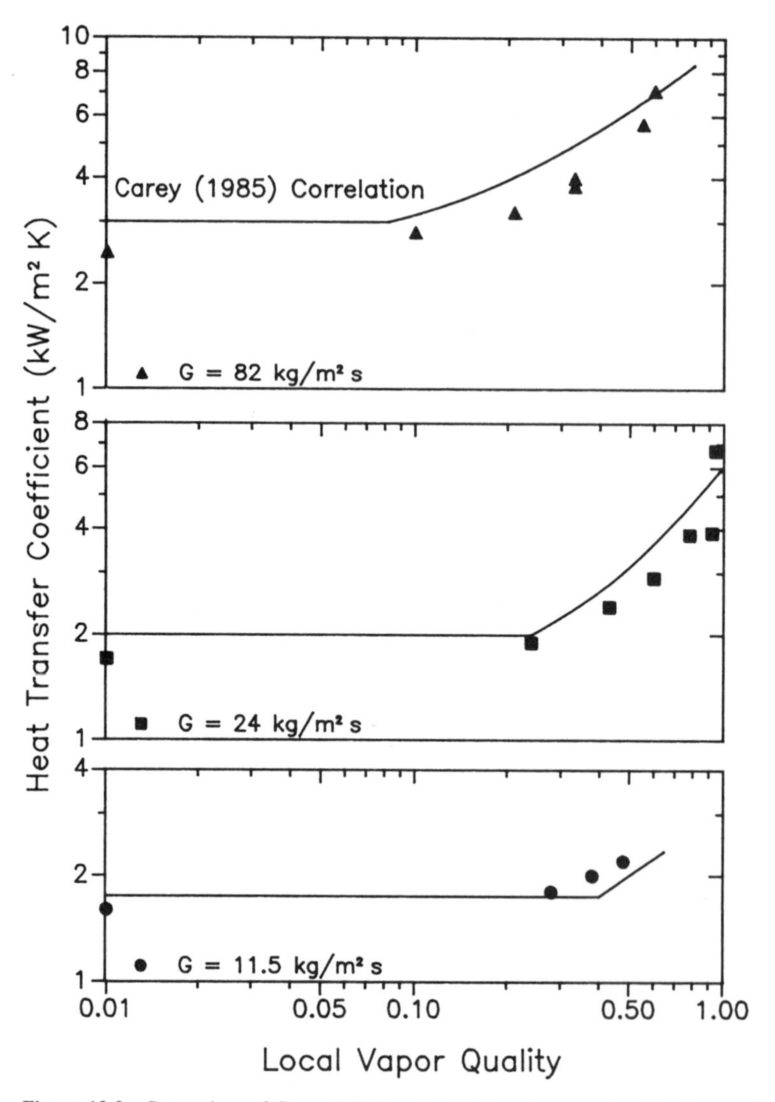

Figure 13-9 Comparison of Carey (1985) method to the nitrogen data of Robertson (1979).

The liquid-only heat transfer coefficient was calculated by relating the modified Reynolds analogy to the Colburn j factor for a partially heated channel to obtain the expression

$$\mathrm{St_L Pr_L^{2/3}} = A\mathrm{Re_L^{-n}} \tag{13-23}$$

A and n were obtained by fitting the above expression to the single-phase data. The values of A and n were determined to be 0.215 and 0.36, respectively. The Stanton number was defined as

$$\mathrm{St_L} \equiv \frac{\alpha_\mathrm{L}}{G(1-x)c_{P\mathrm{L}}} \tag{13-24}$$

The Carey and Mandrusiak (1986) correlation accurately predicted their upward vertical data for water, methanol, and n-butanol. In Mandrusiak, Carey, and Xu (1986), the method was also shown to give satisfactory results for water and methanol boiling in the same unit in a horizontal position when no dryout occurred. In Carey and Shah (1988), however, this method was shown to be unsatisfactory for the R-11 data of Robertson and Lovegrove (1980) and the nitrogen data of Robertson (1979), for which pressures ranged from 2 to 7 bar and the fin geometry was much different. For these data, the following curve gave the best results in Eq. (13-20):

$$F = \left(1 + \frac{3.53}{X_\mathrm{tt}^{1/2}} + \frac{1.05}{X_\mathrm{tt}^2}\right)^{1/2} \tag{13-25}$$

Perforated Fin Correlations

Boiling in perforated fin geometries is less complicated than with serrated fins because the liquid film on the channel walls is only interrupted by the perforations. Robertson (1983) proposed a film flow model based on his data for nitrogen evaporating in a vertical perforated fin test unit. His analysis was similar to that developed for annular flow in round tubes. His correlating expression is given as

$$\frac{\alpha}{\alpha_\mathrm{L}} = 7.0X^{-0.36} \tag{13-26}$$

where the Martinelli parameter is

$$X = \left[(dP/dz)_\mathrm{L}/(dP/dz)_\mathrm{v}\right]^{1/2} \tag{13-27}$$

The liquid-only heat transfer coefficient for perforated fins in the above correlation is calculated with a suitable single-phase perforated fin correlation. (In Robertson [1979], the liquid-only heat transfer coefficients for the perforated fin were only about half those of serrated fins of the same dimensions but were similar to those of plain fins.)

Robertson and Wadekar (1988) demonstrated that the original Chen (1963) correlation for tubes was inadequate for predicting the cyclohexane convective boiling data obtained for their perforated fin test section. It substantially underpredicted the results. At high mass fluxes, however, the data were able to be correlated by using a modified F expression; this was similar to previous work by Carey and co-workers (described above). In addition, Robertson and Wadekar (1988) showed that the data could be described on the basis of the physics of the process, that is by plotting the climbing film Nusselt number against the liquid film Reynolds number, in which the film thickness was estimated from a theoretical model of evaporation from a laminar or turbulent liquid film. Using a "double" velocity profile gave better results than using a "single" asymptotic velocity profile in the film.

Summary

Experimental data and predictive methods are now available for two-phase flow patterns, the onset of nucleate boiling, and boiling heat transfer in plate-fin heat exchangers of various geometries. These methods have not been widely tested against independent sets of data, however. Little has apparently been done to predict two-phase pressure drops, in serrated fin units especially. In addition, the occurrence of dryout in plate-fin heat exchangers must be studied further to develop design guidelines either to avoid it or to predict the lower heat transfer performance at which it occurs, as in the work initiated by Chen and Westwater (1984). The effects of mixtures, both oil contamination of a refrigerant and multicomponent mixtures, on heat transfer must be investigated to widen the range of application of plate-fin designs and to determine whether and how much degradation occurs; this will be similar to the mixture work on enhanced boiling tubes discussed in Chapter Nine. In light of the current state of the art, it is recommended that manufacturers be consulted for the thermal-hydraulic performances of their plate-fin geometries for actual heat exchanger design.

13-7 NOMENCLATURE

a	void fraction
A	heat transfer surface area (m^2)
A	empirical constant in Eq. (13-8)
A	empirical constant in Eq. (13-23)
Bo	Bond number
c_{PL}	liquid specific heat [J/(kg·K)]
C	empirical constant in Eq. (13-9)
D	hydraulic diameter of narrow channel (m)
D_h	hydraulic diameter of serrated channel (m)
D_{hp}	hydraulic diameter based on heated perimeter (m)
D_{hs}	hydraulic diameter at small passage at end of fin (m)
F	Chen parameter
g	acceleration due to gravity (m/s^2)
G	total mass flux [kg/(m^2·s)]
Δh_v	latent heat of vaporization (J/kg)
H	fin height (mm)
ΔH	maximum enthalpy change (J/kg)
L	length of heated channel (m)
L	length of serrated fin (m)
$\dot{m}$	mass flow rate (kg/s)
M	molecular weight
n	exponent in Eq. (13-8)
n	exponent in Eq. (13-23)
Nu_D	Nusselt number
P	pressure (N/m^2)

P_{cr} critical pressure (N/m^2)
P_f liquid frictional pressure drop [(N/m^2)/m]
Pr_L liquid Prandtl number
q heat flux (W/m^2)
Q heat duty (W)
Re Reynolds number based on total mass flux
Re_b boiling Reynolds number
Re_{Di} Reynolds number at inlet
Re_L liquid-only Reynolds number
St_L liquid-only Stanton number
t fin thickness (m)
T_{cr} critical temperature (K)
T_{sat} saturation temperature (K)
T_w wall temperature (K)
u_i inlet liquid velocity (m/s)
U overall heat transfer coefficient [$W/(m^2 \cdot K)$]
We Weber number
x vapor quality
X Martinelli parameter defined in Eq. (13-27)
X_{tt} Martinelli parameter
Y distance from entrance of channel (m)
z axial distance along flow channel (m)
α boiling heat transfer coefficient [$W/(m^2 \cdot K)$]
$\overline{\alpha}$ average heat transfer coefficient [$W/(m^2 \cdot K)$]
α_L liquid-only heat transfer coefficient [$W/(m^2 \cdot K)$]
α_{min} minimum heat transfer coefficient [$W/(m^2 \cdot K)$]
γ_1 Wieting friction factor coefficient
θ angle between channel axis and vertical
λ_L liquid thermal conductivity [$W(m \cdot K)$]
μ_L liquid dynamic viscosity [$Kg/(m \cdot s)$]
μ_v vapor dynamic viscosity [$kg/(m \cdot s)$]
ν_L liquid kinematic viscosity (m^2/s)
ν_v vapor kinematic viscosity (m^2/s)
ρ_L liquid density (kg/m^3)
ρ_v vapor density (kg/m^3)
σ surface tension (N/m)
ϕ_L^2 liquid two-phase multiplier
Ω parameter defined in Eq. (13-13)

REFERENCES

ALTEC International. 1986. Brochure: Brazed aluminum plate-fin heat exchangers. November.

Bennett, D. L., and J. C. Chen. 1980. Forced convective boiling in vertical tubes for saturated pure components and binary mixtures. *AIChE J.* 26:454–61.

Borishanskiy, V. M. 1969. Correlation of effect of pressure on the critical heat flux and heat transfer rates using the theory of thermodynamic similarity. In *Problems of heat transfer and hydraulics of two-phase media,* ed. S. S. Kutateladze, 16–37. Oxford: Pergamon Press.

Carey, V. P. 1985. Surface tension effects on convective boiling heat transfer in compact heat exchangers with offset strip fins. *J. Heat Transfer* 107:970–74.

———. and G. D. Mandrusiak. 1986. Annular film-flow boiling of liquids in a partially heated, vertical channel with offset strip fins. *Int. J. Heat Mass Transfer* 29:927–39.

———. and T. Roddy. 1987. Analysis of the heat transfer performance of offset strip fin geometries in a cold plate operating as part of a two-phase thermosyphon. *Proc. ASME-JSME Int. Symp. Cool. Technol. Electron. Equip.* 565–582.

Carey, V. P., and R. K. Shah. 1988. Design of compact and enhanced heat exchangers for liquid-vapor phase-change applications. In *Two-phase flow heat exchangers: Thermal hydraulic fundamentals and design,* ed. S. Kakac, A. E. Bergles, and E. O. Fernandes, NATO ASI series E, vol. 143, 909–68. Dordrecht: Kluwer Academic Publishers.

Chen, J. C. 1963. *Correlation for boiling heat transfer to saturated fluids in convective flow* (ASME paper 63-HT-34).

Chen, C. C. and J. W. Westwater. 1984. Prediction of boiling heat transfer duty in a compact plate-fin heat exchanger using the improved local assumption. *Int. J. Heat Mass Transfer* 24:1907–12.

Collier, J. G., T. D. A. Kennedy, and J. A. Ward. 1974. Thermal design of plate-fin reboilers. In *Cryotech 73 prod. use ind. gases proc.,* 95–100. Guildford, England: IPC Technology Press.

Galezha, V. B., I. P. Usyukin, and K. D. Kan. 1976. Boiling heat transfer in Freons in finned-plate heat exchangers. *Heat Transfer Sov. Res.* 8(3):103–10.

Hewitt, G. F., and D. N. Roberts. 1969. *Studies of two-phase flow patterns by simultaneous x-ray and flash photography* (Atomic Energy Research Establishment, Harwell, England, report AERE-M 2159).

Johannes, C., and J. Mollard. 1972. Nucleate boiling of helium-I in channels simulating the cooling channels of superconducting magnets. *Adv. Cryog. Eng.* 17:332–41.

Lockhart, R. W., and R. C. Martinelli. 1949. Proposed correlation of data for isothermal two-phase, two-component flow in pipes. *Chem. Eng. Progr.* 45:39–48.

Mandrusiak, G. D., and V. P. Carey. 1988. Convective boiling in vertical channels with different offset strip fin geometries: Part I—Data for annular flow boiling, and Part II—Heat transfer with forced convective and nucleate boiling effects. Submitted for publication.

———. and X. Xu. 1986. An experimental study of convective boiling in a partially heated horizontal channel with offset strip fins. *Advances in Heat Exchanger Design,* ed. R. K. Shah and J. T. Pearson, HTD vol. 66, 55–63. New York: American Society of Mechanical Engineers.

———. 1988. An experimental study of convective boiling in a partially heated horizontal channel with offset strip fins. *J. Heat Transfer* 110:229–36.

Panitsidis, R. D., R. D. Gresham, and J. W. Westwater. 1975. Boiling of liquids in a compact plate-fin heat exchanger. *Int. J. Heat Mass Transfer* 18:37–42.

Robertson, J. M. 1979. Boiling heat transfer with liquid nitrogen in brazed-aluminum plate-fin heat exchangers. *AIChE Symp. Ser.* 75(189): 151–64.

———. 1980. Review of boiling, condensation, and other aspects of two-phase flow in plate-fin heat exchangers. In *Compact Heat Exchangers,* HTD vol. 10, 17–27. New York: American Society of Mechanical Engineers.

———. 1982. The correlation of boiling coefficients in plate-fin heat exchanger passages with a film-flow model. *Proc. 7th Int. Heat Transfer Conf.* 6:341–345.

———. 1983. The boiling characteristics of perforated plate-fin channels with liquid nitrogen in upflow. In *Heat Exchangers for two-phase flow applications,* HTD vol. 27, 35–40. New York: American Society of Mechanical Engineers.

———. 1984. The prediction of convective boiling coefficients in serrated plate-fin passages using an interrupted liquid-film flow model. In *Basic aspects of two-phase flow and heat transfer,* HTD vol. 34, 163–71. New York: American Society of Mechanical Engineers.

———. and R. H. Clarke. 1981. The onset of boiling of liquid nitrogen in plate-fin heat exchangers. *AIChE Symp. Ser.* 77(208): 86–95.

———. 1984. Investigations into the onset of convective boiling with liquid nitrogen in plate-fin heat exchanger passages under constant wall temperature boundary conditions. *AIChE Symp. Ser.* 80(236): 98–103.

Robertson, J. M., and P. C. Lovegrove. 1980. *Boiling heat transfer with Freon 11 in brazed-aluminum plate-fin heat exchangers* (ASME paper 80-HT-58).

———. 1983. Boiling heat transfer with Freon 11 (R-11) in brazed-aluminum plate-fin heat exchangers. *J. Heat Transfer* 105:605–10.

Robertson, J. M., and V. V. Wadekar. 1988. Boiling characteristics of cyclohexane in vertical upflow in perforated plate-fin passages. *AIChE Symp. Ser.* 84:120–25.

Shah, M. M. 1976. A new correlation for heat transfer during boiling flow through pipes. *ASHRAE Trans.* 82:66–86.

Shorin, C. N., V. I. Sukhov, S. A. Shevyakova, and V. K. Orlov. 1974. Experimental investigation of heat transfer with the boiling of oxygen in vertical channels during condensation heating. *Int. Chem. Eng.* 14(3):517–21.

Taitel, Y., and A. E. Dukler. 1976. A model for predicting flow regime transitions in horizontal and near-horizontal gas-liquid flow. *AIChE J.* 22:47–55.

Westwater, J. W. 1983. Boiling heat transfer in compact and finned heat exchangers. *Adv. Two-Phase Flow Heat Transfer* NATO ASI series E 2(64):827–57.

Wieting, A. R. 1975. Empirical correlations for heat transfer and flow friction characteristics of rectangular offset-fin plate-fin heat exchangers. *J. Heat Transfer* 97:488–90.

Yilmaz, S., and J. W. Westwater. 1980. Effect of velocity on heat transfer to boiling Freon-113. *J. Heat Transfer* 102:26–31.

Yung, D., J. J. Lorenz, and C. Panchal. 1980. Convective vaporization and condensation in serrated-fin channels. In *Heat transfer in ocean thermal energy conversion systems,* HTD vol. 12, 29–37. New York: American Society of Mechanical Engineers.

INDEX

Acceleration pressure gradient, 248, 251
Acetone, 147–148, 177–179, 189, 192–195
Additives, 40
Alcohols, 144–147
Air-conditioning coils, 71–72, 312–313
Air separation plants, 271–273, 325
Annular flow, 18, 22, 223, 237, 327
Application guide, 268, 277
Attached promoters, 47–50, 111–117
Augmentation curve, 299

Bent-over fins, 52, 104–106
Benzene, 87–88, 142, 182
Boiling correlations (nucleate):
  Gewa-T tube, 162–169
  High Flux tube, 155–160
  Low-finned tubes, 154–155
  Mixtures, 185–186
  Modified low-finned tubes, 161–170
  Plain tubes, 11–12, 23, 126
  Porous coatings, 155–161
  Thermoexcel-E, 169–170
Boiling correlations (plate fin), 340–347
Boiling curve, 5–7, 111–112
Boiling nucleation:
  Convective boiling, 19, 88–89, 329–332
  Heterogeneous, 30, 34
  Historical developments, 30–36
  Homogeneous, 30
  Mixtures, 180–183
  Photographs in channels, 82, 84, 85, 97, 100–102
  Plate-fin units, 329–332
  Pure fluids, 83–88
  Supersaturation effect, 182–183
  Temperature gradient effect, 33
Boiling range, 176–177
Boiling site density, 29, 36, 108–110
Boiling suppression factor, 21, 23
Bubble departure diameter, 11
Bubbly flow, 18, 22, 223, 237, 327
Bundle boiling correlation, 15–16
Bundle boiling factor, 15–16, 255–258, 285–288, 308, 315–317
Bundle correction factor, 16–17
Bundle diameter, 270, 283–284
Bundle fabrication, 289
n-Butane, 39

Chen correlation, 20–21
Circulation rates in reboilers, 284–285
Compound enhancement, 40
Computer programs, 282–283
Condensation on vertical fluted tubes, 313

Contact angles:
  Pure fluids, 86–87
  Macroscopic, 32
  Microscopic, 32
  Minimum, 31
  Mixtures, 180–181
Critical heat flux:
  Correlations, 253–254
  Enhanced tubes and inserts, 253–254
  Smooth tubes, 21–22
  (*see also* peak nucleate heat flux)
CSBS surface, 122

Dendritic surface, 43, 87
Design considerations:
  Mechanical, 269–271, 288–289
  Thermal, 2, 280–288
Dielectrics, 148–149
Dryout, 18

ECR-40, 69–70
Electroplating processes, 42–44
Electrostatic fields, 40
Enhanced boiling tubes:
  Corrugated, 74–75, 227
  Cruciform, 220–223
  Doubly enhanced, 72–75
  High-fin, 220–221
  Internally finned, 71–72
  Low-finned, 64–66
  Microfinned, 206
  Modified low-finned, 66–70
  Porous layer-coated, 44
  Spirally fluted, 74, 231
Enhanced classifications, 39–40
Entrainment:
  Annular flow, 18, 22, 218
  Exit pipes, 283
  Swirl flow, 218
Ethanol, 43, 87, 88, 279
Ethylene, 273–275, 313–315

FC-72, 88
Film boiling, 13–14
Film temperature, 14
Fin efficiency, 152–154
Flame spraying, 42–45
Flooding mechanism, 110–111
Flow patterns:
  Enhanced tubes, 237–238, 242, 243
  Plate-fin units, 326–329
Fluid vibrations, 40
Fouling, 282, 289–297, 309, 312, 325

Gap size of Gewa-T tube, 102–104
Gewa-T tube, 68
Gewa-TX tube, 68

Hastelloy, 51, 66
Heat exchangers:
  BTX, 278
  Chillers, 306–312
  Chiller-condensers, 269, 272, 275
  Configurations, 268–271
  Internal reboilers, 270, 315–318
  LEFU, 276–279, 292–293
  LNG, 275–276
  Operational limits, 283–284
  Optimization, 282
  OTEC evaporator, 312
  Plate-fin, 323–325
  Prototypes, 308–310
  Seawater evaporators, 280
  Shell sizing, 270
  TEMA types for enhanced exchangers, 269–270
  Thermal constraints, 283–284
  Thermosyphon reboilers, 268–270
  Tube sizing, 270
Heat pumping, 278
Heat transfer mechanisms:
  Bubble agitation, 92
  Capillary evaporation, 94–95, 208–209
  Convection, 96–98
  Enhanced boiling, 93–117
  External convection, 98
  External evaporation, 94
  Laminar convection, 96
  Microlayer evaporation, 7
  Plain surfaces, 91–93
  Thermal boundary layer stripping, 7
  Thin film evaporation, 94, 156
  Vaporization, 92, 94
  Vapor-liquid exchange, 92
Heat transfer regimes, 5–6
Helium-I, 139–140
High Flux tube, 71

Impingement plate, 289
Inner Fin tube, 77
In-tube boiling correlations:
  Corrugated, 249

In-tube boiling correlations (*Cont.*):
- Cruciform, 233
- Internally finned, 246–248
- High-fin, 223
- Porous layer-coated, 252–253
- Star inserts, 251–252
- Twisted tapes, 249–251

Iso-propanol, 144–145

Korotex-II, 43

Latent heat:
- Corrected latent heat, 14
- Differential, 177–180
- Flux, 108–109
- Integral, 180

LNG plants, 275–276
Lockhart-Martinelli parameter, 20
Low-finned tube, 65

Maximum bundle heat flux, 16–17, 284
Maximum heat flux (see critical heat flux)
Maximum heat transfer coefficient, 156–158
Maximum wall superheat, 297
Mechanically aided enhancement, 40
Metals:
- Doubly enhanced tubes, 73–75
- Modified externally finned tubes, 67–70
- Low-finned tubes, 51, 65–66
- Porous coated layers, 70
- Star inserts, 75–76
- Twisted tapes, 76

Metalworking process, 66–67
Minimum heat flux, 10
Mist flow, 18, 223, 237
Mixture boiling:
- Convectional surfaces, 183–186
- Correlation, 185–186
- Effect of physical properties, 177–180
- Enhanced boiling, 186–195
- Gewa-TX, 190–191
- High Flux, 186–190
- Hydrocarbon mixtures, 186–187, 191
- Low-finned tubes, 190–192
- Mixture effects, 192–195
- Nucleation, 180–183
- Refrigerant-oil mixtures (see refrigerant-oil mixtures)

Molar free energy of formation, 30

Natural convection:
- Single tube, 10
- Tube bundle, 15–16
- Nitrogen, 138–139, 271, 331, 338
- Nonwetting coatings, 45–47
- Nozzle placement, 281, 283, 289, 306–307
- Nucleation (see Boiling nucleation)
- *n*-Pentane, 36–37

Onset of nucleate boiling, 88–89
Orientation of surface, 119–121
Oxygen, 42, 57, 271

Particle diameter, 114–116
Particle packing arrangements, 115–116
Peak nucleate heat flux, 7, 8–10, 35–36, 111, 121–122, 284
Perforated fins, 329
Perforated cover plates, 106–111
Phase equilibria, 175–177
Pitch of tube bundles, 259, 288
Plasma spraying process, 44
Plate-fin types, 326–329
Porosity, 115
Porous coatings, 40–45, 70
Porous layer thickness, 114–117
Porous welds, 42
Propane, 39, 142–144
Propylene, 43, 45, 143, 278–279

Radiant heat transfer, 14
Reboilers, 315–321
Reduced pressure, 117–119
Reentrant cavities, 34
Reentrant channels, 81–86
Refrigerants, convective boiling comparisons:
- R-11, 232, 237
- R-12, 58, 229
- R-22, 56, 216, 220, 221, 225, 241–242, 246
- R-113, 226, 234
- R-114, 232

Refrigerants, pool boiling comparisons:
- R-11, 128–132
- R-12, 132–133
- R-22, 132–134
- R-113, 134–135
- R-114, 136–137

Refrigerant-oil mixtures:
- Enhanced tubes, 196–201, 290
- In-tube boiling, 241–246

Refrigerant-oil mixtures (*Cont.*):
  Plain tubes, 196, 243
Ribbed tube, 72–73

Sandblasting, 29
Screw tube, 220–221
Sensible heat flux, 109
Serrated fin, 328–329
Severity factor, 227–228
Shrouds, 49
Slug flow, 18, 22, 223, 237, 327
Spline inserts, 76
Star insert, 75–77, 236
S/T Trufin, 65
S/T Turbo-Chil, 73–74
Subcooled boiling, 12–13, 119, 281, 283
Superheat:
  Maximum bundle, 297
  Minimum, 78
  Nucleation, 8-9, 33, 156
  Onset of nucleate boiling, 19
  Subcooled boiling, 12
  Wall, 12, 17, 157
Surface orientation, 119–121
Surface roughness, 29, 36–38
Surface tension devices, 40

Taylor instability, 10
Teflon:
  Layers, 47
  Spots, 46–47
Temperature approach, 283
Thermal equilibrium, 8
Thermoexcel-E, 66–68, 73
Thermofin, 71–72, 75
Thin film evaporation, 7, 156, 158
Transition boiling, 5–7
Treated surfaces 39, 40–50
Tube costs, 298
Tube layout, 259, 270, 294
Tube selection, 298–301
Turbo-B, 68–69
Two-phase flow instabilities, 239, 283
Two-phase friction factor, 248–251
Two-phase pressure drop, 212, 214, 219, 226–230, 237, 341

U-bends, 288

Vapor disengagement, 283
Vapor distributor, 306–307
Vapor recompression, 278
Vapor trapping, 31–36
Visualization studies, 79–83, 161

Waste heat boilers, 280
Water chillers, 306–312
Water chiller tube, 76
Water, nucleate boiling comparison, 127–128
Wavy flow, 22
Wetted surface area of porous layers, 98
Wicks, 49–50

p-Xylene, 141–142

Zirconium, 51, 66